STUDIES IN GEOPHYSICS

Energy and Climate

Geophysics Study Committee
Geophysics Research Board
Assembly of Mathematical and Physical Sciences
National Research Council

NATIONAL ACADEMY OF SCIENCES
Washington, D.C.   1977

NOTICE: The project that is the subject of this report was approved by the Governing Board of the National Research Council, whose members are drawn from the Councils of the National Academy of Sciences, the National Academy of Engineering, and the Institute of Medicine. The members of the Committee responsible for the report were chosen for their special competences and with regard for appropriate balance.

This report has been reviewed by a group other than the authors according to procedures approved by the Report Review Committee consisting of members of the National Academy of Sciences, the National Academy of Engineering, and the Institute of Medicine.

The Geophysics Study Committee is pleased to acknowledge the support of the National Science Foundation, the U.S. Geological Survey, the U.S. Energy Research and Development Administration, the National Oceanic and Atmospheric Administration, the Defense Advanced Research Projects Agency, and the National Aeronautics and Space Administration.

Library of Congress Catalog Card Number 77-83238

International Standard Book Number 0-309-02636-9

Available from:

Printing and Publishing Office
National Academy of Sciences
2101 Constitution Avenue, N.W.
Washington, D.C. 20418

Printed in the United States of America

Geophysics Research Board

HERBERT FRIEDMAN, Naval Research Laboratory, *Chairman*
PHILIP H. ABELSON, Carnegie Institution of Washington
ARTHUR G. ANDERSON, International Business Machines Corporation
THOMAS C. ATCHISON, JR., U.S. Bureau of Mines
HUBERT L. BARNES, Pennsylvania State University
GEORGE S. BENTON, The Johns Hopkins University
D. ALLAN BROMLEY, Yale University
BERNARD F. BURKE, Massachusetts Institute of Technology
A. G. W. CAMERON, Harvard College Observatory
RICHARD K. COOK, National Bureau of Standards
THOMAS M. DONAHUE, University of Michigan
CHARLES L. DRAKE, Dartmouth College
ALFRED G. FISCHER, Princeton University
JOHN V. EVANS, Massachusetts Institute of Technology
J. FREEMAN GILBERT, University of California, San Diego
THOMAS O. HAIG, University of Wisconsin
WILLIAM M. KAULA, University of California, Los Angeles
WALTER B. LANGBEIN, U.S. Geological Survey, retired
THOMAS F. MALONE, Butler University
ARTHUR E. MAXWELL, Woods Hole Oceanographic Institution
GORDON A. NEWKIRK, JR., National Center for Atmospheric Research

JOHN S. NISBET, Pennsylvania State University
HUGH ODISHAW, University of Arizona
JACK E. OLIVER, Cornell University
VERNER E. SUOMI, University of Wisconsin
FERRIS WEBSTER, Woods Hole Oceanographic Institution
CHARLES A. WHITTEN, National Oceanic and Atmospheric Administration, retired
JAMES H. ZUMBERGE, Southern Methodist University

Assembly of Mathematical and Physical Sciences Liaison Representatives

PRESTON CLOUD, U.S. Geological Survey; University of California, Santa Barbara
JOSEPH W. CHAMBERLAIN, Rice University
ROBERT B. LEIGHTON, California Institute of Technology

Geophysics Study Committee

PHILIP H. ABELSON, Carnegie Institution of Washington, *Cochairman*
THOMAS F. MALONE, Holcomb Research Institute, *Cochairman*
LOUIS J. BATTAN, University of Arizona
CHARLES L. DRAKE, Dartmouth College
RICHARD M. GOODY, Harvard University
FRANCIS S. JOHNSON, University of Texas at Dallas
WALTER B. LANGBEIN, U.S. Geological Survey, retired
HUGH ODISHAW, University of Arizona

NRC Staff
PEMBROKE J. HART
DONALD C. SHAPERO

Liaison Representatives
JAMES R. BALSLEY, U.S. Geological Survey
EUGENE W. BIERLY, National Science Foundation
GEORGE A. KOLSTAD, U.S. Energy Research and Development Administration
CARL F. ROMNEY, Defense Advanced Research Projects Agency
WALTER TELESETSKY, National Oceanic and Atmospheric Administration
FRANCIS L. WILLIAMS, National Aeronautics and Space Administration

Panel on Energy and Climate

ROGER R. REVELLE, University of California, San Diego; and Harvard University, *Chairman*
ROBERT B. BACASTOW, University of California, San Diego
D. JAMES BAKER, JR., University of Washington
CHARLES L. HOSLER, JR., Pennsylvania State University
CHARLES D. KEELING, University of California, San Diego
HANS H. LANDSBERG, Resources for the Future, Inc.
HELMUT E. LANDSBERG, University of Maryland
RALPH A. LLEWELLYN, Indiana State University
J. MURRAY MITCHELL, JR., National Oceanic and Atmospheric Administration
WALTER H. MUNK, University of California, San Diego
HARRY PERRY, Resources for the Future, Inc.
GEORGE D. ROBINSON, Center for Environment and Man, Inc.
JOSEPH SMAGORINSKY, Geophysical Fluid Dynamics Laboratory, National Oceanic and Atmospheric Administration; and Princeton University
VERNER E. SUOMI, University of Wisconsin
WARREN M. WASHINGTON, National Center for Atmospheric Research

Foreword

It has become increasingly apparent in recent years that human capacity to perturb inadvertently the global environment has outstripped our ability to anticipate the nature and extent of the impact. It is time to redress that imbalance.

The world's climate could be perturbed in a number of ways, for example, through excessive destruction of forests and grasslands, thermonuclear war, and through our procedures for satisfying energy demands.

Examination of the possible long-term effects of energy use is particularly timely. With the end of the oil age in sight, we must make long-term decisions as to future energy policies. One lesson we have been learning is that the time required for transition from one major source to another is several decades. We cannot make major mistakes and expect to rectify them quickly. Thus it is important to examine carefully the potential long-term effects of our energy policies.

This report is concerned with the technical considerations that suggest possible impact on the climate of a growing world population and an increasing per capita use of energy. It is intended as a preliminary step in a process, which will require a number of years to complete, aimed at placing in the hands of policymakers credible information on the most likely climatic consequences of major dependence on fossil fuels as a source of energy for an increasingly industrialized society. Even at this early and still somewhat uncertain stage, the implications warrant prompt attention.

The consequences of using fossil fuels as a principal source of energy over the

Foreword

next few centuries comprise part of a family of assessments that should be made to consider the environmental impact of the most attractive alternative sources of energy. Potential difficulties *per se* in the use of fossil fuels as a principal source of energy should *not* be used as an argument for turning to any specific alternative source. Policy decisions with respect to the most desirable source of energy should be based on satisfactory information on the long-term environmental impact of each source and the impact of several combinations of energy sources.

The results of the present study should lead neither to panic nor to complacency. They should, however, engender a lively sense of urgency in getting on with the work of illuminating the issues that have been identified and resolving the scientific uncertainties that remain. Because the time horizon for both consequences and action extends well beyond usual boundaries, it is timely that attention be directed to research needs and to the anticipation of possible societal decision making now. The principal conclusion of this study is that the primary limiting factor on energy production from fossil fuels over the next few centuries may turn out to be the climatic effects of the release of carbon dioxide. This conclusion follows from:

- estimates of energy consumption and probable sources of energy, which, in view of the fact that they reach a century or more into the future, should be treated as a possible scenario rather than as a firm projection;
- a comparative analysis of the likely effects of the release of actual heat and particulates, both globally and locally;
- an examination of several models for the partitioning of carbon dioxide among the oceans, the atmosphere, and the biosphere;
- a review of the models that transform increased carbon dioxide in the atmosphere into an associated rise in global temperatures.

A number of comments are in order on each of the considerations that support this principal conclusion.

First, it seems unlikely that the scenario for energy consumption would actually be realized. World population growth might be curbed. Countries such as the Soviet Union that possess much of the world's coal reserves may be unwilling to serve as energy sources for others. An aroused world society might refuse to accept a drastic and unsatisfactory climatic change. To provide insurance that such a change will be averted will require a carefully planned international program and a fine sense of timing on the part of decision makers.

Second, it should be clear that local climatic effects of energy parks, of the release of particulates, and of sensible heat cannot be ignored, but they are of lesser urgency than the matter of carbon dioxide impact.

It will be noted that there are differences in the quantitative results of models developed by Revelle and Munk, by Keeling and Bacastow, and by others for partitioning carbon dioxide among the atmosphere, the oceans, and the biosphere. What is important is *not* that there are differences but that the span of agreement embraces a fourfold to eightfold increase in atmospheric carbon dioxide in the latter part of the twenty-second century. Our best understanding of the relation between an increase in carbon dioxide in the atmosphere and change in global temperature suggests a corresponding increase in average world temperature of more than 6 °C, with polar temperature increases of as much as three times this figure. This would exceed by far the temperature fluctuations of the past several thousand years and would very likely, along the way, have a highly significant impact on global precipitation.

Finally, it must be recognized that estimates obtained from recently developed ocean–atmosphere models of the relationship between increased atmospheric carbon

Foreword

dioxide and average temperature, humidity, precipitation, and other climatic variables are at an early stage and are just as likely to be too low as to be too high.

To reduce uncertainties and to assess the seriousness of the matter, a well-coordinated program of research that is profoundly interdisciplinary in character, and strongly international in scope, will be required. This research, much of which is outlined in the present report, should extend beyond scientific and technical considerations to include the complex factors and institutional innovations that will enable the nations of the world to act with wisdom and in concert before irreversible changes in climate are initiated. If the preliminary estimates of climate change in the latter part of the twenty-second century are validated, a reassessment of global energy policy must be started promptly because, long before that destined date, there will have been major climatic impacts all over the world.

One final thought. This report has been addressed to what might be the climatic impact over the next century or two of a major dependence on fossil fuels. In the light of a rapidly expanding knowledge and interest in natural climatic change, perhaps the question that should be addressed soon is, "What *should* the atmospheric carbon dioxide content be over the next century or two to achieve an optimum global climate?" Sooner or later, we are likely to be confronted by that issue.

Philip H. Abelson
Thomas F. Malone
Cochairmen
Geophysics Study Committee

Preface

Early in 1974, the Geophysics Research Board completed a plan, subsequently approved by the Committee on Science and Public Policy of the National Academy of Sciences, for a series of studies to be carried out on various subjects related to geophysics. The Geophysics Study Committee was established to provide guidance in the conduct of the studies.

One purpose of the studies is to provide assessments from the scientific community to aid policymakers in decisions on societal problems that involve geophysics. An important part of such an assessment is an evaluation of the adequacy of present geophysical knowledge and the appropriateness of present research programs to provide information required for those decisions. When appropriate, the implications of this evaluation for research strategy may be set forth. Some of the studies place more emphasis on assessing the present status of a field of geophysics and identifying the most promising directions for future research. Topics of studies for which reports are currently in preparation include geophysical predictions and the impact of technology on geophysics. Topics of recently published studies include upper-atmosphere geophysics, water and climate, and geophysics of estuaries.

Each study is developed through meetings of the panel of authors and presentation of papers at a suitable public forum that provides an opportunity for discussion. In completing final drafts of their papers, the authors have the benefit of this discussion as well as the comments of selected scientific referees. Responsibility for the individual essays rests with the corresponding authors.

Preface

Most of the essays in this volume were presented in preliminary form in a symposium at an American Geophysical Union meeting that took place in San Francisco in December 1974. They treat the question of possible constraints placed on energy use by the danger of climatic change.

The introductory chapter provides an overview of the study, summarizing the highlights of the essays and formulating conclusions and recommendations. In preparing it, the Chairman of the panel had the benefit of meetings and discussions that took place at the symposium and the comments of the panel of authors and selected referees. Responsibility for its content rests with the Geophysics Study Committee and the Chairman of the panel.

Contents

OVERVIEW AND RECOMMENDATIONS		1
I. THE CONTEXT		33
1. Projected World Energy Consumption *Harry Perry and Hans H. Landsberg*		35
2. The Changing Climate *J. Murray Mitchell, Jr.*		51
II. EFFLUENTS OF ENERGY PRODUCTION		59
3. Effluents of Energy Production: Particulates *George D. Robinson*		61
4. Impact of Industrial Gases on Climate *Charles D. Keeling and Robert B. Bacastow*		72
5. The Effect of Localized Man-Made Heat and Moisture Sources in Mesoscale Weather Modification *Charles L. Hosler and Helmut E. Landsberg*		96

6. Regional and Global Aspects
   Ralph A. Llewellyn and Warren M. Washington — 106

III. MONITORING AND MODELING — 119

7. Ocean Dynamics and Energy Transfer: Some Examples of Climatic Effects
   D. James Baker, Jr. — 121

8. The Need for Climate Monitoring
   Verner E. Suomi — 128

9. Modeling and Predictability
   Joseph Smagorinsky — 133

10. The Carbon Dioxide Cycle and the Biosphere
    Roger Revelle and Walter Munk — 140

Energy
and
Climate

Overview and Recommendations

INTRODUCTION

Worldwide industrial civilization may face a major decision over the next few decades—whether to continue reliance on fossil fuels as principal sources of energy or to invest the research and engineering effort, and the capital, that will make it possible to substitute other energy sources for fossil fuels within the next 50 years. The second alternative presents many difficulties, but the possible climatic consequences of reliance on fossil fuels for another one or two centuries may be so severe as to leave no other choice.

A decision that must be made 50 years from now ordinarily would not be of much social or political concern today, but the development of the scientific and technical bases for this decision will require several decades of lead time and an unprecedented effort. No energy sources alternative to fossil fuels are currently satisfactory for universal use, and, in any case, conversion to other sources would require many decades. Similarly, finding ways to make reliable estimates of the climatic changes that may result from continued use of fossil fuels could very well require decades.

The climatic questions center around the increase in atmospheric carbon dioxide that might result from continuing and increasing use of fossil fuels. Four questions are crucial:

1. What concentrations of carbon dioxide can be expected in the atmosphere at different times in the future, for given rates of combustion of fossil fuels?

Overview and Recommendations

2. What climatic changes might result from the increased atmospheric carbon dioxide?

3. What would be the consequences of such climatic changes for human societies and for the natural environment?

4. What, if any, countervailing human actions could diminish the climatic changes or mitigate their consequences?

NATURE OF THE PROBLEM

Three by-products of energy production and consumption—heat, particulate matter, and gases—were recognized at the start of the work of the Panel as having the potential for inadvertent modification of global climate. It has been known for some time that cities create their own microclimate (see Chapters 5 and 6). At first, the Panel speculated that increasing urbanization, large power-generating compounds (power parks), and similar developments might, by their output of heat and particles, disturb rainfall or influence other meteorological phenomena on a global scale. However, our study showed that the simple combustion product carbon dioxide has the greatest apparent potential for disturbing global climate over the next few centuries (see Chapters 4 and 10).

Carbon dioxide, although virtually transparent to shortwave solar radiation (visible light), strongly absorbs long-wave radiation (heat) at certain wavelengths where other atmospheric gases are transparent. In the atmosphere, it impedes radiation of heat from the earth's surface into space. An increase in carbon dioxide concentration in the atmosphere could disturb the balance between incoming solar radiation and the radiation of heat from the earth into space with a resulting increase in the temperature of the lower atmosphere. Because glass in a greenhouse traps the sun's heat, although mainly by preventing convection, this phenomenon has come to be known as the greenhouse effect.

In emphasizing questions related to increased atmospheric carbon dioxide we do not imply that serious consequences might not arise also from an increase in the load of particulates in the atmosphere or the growth of large "hot spots" resulting from the uneven distribution of human energy use. It is clearly possible, although expensive, to control the level of atmospheric particulates produced by human activity, and there are other reasons for doing so than the possible effects of high particulate concentrations on climate (see Chapter 3). Present climatic models are not adequate to predict reliably possible large-scale climatic changes resulting from the uneven geographic distribution of heat released by human energy use. But the greater understanding of climate required to answer questions about the effect of carbon dioxide could make it possible to give useful estimates of the effects of uneven heat releases. Even a future world population of ten billion people, with a per capita energy use several times greater than at present, would release an amount of heat equivalent to only one thousandth of the global net radiation received from the sun. The short residence times of tropospheric aerosols limit the threat that they pose because the atmosphere can be cleansed of them in a matter of weeks.

The average global temperature is only one of a constellation of dynamically related variables that, taken together, describe climate. Others include statistical properties of temperature, cloudiness, precipitation, and wind. The possibility that a moderate change in one of these variables could lead to a major shift in global climate cannot be ruled out. Historical records and indirect indices of past climates do indeed show marked shifts in temperature, precipitation, and ice volume. About 60 million years ago, the warm Mesozoic era ended and a gradual cooling began, leading to the present glacial age. The last 2 million years have been characterized by ice ages relieved by warm interglacial periods. The most recent ice age, during which average midlatitude temperatures were 5 to 10°C below those we presently enjoy, ended about 10,000 years ago.

Some of the basic processes governing climatic change are poorly understood. It is not known whether climate changes occur in steplike transitions from one dynamically stable steady state to another or whether they represent a more gradual passage through a continuum of climate states. Either kind of change could be induced by variations in external parameters, such as the amount of radiation coming from the sun, or by spontaneous internal redistributions of energy within the physical components of the climate system (see Chapter 2). If changes in climate are steplike, then a disturbance of climate such as would be produced by a large increase in carbon dioxide production would be especially worrisome; a slow change might be the forerunner of a comparatively abrupt transition to a new climatic regime. If, on the other hand, changes in climate are gradual, then the effects of increasing carbon dioxide in the atmosphere would build steadily to produce a more gradual global shift in climate. In either case, agricultural belts would be shifted by changing seasonal precipitation and temperature patterns. For some countries with marginal agriculture, the impact on food production could be severe. For this and other reasons, the prospect of a man-made modification of global climate must be taken seriously.

If the potential for climatic change discussed in this report is further substantiated, then it may be necessary to reverse the trend in consumption of fossil fuels. Alternatively, carbon dioxide emissions will somehow have to be controlled or compensated for (no practical means of doing so seem to be readily at hand). In the face of so much uncertainty regarding climatic change, it might be argued that the wisest attitude would be laissez-faire. Unfortunately, it will take a millennium for the effects of a century of use of fossil fuels to dissipate. If the decision is postponed until the impact of man-made climate changes has been felt, then, for all practical purposes, the die will already have been cast.

MAGNITUDE OF THE PROBLEM

Man's use of energy has increased manyfold since the industrial revolution, yet large-scale effects on climate are not readily apparent. One should, therefore, look ahead for significant relationships. Thus estimates of future world population, future energy uses, and future sources of energy are central to an assessment of future climatic impact. Harry Perry and Hans H. Landsberg have undertaken to produce such estimates,* which, as they note, are intended simply as plausible springboards for analysis and discussion—not as forecasts. Nevertheless, the model they present sets forth vividly the implications of growing population and a continuing demand for energy.

They envision a world population of some 10 billion by the latter part of the next century and total energy consumption over 5 times current levels. Perhaps surprisingly, all this energy could be provided by fossil fuels, mainly coal. On this basis, annual heat and carbon dioxide production would also be over 5 times current levels, while annual production of particles (because of the need to use dirtier fuels) could be perhaps 20 times current values. Thus considerable man-made heat would be released into the environment, but it would still be only a small fraction of the natural energy flows on a global or regional scale; local concentrations, however, could be much greater. While particulate *production* might be very high, there is no reason to expect that *release* of particulates to the environment would be correspondingly high. On the contrary, there is every reason to suppose that present

---

*These estimates are discussed in Appendix A at the end of this chapter and in Chapter 1.

means of controlling particle emissions will be improved to a high degree. Certainly a twentyfold increase in the emission of particles would be intolerable because of hazards to human health.

Perry and Landsberg have calculated that in 1973, energy use was 7.6 billion metric tons of coal equivalent or 5.8×10^{16} kilocalories. This is about 0.01 percent of the incoming solar radiation. Their figures for 2075 indicate that the total energy used by human beings may correspond by then to about 0.1 percent of the incoming solar energy. On a worldwide basis, the climatic effect of the added heat would be small. On a local scale, however, it might be significant. Over Japan, for example, the heat released by human energy use could be nearly 2.6 percent of the solar radiation absorbed near the earth's surface, and over Western Europe it could be about 0.6 percent. Even with a population of 20 billion people and a per capita energy demand 10 times the present world average (twice that of the United States in 1975) the total energy release would be 400 Gt of coal equivalent, or about 0.3 percent of the absorbed solar radiation. Current general circulation models indicate that, if the heat release were uniformly distributed on the earth's surface, the corresponding increase in average surface temperature would be about 0.6 °C, and at latitudes above 50° perhaps 2-3 °C.

According to the data and estimates in Chapters 4 and 10, somewhat less than half of the carbon dioxide released by man since the industrial revolution has remained in the atmosphere. During that time, about a 13 percent rise in atmospheric concentration of carbon dioxide has taken place. Most of the remainder is inferred to have been taken up by the oceans and by the terrestrial biosphere. One can estimate the amount of carbon dioxide that may be released through the middle of the next century and, by the use of models of the carbon cycle, the amount that may be expected to remain in the atmosphere. It is not implausible that *the peak atmospheric concentration occurring in A.D. 2150 to A.D. 2200 might be four to eight times the preindustrial level.* Moreover, concentrations much higher than today's may persist for many centuries thereafter.

Manabe and Wetherald (1975) have constructed a three-dimensional climatic model of the general circulation of the atmosphere that simulates the effects of a doubling in atmospheric carbon dioxide. This model, although recognized to be imperfect in a number of significant ways, is the most complete one yet devised. For a doubling of carbon dioxide in the atmosphere, the model predicts about a 2-3 °C rise in the average temperature of the lower atmosphere at middle latitudes and a 7 percent increase in average precipitation. The temperature rise is greater by a factor of 3 or 4 in polar regions (see Chapter 9). For each further doubling of carbon dioxide, an additional 2-3 °C increase in air temperature is inferred. *The increase in carbon dioxide anticipated for A.D. 2150 to A.D. 2200 might lead to an increase in global mean air temperature of more than 6 °C*—comparable with the difference in temperature between the present and the warm Mesozoic climate of 70 million to 100 million years ago.

CONCLUSIONS

- It does not now appear that the direct generation of heat from the production and consumption of energy over the next few centuries will cause a rise of more than 0.5 °C in global average air temperature, although it may have substantial effects on local climates. If the correspondingly increased particulate emissions are properly controlled, there should be little global effect on climate from an increased atmospheric burden of aerosols.
- *The climatic effects of carbon dioxide release may be the primary limiting*

Overview and Recommendations

factor on energy production from fossil fuels over the next few centuries. The prospect of damaging climatic changes may thus be the stimulus for greater efforts at conservation and a more rapid transition to alternate energy sources than is justified by economic considerations alone. The potential effect of carbon dioxide on climate could be exacerbated by fluorocarbons, nitrous oxide, and other industrial gases. The natural variability of climate could increase or reduce the impact of such man-made effects.

■ There are profound uncertainties regarding the carbon cycle, climate, and their interdependence. *These uncertainties can be resolved only by a well-coordinated effort of extraordinarily interdisciplinary character.* The focus for such an effort is not provided by any existing institutional mechanisms.

RECOMMENDATIONS

■ The possibility of modification of the world's climate by carbon dioxide released in the production of energy from fossil fuels should be given serious prompt consideration by concerned national and international organizations and agencies. *Two kinds of action are needed: organization of a comprehensive worldwide research program and new institutional arrangements.*

■ A worldwide comprehensive research program should be undertaken. It should include studies of the carbon cycle, climate, future population changes, and energy demands and ways to mitigate the effects of climatic change on world food production.

1. *Research on climate.* The development and verification of climatic models should continue. Adequate means to monitor climate should be provided both to verify models and to give warning of climatic change. Further studies of the sensitivity of climate to disturbances in the radiation balance should be carried out using advanced climate models. Paleoclimatology should be pursued in the interest of understanding past climate changes as well as to provide data for verification of models, both of climate and of biogeochemical cycles.

2. *Studies of world population and energy demands.* These studies should include estimates of the contribution to our energy requirements that can be made by renewable resources. Investigation of energy sources that do not release waste gases and particles and minimize the release of heat should be intensified. The problems associated with energy supply and demand call for the further study of conservation measures and their prompt implementation.

3. *Food.* In view of the prospective impact of man's activities on climate as well as the natural variability of climate, greater attention must be given to agriculture and water resources from the point of view of mitigating the effects of climatic change.

4. *Carbon dioxide and the atmosphere–ocean–biosphere system.* A better understanding of the partitioning of carbon among the biosphere, oceans, and atmosphere is *essential* and might be obtained by the following measures:

(a) Measurements of variations with time in the ratio of the two stable isotopes of carbon (^{13}C and ^{12}C) in the atmosphere are needed to determine the *net* flux of carbon between the atmosphere and the biosphere. Past variations in this ratio can be obtained from measurements of these two isotopes in tree-ring sequences from trees in isolated locations—as distant as possible from major biological and industrial sources of carbon dioxide. Changes in the ^{13}C-^{12}C ratio are likely to be small relative to random measurement errors, and consequently many measurements are needed over a wide range of geographic locations.

(b) Better estimates should be made of the area of land annually cleared for

Overview and Recommendations

agricultural and other purposes. From 1972 onward, these estimates can be obtained from earth-resources satellite data. To obtain estimates for the period before 1972, a historical study of the growth of cultivated areas in all continents from the early part of the nineteenth century should be made.

(c) Attempts should be made to estimate changes in forest biomass throughout the world, particularly in tropical and subtropical areas. The principal component of this biomass consists of wood in living trees; measurements of differences in the thickness of successive tree rings might indicate changes in the rate of net primary production of trees, at least in temperate latitudes. Many tree-ring sequences (of the order of several thousand) would be necessary for an adequate sample. Efforts should also be made to estimate time variations in the quantity of foliage and other organs of trees that participate actively in photosynthesis and in the rate of fall of "litter" (dead leaves and branches) from trees.

(d) Better estimates are needed of the fraction of soil humus from which carbon dioxide is released to the atmosphere. Changes in this quantity should be determined in agricultural and other cleared areas. Surveys of the present distribution of humus in soils throughout the world are needed to serve as a baseline for comparison with future measurements.

(e) Intercomparable mean monthly values of the partial pressure of atmospheric carbon dioxide should be obtained from continuous measurements at a number of carefully selected stations in different latitudes in both the northern and southern hemispheres. One of the principal objectives of such a network of stations would be to study year-to-year variations in the airborne fraction of carbon dioxide released to the atmosphere by fossil-fuel combustion and land clearing. These variations appear to be related to fluctuations in carbon dioxide uptake and release by near-surface ocean waters, and their elucidation should give us greater insight into oceanic processes affecting the partitioning of carbon dioxide between the ocean and the atmosphere.

(f) Further insight into these processes might be gained by time sequences of measurements of the total carbon dioxide content and partial pressure of free carbon dioxide at a global network of observing stations in surface and subsurface ocean waters. These quantities vary widely with local biological and other ocean processes, and hence it may not be possible to make useful interpretations of the measurements from the standpoint of the global carbon dioxide problem. Further analysis of the desirability of this type of measurement should be undertaken.

(g) Better estimates are needed of the quantity of carbon dioxide released by fossil-fuel combustion. International statistics on the consumption of fossil fuels should be supplemented by estimates of the carbon content of the fuels consumed each year. The present uncertainty in the amount of carbon dioxide released is of the order of 10 to 15 percent because the estimates of fuel consumption are given in terms of fuel energy rather than carbon content.

(h) A series of measurements of the dispersal of tritium from atmospheric nuclear-weapons tests in subsurface ocean waters should be made at approximately five-year intervals following the lines of the GEOSECS expeditions of the early 1970's. Such measurements of time variations in the oceanic distribution of tritium appear to be the most satisfactory experimental means of studying "stirring" processes (advection, convection, and turbulent mixing) in the upper thousand or so meters of the ocean waters. These processes are critically important in estimating partitioning of fossil-fuel carbon dioxide between the ocean and the atmosphere.

(i) In principle, an independent check on oceanic stirring processes could be obtained if the Suess effect (the decrease in the radiocarbon content of the atmosphere from the beginning of the nineteenth century to 1950 resulting from injection of ^{14}C-free fossil-fuel carbon dioxide into the atmosphere) were more accurately

Overview and Recommendations

known. The present uncertainty in the Suess effect is about ±25 percent. Many more measurements of ^{14}C in tree rings covering the time span from 1800 to 1950 in trees from carefully selected locations are needed.

(j) The following kinds of observations should be given consideration for future studies but with a lower priority than the above recommendations relating to carbon dioxide.

(*i*) Knowledge of rates of interchange between the interstitial waters of calcareous sediments and the bottom waters overlying the sediments would allow better estimates of probable rates of solution of calcium carbonate and the corresponding increase in the capacity of the ocean to absorb carbon dioxide.

(*ii*) More data on the distribution of aragonite (the more soluble of the two crystalline forms of calcium carbonate) in shallow and deep calcareous sediments are needed for better estimates of the possible solution of calcium carbonate.

(*iii*) Whether solution of calcium carbonate has actually occurred and if so to what extent can be determined directly by measurement of changes in the alkalinity of seawater. New methods of measuring the alkalinity have a precision of about 1 part in 10,000, corresponding to a change in atmospheric carbon dioxide of about 0.1 percent.

(*iv*) Carbon dioxide could be extracted from the atmosphere–subsurface ocean system if the rate of fallout of particulate organic matter from the subsurface layer into the deep ocean waters could be accelerated. This might be possible if the photosynthetic production of organic matter in the near-surface ocean waters could be increased. Photosynthetic production in these waters appears to be determined by the supply of dissolved phosphorus and nitrogen compounds. In the future, it might be possible to disperse large quantities of industrially produced phosphorus and nitrogen over extended areas in the oceans, at a cost that would be relatively small compared with the total cost of carbon dioxide-producing fossil fuels. The effectiveness of this tactic can probably be determined by comparative measurements of the rate of fallout of organic matter in both presently highly productive and unproductive ocean areas. In principle, fertilization of ocean waters with 10 million tons of phosphorus would produce fallout of about 300 million tons of organic carbon.

■ All the foregoing recommendations for research relate to global concerns and therefore the cooperation of such international agencies as the World Meteorological Organization, the Intergovernmental Oceanographic Commission, and the International Council of Scientific Unions should be sought in responding to them. The necessary research will be expensive in terms of scientific and technical manpower and research facilities and must by continued over many years. Consequently, it must be fostered and supported by governments. A high degree of international governmental cooperation is called for because of the need for a worldwide set of measurements and network of observing stations.

Consideration should be given to the establishment at the national level of a mechanism to weave together the interests and capabilities of the scientific community and the various agencies of the federal government in dealing with climate-related problems. Solutions to those problems will involve coordination of research in many scientific disciplines and are likely to require adjustments in national policy or the formulation of new legislation. Such a mechanism might be embodied in a Climatic Council with the following functions:

1. To serve as the focal point within the United States for the development of a global research and action program.

2. To coordinate activities that cross disciplinary, institutional, and organizational boundaries.

Overview and Recommendations

3. To serve as a link with international mechanisms that guide international research and action programs concerned with carrying out aspects of the above recommendations.

POTENTIAL CONSEQUENCES OF CLIMATIC CHANGES

WARMING OF THE OCEAN WATERS

A warmer atmosphere could not persist without a comparable warming of the upper layers of the oceans. The effects would be several: a reduction in the amount of sea ice, a rise in sea level, a flux of carbon dioxide from the ocean to the atmosphere, a reduction in the vertical stirring of the oceans, and a poleward shift in marine ecosystems including the fish population. Reduction of sea ice should decrease albedo and lead to further warming, higher precipitation, and possibly a new build-up of snow and ice poleward.

An increase by 5°C in the average temperature of the top 1000 m of ocean water would raise sea level by about 1 m, because of the expansion of water volume. Such an increase in ocean temperature would raise the partial pressure of carbon dioxide in the water by about 30 percent. After equilibration with the atmosphere, which would probably take place within a few years, the carbon dioxide content of the air would be increased by about 17 percent.

The expected polar warming would affect the rate of ventilation of all subsurface waters. A "lid" of relatively warm water would be formed over the colder deep waters, thereby increasing the vertical density stratification of the oceans. This would inhibit vertical mixing and stirring processes, which would in turn reduce the rate of nutrient supply to the near surface ocean waters and hence the productivity of marine plants. The quantity of dead organic carbon falling from the surface layers to the deep water would be lower, and consequently the rate of uptake of carbon dioxide by the deep sea would also be lower. According to our models, atmospheric and presumably oceanic near-surface temperatures would increase much more at high than at low latitudes. Circulation of the deep waters and vertical exchange between the deep and near-surface waters might be profoundly altered by reduction or even cessation of present vertical convection and formation of deep water in the north Atlantic.

Experience in previous periods of ocean warming indicates that the area of sea ice would be substantially reduced, probably so much that both the Northwest and Northeast Passages would be open for shipping throughout most of the year.

During earlier decades of this century, the slight rise in average air temperature over the northern hemisphere, and the concomitant warming of surface-water layers of the ocean, brought about a marked shift in the locations of certain commercially important fish populations, notably the North Atlantic cod. Much of the cod fishery shifted to the waters off Greenland and north of Iceland. Hence an abnormally large warming of the atmosphere would surely have significant effects in the geographic extent and location of important commercial fisheries. Because different marine organisms respond differently to temperature changes, marine ecosystems might be seriously disrupted.

EFFECTS ON THE POLAR ICE CAPS

In the present state of understanding, it is impossible to forecast what might happen to the Greenland and Antarctic ice caps as a result of a rise of several degrees in global average air temperature. It is likely that the temperature in the Antarctic would still remain below the freezing point, so that melting at or near the surface of the ice probably would not occur. On the other hand, a substantial change in climate might greatly increase the annual snowfall in Antarctica and Greenland, resulting in a substantial increase in the thickness of the ice. This, in turn, would increase horizontal stresses at the base of the ice caps, which might result in surges or slides of ice masses into the sea. If these surges resulted in destruction of the West Antarctic ice cap, there might be a corresponding rise in sea level of about 5 m within 300 years (Hughes, 1974, 1977).

EFFECTS ON WORLDWIDE AGRICULTURE

Far-reaching consequences of a large increase in atmospheric carbon dioxide would be felt in agriculture—mankind's basic industry. With our present level of understanding, we cannot specify these consequences completely, let alone quantitatively. We can only suggest some of the possible effects. A few of these would be beneficial; others would be disruptive. Five factors must be considered: the effect of high levels of carbon dioxide on plant metabolism; higher annual average temperatures; spatial shifts in agroclimatic regions, especially in the precipitation patterns in different regions; the possibilities of greater or less variability from year to year in different regions; and the effects of possible increased cloudiness on the growth of crops.

Effects on Photosynthesis

Both theory and experiment show that raising the carbon dioxide content of the air in contact with plants increases the photosynthetic production of organic matter, provided other requirements for plant growth—nutrients, water, and sunshine—are present in abundance and the plants are not under stresses caused by too low or too high temperatures, soil acidity or alkalinity, lack of oxygen in the root zone, diseases, or other factors. With modern farming technology it is possible to provide adequate supplies of water and major and minor nutrients and to eliminate most causes of stress. Atmospheric carbon dioxide, incoming solar radiation, and the genetic potential of crop plants can then become the limiting factors of agricultural production.

Under normal farming conditions, the net photosynthetic product, that is, the organic matter remaining after the plant has used some of its own product in respiration, will not increase so rapidly as the atmospheric carbon dioxide (Waggoner, 1969). For the terrestrial biosphere as a whole, we have estimated that the factor of proportionality (designated as β) is about 30 percent, but it can be much higher for agricultural crop plants. Future agronomic and genetic research might bring this factor close to 1 (Waggoner, 1969; Hardman and Brun, 1971).

Other changes that might be brought about by higher atmospheric carbon dioxide would work in an opposite direction. If average air temperatures increase significantly, plant respiration is also likely to increase (Waggoner, 1969; Botkin et al., 1973). The net photosynthetic product may be reduced even though gross photosynthesis is raised.

If average cloudiness (the proportion of the land area covered by clouds) rises, the quantity of incoming solar radiation will be lowered, and the energy available to crop plants for photosynthesis will diminish. Excessive cloudiness during the monsoon season in India and Bangladesh already limits crop yields, compared with those obtained on the same farms in the sunnier months of October through March.

Overview and Recommendations

Poleward Shift of Agroclimate Zones

A rise in average annual global air temperature, increasing toward higher latitudes as predicted in the model of Manabe and Wetherald (1975) would result in a general poleward movement of agroclimatic zones. At higher latitudes, there would be a longer frost-free growing season than at present, and the boundaries of cultivation could be extended northward in the northern hemisphere. At the same time, summer temperatures might become too high for optimum productivity of the crops presently grown at middle latitudes, such as corn and soy beans in Iowa, Illinois, Indiana, and Missouri, and it might be necessary to shift the Corn Belt toward the north. But the acid podzol soils over large areas in these higher latitudes are badly leached, and extensive and expensive soil amendments would be required even to approach the yields now obtained in the remarkable soils of the Corn Belt.

The model predicts a global rise in average annual precipitation, which at first glance would seem to benefit agriculture. But the accompanying higher temperatures would also increase evapotranspiration from cultivated lands, and some part, perhaps all, of the benefits from the additional water supply would be lost. In some regions, higher evapotranspiration might exceed the increase in precipitation; in others, the reverse might be true (Manabe and Wetherald, 1975).

In general, the most serious effects on agriculture would arise not from changes in global average conditions but from shifts in the location of climatic regions and changes in the relationships of temperature, evapotranspiration, water supply, cloudiness, and radiation balance within regions. Present cropping patterns, crop varieties, and farming technology in different climatic regions are based on cumulative experience over many years in the selection of appropriate crop species and varieties for each region and in adapting both the plants and their physical environment to each other in as nearly an optimal fashion as possible. These adaptations have remained fairly satisfactory over the relatively narrow range of climatic changes that have occurred in the historic past. But large changes in climatic relationships within regions such as might be brought about by a doubling or quadrupling of atmospheric carbon dioxide would almost certainly exceed the adaptive capacity of presently grown crop varieties.

The regional changes in temperature–precipitation relationships that can accompany even comparatively small excursions in global average temperatures are illustrated by paleoclimatic studies. For example, during the so-called climatic optimum of several thousand years ago, when the average temperature was perhaps 1.5°C higher than at present, precipitation probably increased over southern Europe, northern Africa, southern India, and eastern China, while over large parts of the United States, Canada, and Scandinavia, the climate was drier (Kellogg, 1977).

It cannot be expected that regional climatic changes resulting from a large rise in atmospheric carbon dioxide would be simply an exaggerated replica of past changes. Both the seasonal and latitudinal effects of added carbon dioxide should be different from, for example, the effects of a global change in incoming solar radiation. Since both water vapor and carbon dioxide absorb and reradiate infrared energy, the effect of added carbon dioxide will be relatively more important in the dry air of high latitudes, in the upper troposphere, and in the stratosphere than in the moist air of the tropics. Similarly, because absolute humidity in winter is less than in summer, the effects of added carbon dioxide will be relatively more significant in the winter months. These latitudinal and altitudinal differences in the role of carbon dioxide are taken into account in the model of Manabe and Wetherald (1975); and they must form an integral part of future three-dimensional dynamic models that attempt to specify regional climatic changes in temperature, precipitation, evapotranspiration, and cloudiness.

Overview and Recommendations

Studies of oxygen and carbon isotope ratios in deep-sea cores suggest that the higher temperatures of the climatic optimum may have been due to a temporary increase in atmospheric carbon dioxide resulting from the changes in ocean circulation that followed the melting of the ice cap. If this suggestion can be substantiated, further paleoclimatic studies of seasonal differences in temperature-precipitation relationships during the climatic optimum should provide valuable insights into the future effects of increased atmospheric carbon dioxide (Berger and Garner, 1975; Berger and Killingley, 1977).

Serious Effects in Arid and Semiarid Regions

The most serious effects of possible future climatic changes could be felt along the boundaries of the arid and semiarid regions in both hemispheres. These are the zones of atmospheric subsidence where precipitation is scanty and highly variable: the southwestern United States and northern Mexico; the belt of relatively dry lands extending from southern Europe and northern Africa (including the Sahara), eastward across Arabia and south Asia to Pakistan and northwestern India; northeastern Brazil, northern Chile and southern Peru, western Argentina, southwest Africa, and Australia. We need to be able to estimate whether these belts of aridity and semiaridity will move toward or away from the poles and whether they will expand or contract in area.

Short-term variations in climate, especially in rainfall, often persisting for several years, are characteristic of the semiarid regions. The human disasters caused by drought in these regions are familiar and dramatic. But the variations are also closely related to destruction of the resource base through processes of desertification—wind and water erosion and sedimentation—which make large areas unfit for agriculture or pasturage, deterioration in soil and groundwater quality through salinization, and destruction of natural vegetation and its replacement by plants unfit for grazing.

The effects of relatively short-term climatic variations in the semiarid zone are worsened by the seemingly inevitable behavior patterns of farmers and pastoralists, who, during relatively wet periods, expand their cultivated areas and drive their livestock into marginal lands beyond the region's carrying capacity. During dry periods, these marginal lands, with their natural protective cover destroyed by cultivation and grazing, are eroded at rates exceeding by orders of magnitude those that had prevailed before.

From this discussion of the human consequences of a marked climatic change that might be brought about by the addition of large quantities of carbon dioxide to the atmosphere, it may be concluded that world society could probably adjust itself, given sufficient time and a sufficient degree of international cooperation. But over shorter times, the effects might be adverse, perhaps even catastrophic.

COUNTERVAILING MEASURES

Two kinds of countervailing measure against the possible climatic effects of added carbon dioxide can be visualized: measures to reduce possible climatic changes themselves and measures to reduce the impact on human affairs. In the first category, it is possible to conceive of ways of reversing the changes in the radiation balance of the earth that might result from added carbon dioxide or ways of removing the added carbon dioxide from the air. In the second category, we are concerned primarily with ways to increase the robustness and flexibility of the world's food-supply systems. We shall consider these first because they are less problematic and largely within the range of present technology.

Overview and Recommendations

Improving the Food Supply System

From an agricultural point of view, arid and semiarid lands are defined as regions where water is not available in adequate quantities for crop production. Irrigation is the classical and still the most practical remedy. In this process, water is transported from mountainous and hilly regions or humid lowland areas, where the amount exceeds that which can be used for agriculture, to the arid or semiarid regions. Because the water supply is usually widely variable from season to season and from year to year, the water is commonly stored during wet periods, in surface reservoirs or underground, for use during the dry times when it is needed.

The stabilization of water supply provided by storage of irrigation water, and the increased quantities made available where they are needed by transporting the water, are the essential basis of modern, high-yielding agriculture, particularly in semiarid zones and in subtropical latitudes. To be of permanent benefit, irrigation development must nearly always be accompanied by parallel development of drainage facilities. Combined development of irrigation and drainage requires large capital investments of the order of $500 to $1000 per hectare (1 hectare = 0.01 km^2). For example, the cost of full development of irrigation of the Ganges Plain in India, which contains over 50 million hectares of irrigable land, has been estimated to be close to $50 billion. Such development could lead to an annual increase in crop production of several hundred million tons of food grain, worth $20 billion to $40 billion (Revelle, 1976).

It will be of particular importance to counter effects of possible increases in short-term climatic variation. For this purpose, large reservoirs are needed, either on the surface behind dams or in underground aquifers. Wherever feasible, underground storage is preferable, because it is cheaper per unit volume of stored water; water losses by evaporation are small; and the volume stored can be large enough to provide stable supplies even in the face of the persistent drought periods with time constants of a decade. But in some situations, no practical reservoir could provide long-term protection against climate change.

Equally careful research, planning, and investment are needed to develop methods for conserving water. With present methods of farm-water management in less-developed countries, only about a third of irrigation water supplies are beneficially used. Large savings in water could be obtained by improved water management and in many cases by adoption of new irrigation practices. Even more useful in most areas will be the introduction of water-saving crops; for example, crops grown during the season of minimum evapotranspiration; crop varieties with the shortest possible growing season; and row crops such as peanuts and sugarbeets where evapotranspiration is reduced by the presence of clean, bare ground between the planted rows. The best way to save water, however, is through the use of high-yielding crop varieties. Hardly any more water is required to irrigate a variety of wheat or corn yielding three or four tons per acre than is required for a variety yielding less than one ton per acre.

The impact on world and regional food supplies of short-term variations in climate can be greatly lessened by maintaining food reserves. On a worldwide basis, under the present climatic regime, such reserves should represent about 5 percent of average annual production; that is, the excess or deficit over periods of several years of food grain production with respect to demand is about 5 percent of the average annual production. The primary purpose of food reserves of this magnitude would be to stabilize prices of food staples for both farmers and consumers. Because food is an essential requirement for human life, the demand for food is extremely inelastic in relation to price; food supplies cannot be rapidly increased in response to rising prices. Experience shows that food prices may rise or fall by several hundred percent with a deficit or surplus of only a few percent in supply in relation to demand.

Counteracting a Changed Radiation Balance

One way to counteract the climatic effects of added carbon dioxide in the air would be to increase the albedo, or reflectivity, of the earth and thereby to reduce the incoming solar radiation. No practical, plausible, and reliable means to accomplish such an increase seem to be at hand.

One means to accomplish this that has been discussed might be to spread small reflecting particles over large areas of the ocean surface. To reduce the cost and increase the effectiveness of this measure, the particles should have a density very close to that of seawater and should be chemically stable over periods of many months. One material that has been suggested is very thin platelets of latex. If these platelets had a thickness of 0.01 mm, about 10 tons would be required per square kilometer or 50 million tons per year to cover 5 million square kilometers—1 percent of the earth's surface. At $100 per ton, the cost would be $5 billion per year—about 0.2 percent of the world's annual expenditure for fossil-fuel energy anticipated during the next century (PSAC, 1965).

The disadvantages of such a scheme are obvious and may be insuperable. The material might eventually pile up on the coastlines of the world, with unacceptable environmental consequences, and the effect on fisheries might be disastrous.

A completely different kind of countervailing measure might be to store the added carbon dioxide in the biosphere. The present biospheric organic carbon is about four times the carbon in the atmosphere. Perhaps a fourth of this mass is in the roots, trunks, branches, and leaves of living trees. Most of the remainder is in the soil humus or in dead organic matter in lakes, marshes, and wetlands.

Forests now cover about 50 million square kilometers, a third of the land surface of the earth. Doubling this area or doubling the mass of living trees in the existing forests would permit storage of 700 Gt (1 gigaton = 10^9 metric tons) of carbon—about a seventh of the carbon in fossil fuels but between a third and a sixth of the carbon that might otherwise be added to the atmosphere by fossil-fuel combustion. Such an increase in forest mass would have a significant mitigating effect, but it would be extremely difficult to accomplish, even on the hundred-year time scale we are considering. Major changes in land use on a worldwide basis, and consequently in world political and social organization, would be required. With the continuing growth in human populations and economic production, and the resulting needs for expanded food, fuel, and timber production, present trends are in exactly the opposite direction, since forests are being cut down for fuel and lumber and land is being cleared for agriculture.

Without human intervention, increased photosynthesis due to rising atmospheric carbon dioxide will probably cause forests and soil detritus to grow, perhaps by over 1000 Gt in the next 200 years (see Chapter 10).

It has been suggested that if wood grown in fertilized and irrigated forest plantations were cut and preserved against decay, significant amounts of carbon dioxide could be removed from the atmosphere, counteracting to some extent the increase of atmospheric carbon dioxide from fossil-fuel combustion. But it seems obvious that if large quantities of organic material were to be grown and collected, it would make more sense to use the material as an alternative source of energy to fossil fuels. If this were done, carbon dioxide would simply be recycled between the atmosphere and the biosphere, and the net addition of carbon dioxide to the atmosphere from fossil-fuel combustion would be reduced by the amount of biomass energy substituted for coal, oil, and natural gas.

Overview and Recommendations

It appears that any attempt to reduce the climatic impact of additional carbon dioxide in the atmosphere would be formidably difficult, especially since the effort would need to be continued over the next millennium and might even have unacceptable consequences. On the other hand, mitigation of the climatic effects on human affairs might be possible and even desirable from points of view other than that of climatic change; but it would require planning, research, and investment of international scope on an unprecedented scale.

It may well be the case that increasing reliance on renewable resources, with the concomitant reduction in the carbon dioxide burden in the atmosphere, will emerge as a more practical alternative to these countervailing measures.

NATURE AND LIMITATIONS OF MODELS

THE CARBON CYCLE

During the last 110 years, the carbon dioxide content of the air (expressed as weight of carbon) has increased by 72 to 83 Gt, or 11.5 to 13.5 percent. During the same period, 127 Gt of carbon in fossil fuels and in limestone have been converted to carbon dioxide and released to the atmosphere (2 percent of this amount came from cement manufacturing and the remaining 98 percent from combustion of fossil fuels). Volcanoes may have added around 4 Gt, less than 3 percent of the "anthropogenic" carbon. Rock weathering has probably subtracted an amount equal to the volcanic flux. The clearing of forests, woodland, savannas, and grasslands for farming and other human modification of land biota and soils throughout the world may have brought about a further net release of roughly 70 Gt of carbon, as carbon dioxide, into the atmosphere.

The excess of carbon dioxide released over that remaining in the atmosphere has undoubtedly been absorbed in the subsurface water layers of the ocean and in the pool of organic matter on the land (about 70 percent of this organic pool, around 2000 Gt, consists of dead organic material—mostly soil humus—and roughly 30 percent is in the trunks, stems, roots, branches, and leaves of living plants). Our calculations indicate that about 40 percent of the carbon dioxide released to the air has been absorbed in the land organic pool, about 20 percent has been absorbed by the oceans, and about 40 percent remains in the air. General discussion and review of these questions are contained in Woodwell and Pecan (1973), Wilson and Matthews (1970), Matthews *et al.* (1971), Dryssen and Jagner (1972), Proceedings of the Dahlem Workshop (1977), Proceedings of the ERDA Workshop (1977), Proceedings of the SCOPE Workshop (1977), and Baes *et al.* (1976).

If fossil fuels remain the principal energy source for world society during the next hundred years, our estimates indicate that around 2500 Gt of carbon, as carbon dioxide, could be released to the air by 2090, about 20 times as much as the amount produced to date by fossil-fuel combustion and over 4 times the preindustrial content of the atmosphere. More than half of this amount would probably remain in the air. At first thought, this statement appears paradoxical, because the ocean contains nearly 60 times as much carbon as the air, and the terrestrial organic pool nearly 4 times as much, and it might be supposed that the division of added carbon dioxide between the atmosphere, the oceans, and the land organic pool would be in proportion to the amounts already present. But the amount of carbon dioxide that the ocean is able to take up is limited by the small amounts of carbonate ion in seawater and by the low solubility of free carbon dioxide. Similarly, uptake of carbon in the biosphere is limited by the balance between photosynthetic production and oxidation of organic matter.

Because of the highly stratified nature of the oceans, the exchange between surface waters and deep waters is very slow. Hence, as long as the rate of addition of fossil-fuel carbon dioxide continues to increase exponentially, only a small fraction of the total volume of the oceans can serve as an absorber of an important fraction of the added carbon dioxide. Keeling and Bacastow (see Chapter 4) have calculated that this slow turnover of the oceans, combined with the relatively low carbonate-ion concentration of surface seawater, could result in about 80 percent of the carbon dioxide added during the next century being retained in the air. If so, the concentration of carbon dioxide in the atmosphere in the twenty-second century might be 7 times the preindustrial value. As they point out and as we discuss below, considerable uncertainty attends this figure.

DEEP CIRCULATION OF OCEANS

The real nature of the intermediate and deep circulation of the ocean (see Chapters 4 and 7) and the location and size of the regions where the deep waters come in contact with the surface are poorly understood at present. The variations from year to year in the proportion of fossil-fuel carbon dioxide removed from the air, shown in measurements made at both Mauna Loa and the South Pole, suggest that the interchanges between the intermediate water and the surface occur in irregular pulses. No very reliable estimate can be made of how the rates of interchange might be modified by changes in air and surface ocean temperatures or in circulation patterns accompanying possible climatic changes.

Because of the relatively large carbon dioxide content of the oceans and the large variability in the partial pressure of carbon dioxide in the waters near the ocean surface, the fraction of the added carbon that has been absorbed in the sea cannot be measured directly. But it can be estimated by comparison with the distribution in the upper ocean layers of tritium and other trace substances produced in nuclear-weapons tests during the two decades after World War II, principally in 1962–1963. Measurements of tritium in the upper 500 to 700 m of the ocean were begun in 1964–1965 by Östlund and co-workers in the Atlantic (Östlund *et al.*, 1969) and by Suess and co-workers in the Pacific (Michel and Suess, 1975). These have been extended by Östlund on nearly 1000 samples collected on the GEOSECS expeditions (Östlund *et al.*, 1974; Östlund, Institute of Marine Science, Miami, personal communication, 1977).

These measurements show that tritium has been mixed down to an average depth of about 360 m over the last 12 years. Apparently the tritium is carried downward partly by advection along surfaces of equal density, partly by sinking of cold surface waters in high latitudes of the Atlantic, and partly by vertical mixing. The observed downward diffusion of tritium can be used to estimate the time constant for diffusion of material added to the well-mixed surface layer of the sea into the main mass of the ocean (see Chapter 10). Since the concentration of tritium in the 100-m-thick mixed surface layer has diminished by a factor of $1/e$ in about 12 years, the time constant for diffusion into the deeper waters, with a volume 40 times larger than that of the mixed layer, should be at least 500 years.

Because of the buffer mechanism of seawater, a 10 percent increase in atmospheric carbon dioxide corresponds at equilibrium to a 1.1 percent increase in the carbon dioxide content of surface ocean waters (see Chapter 4). The mixed surface layer contains approximately two moles of carbon dioxide per cubic meter, or 864 Gt of carbon in a 100-m-thick layer. Hence, for a 13 percent increase in atmospheric carbon dioxide from 1860 to 1973, the carbon dioxide content of the mixed layer should have increased by 12.5 Gt. If the carbon dioxide has diffused downward to an average depth of 360 m, the total added to the ocean would be 45 Gt of carbon—

about 22 percent of the carbon added to the atmosphere by fuel combustion and land clearing.

Carbon dioxide from fossil fuels contains no ^{14}C and hence the proportion of ^{14}C in the atmosphere prior to the nuclear tests of the post-World War II period should have been less than that of 100 years ago. The magnitude of this phenomenon—the Suess effect—is about 2 ± 0.5 percent for A.D. 1950, indicating that the atmospheric carbon dioxide has mixed with biosphere and ocean reservoirs, which are some 5 times larger than the atmospheric reservoir. If the part of the biosphere that exchanges with the atmosphere during 29 years (the average age of carbon added from fossil-fuel combustion) is 1.8 times the atmospheric reservoir, the exchanging part of the ocean must be 3.2 times the size of the atmospheric reservoir, corresponding to a water layer about 230 m thick. However, consideration of the exchange rates of carbon dioxide between the oceans and the atmosphere suggests that the mean depth of complete isotopic mixing calculated from the Suess effect may be less than the mean depth of diffusion of fossil-fuel carbon dioxide. According to Broecker (1977), the chemical equilibration time between the ocean and the atmosphere is smaller than the isotopic equilibration time. Thus the apparent isotopic equilibration depth of 230 m calculated from the Suess effect implies an equivalent chemical depth of perhaps 400 m. This would allow absorption of about 25 percent of the carbon dioxide added to the atmosphere from land clearing and fossil-fuel combustion.

From either method of estimating the absorption of carbon dioxide in the oceans we are left with the conclusion that a large fraction must have been taken up by the biosphere. Bacastow and Keeling (1973) also reached this conclusion using a time-dependent model. This requires that the rate of net primary production by photosynthesis should have increased relative to the rate of oxidation by animals, microorganisms, and fires.

PROCESSES IN THE BIOSPHERE

The portion of the terrestrial biosphere that exchanges carbon dioxide with the atmosphere consists of two components: the biomass of living plants and animals, mostly the trunks, branches, roots, and leaves of trees, and litter and soil organic matter (humus). Whittaker and Likens (1975) have compiled and evaluated data on the biomass (see Table 10.1). They conclude that 90 percent of the total of around 830 Gt is in the world's forests, which cover nearly 50 million km². Tropical forests, with an area of 24.5 km², contain more than half of the total. Woodland and shrubland, savannas, grasslands, desert and semidesert scrub, swamps and marshes, and cultivated land together contain only 84 Gt or 10 percent of the total biomass, although they cover nearly 75 million km². Net primary production of organic matter (photosynthesis minus plant respiration) is more evenly divided: forests produce 33 Gt of carbon per year, and all other vegetation produces nearly 20 Gt. These estimates correspond to an average efficiency of photosynthetic conversion of solar energy on the earth's land surface of about 0.1 percent. In a steady state, net primary production must be balanced by the oxidative activities of animals and microorganisms (heterotrophic respiration) and fires. Thus about 8 percent of the carbon dioxide content of the atmosphere is turned over each year by terrestrial biological activities.

Bohn (1976) has recently estimated from the World Soil Map of FAO-UNESCO (1971) and other sources that the content of organic carbon in the world's soils in somewhat less than 3000 Gt, about three times the previously accepted value (Baes *et al.*, 1976). Of this total, about 860 Gt is in peaty materials (dystric and gelic histosols in the FAO nomenclature), covering 4.3 million km², which presumably exchange carbon very slowly, if at all, with the atmosphere. The remaining approximately 2000 Gt of carbon in soil organic matter can be assumed to lose carbon to the atmosphere by oxidation at about the same rate as carbon is added by the accumulation of dead plant material.

The world's soils contain on the average 18 kg/m^2 of exchangeable soil organic matter, or 180 tons per hectare. There is a wide variation in different soil types from 40 to 600 tons per hectare.

The average residence time of carbon in the terrestrial biosphere is simply the total mass of organic matter, about 2800 Gt, divided by 53 Gt per year (the rate of heterotrophic respiration plus fires), or 53 years. Variations in global temperature and precipitation should bring about short-term variations in the rates of photosynthesis and oxidation of organic matter in the terrestrial biosphere, but over periods of ten years or more these changes should be small and tend to balance out. They are likely to be considerably less than the changes brought about by the secular increase in atmospheric carbon dioxide or by human activities such as clearing of forests, reforestation, and destruction of soil humus.

The quantity of organic carbon in the oceanic biosphere is estimated to be about half as large as that of the terrestrial biosphere, 1650 ± 500 Gt, but with the important difference that almost all of it consists of dead organic matter in small suspended particles and dissolved substances distributed fairly evenly throughout the ocean waters. Carbon in the living marine biomass, almost all in the top 100 m of the ocean, is approximately 0.1 percent of the total marine organic carbon (Whittaker and Likens, 1975). The rate of primary production (photosynthesis minus plant respiration) in the ocean is estimated at 20-25 Gt per year (Whittaker and Likens, 1975), probably about half that of the terrestrial biosphere. This rate is believed to be determined by the rate at which the essential plant nutrients, phosphorus, and nitrogen compounds are brought to the sunlit surface waters by vertical motions within the oceans.

Broecker (1974) estimates that about 1.7 Gt per year, 8 percent of the organic matter produced by photosynthesis in the subsurface ocean waters, sinks into the depths, where most of it is oxidized to carbon dioxide and water. This process is balanced by a slow upwelling of deep waters with higher carbon dioxide content into the surface layers. The deep ocean waters contain about 10 percent more carbon dioxide than they would if they were at equilibrium with the partial pressure of carbon dioxide in the present atmosphere. The combination of biological and gravitational processes can be thought of as a pump that maintains a high carbon dioxide content in the deep water and a low content in the surface waters and in the atmosphere. If the pump ceased to act, the atmospheric carbon dioxide would eventually increase severalfold. Variations in the effectiveness of the pump could have occurred without detection during the past 100 years and could have caused significant changes in the atmospheric carbon dioxide content.

Several effects of human activities during the past 100 years may have changed the quantity of organic material in the terrestrial biosphere and, correspondingly, the content of carbon dioxide in the atmosphere. Perhaps the most important of these has been the clearing of lands for agriculture, which is taking place with the rapid increases in world population and food demand. Some workers have estimated that about 2 million hectares of forest lands are being cleared each year (or destroyed by continued shortening of the cycle of slash and burn agriculture) in Africa (Persson, 1976); about 6 million hectares in Latin America (FAO, 1976); and perhaps 600 thousand hectares in Southeast Asia (Swedish Royal College of Forestry, 1974). The total area of land cleared for agriculture between 1950 and 1970 has been estimated by Revelle and Munk (see Chapter 10) from data published by the U.S. Department of Agriculture and the Food and Agriculture Organization of the United

Nations, which show annual increases in cultivated areas in 35 countries. During this period, the areas sown to crops in the Soviet Union may have risen by 88 million hectares, and in Africa south of the Sahara, Latin America, and Asia (outside the Soviet Union and China) by 74, 50, and 84 million hectares, respectively. Thus, the average annual increase in crop area for these regions combined was 15 million hectares. Assuming that somewhat less than 30 percent of this area was originally forest, and that approximately 30 tons per hectare of soil humus is oxidized after cultivation has commenced, the total loss of carbon from the biosphere to the atmosphere over 20 years was close to 26 Gt, an average of 1.3 Gt annually.

The loss of carbon dioxide from agricultural clearing during the past 100 years can likewise be estimated from the probable worldwide increase in cultivated land. From 1860 to 1970, the world population rose from somewhat more than 1200 to over 3400 million people, a nearly threefold increase. Agricultural yields per hectare increased much more slowly than food demand arising from the increase in population and rise in income. The increased food demand must have been met mainly by enlarging the area of cultivated land, perhaps from 500 million hectares in 1860 to 1370 million hectares in 1970. Assuming that about 30 percent of the new farm land was originally forest and that 30 tons of carbon dioxide per hectare in the soil humus were oxidized, addition of carbon dioxide to the atmosphere could have been 72 Gt in 110 years. This is less than 3 percent of the mass of organic carbon in the biosphere but nearly two thirds the addition to the atmosphere from fossil-fuel combustion. The total mass of carbon dioxide released by fossil-fuel combustion and land clearing combined could have been over 200 Gt, a third of the initial atmospheric content.

Populations in the less-developed countries have increased roughly exponentially, and agricultural clearing may well have followed the same course (see Chapter 10), with an annual growth rate of about 3 percent from 1860 to the present. The total potential arable land on earth is limited, however, to 2600 million hectares (Revelle, 1976) compared with a presently cultivated area of about 1400 million hectares. Even assuming that all presently uncultivated but potentially arable land is now in forests, the maximum future addition of carbon dioxide from land clearing is only about 240 Gt, less than 5 percent of the total carbon reserve in fossil fuels. Land clearing in the future is likely to follow a logistic curve, similar to that used by Keeling and Bacastow (see Chapter 4) for fossil-fuel carbon dioxide production but with a maximum rate of clearing around 2010, compared with the maximum carbon dioxide production from fossil fuels near the end of the twenty-first century.

Reforestation, the planting of trees in previously cleared areas, will increase the size of the terrestrial biosphere at the expense of atmospheric carbon dioxide. According to Bolin (1977) reforestation has occurred over 30 million to 60 million hectares in China and about 10 million hectares in other less-developed countries, while in the developed countries it is not certain whether there has been an increase or a decrease in the forested area by as much as 12 million hectares. The increase in net primary production, less oxidation and fires, resulting from reforestation may be as much as 2.5 tons per hectare per year; hence, the annual subtraction of carbon from the atmosphere by reforestation may lie between 0.07 and 0.2 Gt.

Bolin (1977) suggests that wood cut for lumber, paper pulp, and especially firewood may exceed net primary production in existing forests. Wood is "the poor man's oil," and in Africa, Latin America, and parts of Asia the average per capita use of firewood (primarily for cooking) is about 0.7 ton per year (see Chapter 10), corresponding to 350 kg of carbon per capita, or 0.875 Gt for the 2500 million people of the poor countries. This is more than the FAO estimate of the harvest of roundwood from the world's forests. But, even accepting this figure, the total wood harvest is only 5 percent of the estimated net primary production in tropical and subtropical forests (see Table 10.1). Thus, it seems unlikely that wood cutting has resulted in a net addition of carbon to the atmosphere.

As G. C. Delwiche of the University of California at Davis (personal communication) has pointed out, several other human activities may result in adding or subtracting carbon from the biosphere or the inorganic components of the soil. Cultivation of calcareous soils in arid and semiarid regions, and expansion of areas of irrigated land, may result in an uptake of carbon by the soil as calcium carbonate or organic matter. Drainage and cultivation of swamps and marshes will liberate organic carbon to the atmosphere as will some improved forest practices. At present, we are unable to estimate the magnitude of these various effects.

M. Stuiver of the University of Washington (personal communication) has estimated that the total amount of carbon dioxide released from the terrestrial biosphere by clearing of forests and other human activities corresponded to 120 Gt of carbon between 1850 and 1950 and possibly to an additional 25 Gt between 1950 and 1970, a total of 145 Gt. This is approximately twice our estimated value. He bases his estimates on an observed decrease with time in the ratio of the two stable carbon isotopes, $^{13}C/^{12}C$, in the atmosphere over the past 120 years, as measured in tree rings. Biosphere carbon, like fossil-fuel carbon, is deficient in ^{13}C, and hence a transfer of carbon from the biosphere and the fossil-fuel reservoir to the atmosphere should result in a lowering of the atmospheric ratio of ^{13}C to ^{12}C. Tree-ring measurements by Stuiver and others suggest a lowering of this ratio by 1.5 per thousand since 1850. This is a small effect and could be largely accounted for by changes in the heights of the sample trees relative to the average height of the leaf canopy in the surrounding forest. (Trees at different relative heights in the forest are bathed in carbon dioxide with a different ^{13}C-^{12}C ratio.) The measurements do not show an exponential decease in $^{13}C/^{12}C$ with time, which should be expected from the accelerating use of fossil fuels and the probable acceleration of land clearing in less-developed countries. Moreover, the changes in the ^{13}C-^{12}C ratio would not occur unless the carbon released from the biosphere had remained in the atmosphere. This requires the carbon dioxide content of the atmosphere in the middle of the nineteenth century to have been only about 268 parts per million, much lower than the most probable value. It is more likely that a large fraction of the released carbon has been reabsorbed in the biosphere.

Many more measurements of possible changes with time in the ^{13}C-^{12}C ratio, using tree-ring sequences from trees in isolated locations, are needed to resolve this question. The ^{13}C-^{12}C ratio should also be measured directly on air samples collected over a wide range of geographic locations at the present time as a baseline for the detection of future changes.

Besides clearing of land for agriculture, the two principal effects of human activities on the biosphere should be an increase in net primary production over oxidation, resulting from the fertilization of the biosphere by carbon dioxide added to the air, and a similar increase resulting from the excess of nitrogen fixation over denitrification.

According to G. C. Delwiche and G. K. Likens of the University of California at Davis (personal communication), the present annual fixation of atmospheric nitrogen from human activities amounts to 68 Mt (40 Mt of nitrogen fertilizer, 10 Mt of other chemical products, and 18 Mt in combustion of fossil fuels). This is 35 percent of world total nitrogen fixation. The remaining 65 percent (127 Mt) is fixed by terrestrial and marine bacteria—including blue-green algae—and in lightning discharges. Denitrification is estimated roughly at 160 Mt annually. Thus, the excess of nitrogen fixation over denitrification is perhaps 30 Mt per year. If this increment of fixed nitrogen were stored in the terrestrial biomass, principally in the wood of forest trees that have a high carbon-to-nitrogen ratio, the carbon added annually to the biosphere would be 2300 Mt. If it were stored in soil humus, the additional carbon would be only 300 Mt. It seems unlikely that the excess fixed nitrogen has had much effect on the marine biosphere, which, as we have indicated, is limited by the flux of phosphorus as well as nitrogen to ocean surface waters.

Any uptake of carbon dioxide by the biosphere requires that the rate of net primary production by photosynthesis should be greater than the rate of oxidation by microorganisms, animals, and fires. It has long been known that photosynthetic production increases when crop plants are bathed in a carbon dioxide-rich atmosphere, provided other factors of production are not limiting. In the natural world, however, plant production is limited by the availability of nitrogen, phosphorus, and minor nutrients; the intensity of solar radiation; the adequacy of the supply of water; and environmental stresses, such as low and high temperatures; in addition to the concentration of atmospheric carbon dioxide. Experimental data indicate that in the absence of other limiting factors, net primary production increases with the logarithm of the increase in carbon dioxide (see Chapter 10 and Bacastow and Keeling, 1973). In the heterogeneous conditions of the natural world, we may introduce a factor β, which takes account of the presence of other limiting factors. This leads to simple mathematical relationships among net primary production, atmospheric carbon dioxide, photosynthesizing biomass, rate of oxidation, and the total weight of carbon in the biosphere (see Chapter 10).

Except for the changes associated with agricultural land clearing, the portion of the terrestrial biomass that carries out photosynthesis may not vary significantly in the future, if, as seems likely, plants now optimally occupy the land surfaces available to them. The size of the terrestrial biosphere is then uniquely determined by the atmospheric carbon dioxide content. Computations by Revelle and Munk (see Chapter 10) based on this assumption indicate that the mass of the biosphere would ultimately increase by 700 Gt or 25 percent, if all fossil fuels are burned over the next one or two centuries, and the mass of carbon dioxide in the ocean and the atmosphere would be correspondingly reduced. The peak concentration of carbon dioxide in the atmosphere near the end of the twenty-second century would be 2900 Gt or about four times the present value. Somewhat similar outcomes would result from any other model in which the size of the photosynthesizing biomass does not vary linearly with the size of the biosphere. Present empirical data on past changes in the terrestrial biosphere and the ecological relationships involved do not provide evidence sufficient to determine a choice among possible models of the terrestrial biosphere. Consequently, a considerable range of uncertainty must be associated with predictions of atmospheric carbon dioxide levels.

We conclude that a principal uncertainty in estimating the likely increase in atmospheric carbon dioxide from future fossil-fuel combustion lies in our present inability to estimate the uptake of carbon by the terrestrial biosphere. Strenuous research efforts and extensive surveys should be undertaken to reduce this uncertainty.

GEOLOGIC SOURCES AND SINKS

Beside absorption of carbon in the oceans and terrestrial biosphere, five geological processes can appreciably affect the carbon dioxide content of the atmosphere over periods of centuries: (1) the flux of carbon dioxide from the earth's interior through volcanos and in other ways, (2) weathering of silicate rocks on land, (3) weathering of carbonate rocks on land and the subsequent redeposition of carbonates on the

Overview and Recommendations

sea floor, (4) burial of organic material in marine and lacustrine sediments, and (5) solution of calcium carbonate in the sediments of the sea floor.

Bowen (1966) estimates that the flux of carbon dioxide from the earth's interior adds about 40 Mt of carbon to the atmosphere each year. This is less than 1 percent of the addition from fossil fuels. It is approximately balanced by the removal of carbon dioxide in the weathering of silicate rocks, which is estimated by Garrels *et al.* (1975) from the average content of dissolved silica in river waters to be 42 Mt of carbon per year.

Weathering of limestones and dolomites removes carbon dioxide from the air in the ratio of one mole per mole of calcium or magnesium. The amount of bicarbonate in river waters resulting from limestone and dolomite weathering is about twice the amount of carbon dioxide removed from the air; half of this quantity is contributed by the rocks themselves.

The total quantity of bicarbonate ions carried to the sea by rivers each year is estimated by Garrels *et al.* (1975) as equivalent to 368 Mt of carbon. Of this amount, 205 Mt is removed from the atmosphere—163 Mt from the weathering of carbonates and 42 Mt from the weathering of silicates. The annual loss of atmospheric CO_2 to river waters is thus about 3 percent of the total added annually at present to the atmosphere by combustion of fossil fuels and land clearing. In a steady state, the carbon from the weathering of limestone and dolomites is ultimately restored to the atmosphere when calcium carbonate is precipitated in the upper layers of the sea and deposited in ocean sediments, releasing an amount of carbon dioxide equal to the amount of carbonate deposited. Broecker (1974) estimates the rate of deposition of carbon as calcium carbonate as 69 to 216 Mt per year, in fair agreement with the calculated quantity removed from the atmosphere by weathering of limestones and dolomites.

Future increases in the amount of carbon dioxide in the atmosphere should increase the rate of weathering of both silicate and carbonate rocks, while calcium carbonate precipitation in the sea should be reduced because of the increased acidity resulting from addition of carbon dioxide to the seawater directly from the air. But as the following considerations indicate, the effects are likely to be small.

For silicate rocks, the rate of weathering could conceivably grow as rapidly as the rate of increase in atmospheric carbon dioxide. A quadrupling of atmospheric CO_2 would then increase the rate of extraction of carbon dioxide from the atmosphere to 170 Mt of carbon per year. This is less than 0.01 percent of the postulated increase in atmospheric CO_2.

With respect to carbonate rock weathering it might be supposed that if present river waters are saturated with calcium carbonate, a quadrupling of atmospheric carbon dioxide would result in a 60 percent increase in the rate of extraction of carbon dioxide from the air, since at equilibrium

$$\frac{(Ca^{2+})(HCO_3^-)^2}{pCO_2 \text{ atm}} = \text{constant}.$$

To a fair approximation,

$$2\, Ca^{2+} = HCO_3^-,$$

and therefore

$$\frac{\Delta HCO_3^-}{HCO_3^-} = \left(\frac{\Delta pCO_2}{pCO_2} + 1\right)^{1/3} - 1.$$

In fact, present river waters are substantially undersaturated with calcium carbonate. Their content of "free" carbon dioxide is on the average ten times higher (and their CO_3^{2-} concentration correspondingly lower) than that at equilibrium with present atmospheric carbon dioxide, while the average calcium content is somewhat lower than the saturation value for water with a CO_2 content at equilibrium with the present pressure of CO_2 in the atmosphere (Garrels *et al.*, 1975). The marked undersaturation of calcium carbonate in present rivers is demonstrated by the close correlation between the contents of calcium and bicarbonate in different rivers and the total dissolved solids. If the rivers were saturated with calcium carbonate, their calcium and bicarbonate concentrations would be relatively constant regardless of the total quantities of dissolved solids.

Garrels *et al.* (1975) examined the content of "free" carbon dioxide in 17 different rivers covering a wide range of latitudes, rock types in the drainage basins, river sizes and stages, and total dissolved solids. The CO_2 pressures were 3–30 times higher than that in the present atmosphere and were nearly independent of all the above river characteristics. The average CO_2 pressure was about the same as that of the interstitial waters of soils. Apparently the CO_2 pressure in the world's rivers depends on microbiological oxidation processes that are relatively independent of the carbon dioxide pressure in the atmosphere.

It seems reasonable to conclude that even with a quadrupling of atmospheric carbon dioxide, the annual loss of carbon in river waters would not exceed 500 Mt, less than 0.03 percent of the additional carbon in the atmosphere.

Organic carbon is deposited in marine sediments at a rate, according to Broecker (1974), about one fourth of that of carbon as calcium carbonate, and this represents a net loss of atmospheric carbon that must, on a geologic time scale, be counterbalanced by other processes. The rate of deposition of organic matter is apparently determined by the amount of phosphorus carried by rivers from the land to the sea, which may be related to the atmospheric carbon dioxide concentration through weathering processes. Photosynthetic reduction of carbon dioxide by plants and the subsequent deposition of a small fraction of the plant remains in marine sediments, leaving the photosynthetically produced oxygen in the air, is probably the chief process that, over geologic eons, has resulted in our planet having an atmosphere with free oxygen. But the present annual rate of deposition is probably not more than 20 to 50 Mt. This is negligible compared with the rate of addition of carbon dioxide to the air.

A process of greater potential significance could be the dissolution of calcium carbonate in marine sediments. The upper water layers of the sea in the tropics and subtropics appear to be markedly oversaturated with respect to both calcite and aragonite, the two crystalline forms of calcium carbonate. Carbonate-rich sediments, including coral reefs and atolls and deposits that may be formed inorganically, are widespread at shallow depths. In the deeper waters of low latitudes and midlatitudes, pelagic sediments containing a high percentage of calcareous skeletons of marine animals and plants occur above a "compensation depth," at which, because of the combined effects of pressure and temperature, the water appears to be about saturated with calcium carbonate. Above the compensation depth, the waters are supersaturated; below it, they are undersaturated, and the fraction of calcium carbonate in pelagic sediments sharply diminishes. The compensation depth is different in different oceans, because of their differing carbonate ion concentrations, being about 4500 and 1000 m for calcite and aragonite, respectively, in the Atlantic (and probably the Indian Ocean), and perhaps as shallow as 500 and 300 m for these two mineral species in the North Pacific north of $10°$ N. South of $10°$ N, the saturation depth for calcite in the Pacific is probably between 3500 and 4000 m, judging by the distribution of calcium carbonate in the pelagic sediments (Sverdrup *et al.*, 1942).

When some of the carbon dioxide produced by fossil-fuel combustion and forest clearing reaches the deep sea, the carbonate ion concentration in the deep water will diminish and consequently the compensation depth will rise. Calcium carbonate on the surface of the sediments should dissolve at depths between the old and new compensation depths. At a depth of the order of 10 cm below the sediment surface (beneath the zone of burrowing organisms), calcium carbonate will be effectively insulated from the overlying waters and will dissolve very slowly, if at all. Between 1 and 10 cm below the sediment surface, the rate of solution of sedimentary calcium carbonate is limited by the rates of exchange between the interstitial water in the sediment and the overlying seawater. Experimental data summarized by Guinasso and Schink (1975) indicate that this rate is several orders of magnitude lower than the rate of molecular diffusion in water. Calculations based on the data suggest that the rate of solution of calcium carbonate, even if the superjacent waters become more acid, is likely to correspond to less than 1 Gt of carbon a year, a relatively small proportion of the anticipated increase in atmospheric carbon. The alkalinity of the seawater, and hence its ability to absorb carbon dioxide, will increase in proportion to the quantity of calcium carbonate dissolved.

If the compensation depth rose to the sea surface, the ultimate result could be, as Keeling and Bacastow show (see Chapter 4), that the carbon dioxide content of the atmosphere in the early twenty-second century would, *ceteris paribus*, be only five times the preindustrial value, instead of eight times if carbonate solution did not occur, and it would drop off to twice the preindustrial value a thousand years from now. However, this would require solution of a layer of shallow-water carbonate sediments several meters thick, which probably could not occur without massive human intervention. Observed mixing rates resulting from the activities of burrowing organisms in shallow water sediments (Guinasso and Schink, 1975; Guinasso, 1976; Sundquist *et al.*, 1977; Goreau, 1977; Berger, 1977) indicate that the diffusion of interstitial waters between these sediments and the superjacent waters is slow, although much more rapid than in deep-sea sediments, and hence the rate of solution of calcium carbonate in undisturbed sediments will result in only a small rate of increase in the capacity of the ocean to absorb carbon dioxide.

Up to the present, this effect is probably negligible. The carbon dioxide added from the air has not penetrated to sufficient depths to raise the compensation depth, except in the North Pacific, where the compensation depth may have become a few meters shallower. But the total quantity of carbon dioxide in the top meter of sediments in the uppermost thousand meters of the entire Pacific Ocean corresponds only to 230 Gt of carbon (Sverdrup *et al.*, 1942), and perhaps 90 percent of this lies south of 10° N where the compensation depth has almost certainly not been affected. An upper limit to the quantity of sedimentary calcium carbonate dissolved up to the present is probably about 1 Gt of carbon, less than one fifth of the addition from fossil fuels in the single year 1973.

We conclude that none of the geologic processes described above would significantly affect the increase in atmospheric carbon dioxide from fossil-fuel combustion that is likely to occur during the next few centuries.

CLIMATE

Any useful prediction of the effect of carbon dioxide added to the air must depend on an understanding of the physical processes that determine climate. In general, we know that both carbon dioxide and water vapor in the air absorb and reradiate infrared radiation. Hence the temperature of the lower air must be at the level where the fraction of the infrared radiation passing outward to space is equal to the total incoming radiation from the sun, less the amount reflected from clouds and from the earth's surface. Because of this greenhouse effect produced by carbon dioxide and water vapor, the average temperature at the earth's surface is about 30 °C higher than it would be in the absence of these two gases.

As mentioned above, Manabe and Wetherald (1975) have calculated the magnitudes of the effects of a doubling of the atmospheric carbon dioxide—it would result in about a 2-3 °C rise in the earth's average air temperature, about a 2 percent increase in relative humidity, and a 7 percent increase in average precipitation. The temperature rise would be greater at high latitudes, about 10 °C at latitude 80° N. Temperatures in the stratosphere would decrease. At a height of 30 km, this decrease would be about 5 °C. The increase in average air temperatures near the ground would be about 4-6° for a fourfold increase in CO_2 and 6-9° for an eightfold increase (see Chapter 9). A global average rise of 6 °C would be comparable with the extreme changes of global temperature believed to have occurred in geologic history of the last billion years.

Although the model used by Manabe and Wetherald is the most sophisticated that can be constructed at present, it does not take account of four climatic feedback mechanisms that could change the calculated results: atmosphere-ocean carbon dioxide interaction, cloud reactions, ocean temperature and heat transport changes, and aerosol reactions.

Atmospheric-Ocean Carbon Dioxide Interactions

If an increase in atmospheric carbon dioxide raises the air temperature over the sea it will also warm the waters near the surface and thereby reduce the solubility of carbon dioxide. Some carbon dioxide will be driven from the sea into the air, or conversely less carbon dioxide will be absorbed. A more important effect will be the increase in the vertical density gradients in the ocean, which may significantly change interactions between the air and deep-ocean circulation.

Cloud Reactions

An increase in the water vapor content of the air could result in an increase in low-level clouds. This might be expressed as a larger area of the earth's surface being covered by clouds and also by changes in the heights of cloud tops, which determine their temperature and hence their radiative properties. An increase in the surface area covered by clouds would result in a proportional increase in albedo—that is, in the reflectivity of the earth. A higher albedo in turn would reduce the percentage of solar radiation reaching the earth's surface and thus would cause a lowering of air temperatures. This could partly compensate for the rise in temperature that would otherwise be produced by the rise in atmospheric carbon dioxide, although Manabe and Wetherald find in some of their numerical experiments that increases in low cloudiness are offset from the standpoint of radiation balance by changes in middle and high cloudiness.

Ocean Temperature and Heat Transport Changes

The Manable-Wetherald model assumes a swamplike ocean with no heat capacity but with the ability to provide water vapor to the atmosphere. Their result would have been completely different if a constant sea-surface temperature had been assumed, implying an ocean of infinite heat capacity. This would greatly diminish the response of the ocean-atmosphere system to changes in the carbon dioxide content or even to changes in solar radiation. In fact, the ocean has a large but finite heat capacity, and it surrounds continental land masses that have very low

Overview and Recommendations

heat capacity. This fact is particularly important in considering seasonal cycles of heating and cooling, because of the long lag time and persistence of changes in ocean temperatures near the surface.

The ocean also transports heat from low latitudes to high latitudes. The quantity of heat transported poleward by the oceans is about equal to that transported by the air. Any model that attempts to predict atmospheric temperature changes, let alone the diffusive mixing in the oceans that determines the amount of carbon dioxide absorbed, must take into account possible changes of the ocean circulation. That is, it must be a coupled ocean-atmosphere model. The construction of such a model is difficult at present, because the main motions of the sea appear to be in eddies of the order of 200 km in diameter (much smaller than the main components of the atmospheric circulation), and satisfactory methods for arriving at even a statistical description of these oceanic motions have not yet been developed. Ways of expressing the effects of these eddies in terms of larger-scale oceanic parameters are needed.

Aerosol Interactions

Aerosol particles absorb and scatter sunlight, and they also absorb and reradiate infrared radiation. The backscattering of solar radiation tends to increase the net albedo, while the absorption and reradiation of sunlight in effect reduce the net albedo. Thus, "dark" particles, which have a high ratio of absorption to backscatter, tend to decrease the albedo over a light surface such as a snow field, a low cloud deck, or desert areas. Aerosols over the ocean, which has a low albedo, tend to increase the net albedo. Modeling the interaction of aerosols such as dust, sea salts, or sulfates cannot be done very well at present (see Chapter 3) because their introduction into the atmosphere is either not coupled to the dynamics of climate (fires, volcanoes) or is coupled in a very complex way (wind-raised dust and sea spray).

OUTLOOK

We now summarize what will be needed to improve our understanding of the phenomena involved in the carbon dioxide problem and indicate some of the elements that should be included in planning to close gaps in our knowledge, so that future decisions regarding the exploitation of energy resources can be made on as sound a basis as possible.

CARBON CYCLE

Further work is needed to understand the carbon cycle that takes into account buffering by the biosphere, rock weathering, and calcium carbonate precipitation and solution in lakes and the oceans. A program of measurement of tracers that indicate the mixing rates of the upper oceans (e.g., tritium), and also of oceanic carbon dioxide content both near the surface and at a series of depths below the surface, should be undertaken. Present efforts to develop better methods of studying the ocean circulation in three dimensions should be broadened and intensified. Research on the biological and chemical processes in ocean banks and reefs, and in the calcareous oozes of the deep seafloor, should be encouraged. Baseline measurements of soil humus and of the other living and dead components of the terrestrial organic pool should be made on a worldwide basis, and these measurements should be repeated at suitable intervals to allow direct estimates of the net uptake or net release of carbon from this reservoir.

Overview and Recommendations

CLIMATE

Clearly, a fundamental understanding of the mechanisms and dynamics of climate is needed. The question of how to meet that need has been addressed in a number of recent reports (U.S. Domestic Council, 1974; Matthews *et al.*, 1971; GPS No. 16, 1975; U.S. Committee for the Global Atmospheric Research Program, 1975). Broadly speaking, these reports agree that modeling is the most promising approach to understanding climate. They document the considerable gains achieved through modeling but point out that much remains to be done. Further progress will require new observations, both to point the way to new advances and to test models, and technological innovation in observational and data-gathering systems, more powerful computers, and perhaps new institutions through which to plan, organize, and execute research programs and to provide access to the results of these programs to the widest possible audience (see Chapter 8).

Problems that will have to be considered in improving our understanding of energy-related aspects of climate include the following.

Ocean-Atmosphere Interaction

The oceans play an important part in storing and transporting heat involved in climate dynamics. Coupled ocean-atmosphere models, including mechanisms of sea-ice formation and its effect on radiation balance and of the exchange of heat between the ocean and the atmosphere, are needed (for further discussion see Ocean Science Committee, 1975).

Radiation Balance

The physical and dynamical mechanisms of cloud formation are not well understood. Cloud formation in turn affects the radiation balance. This is one of the most challenging problems in climatic modeling.

Fluorocarbons and oxides of nitrogen from fertilizers can, like carbon dioxide, cause a greenhouse effect. Global warming due to the use of chlorofluoromethanes and the increase in atmospheric content of nitrous oxide has been predicted (Committee on Impacts of Stratospheric Change, 1976; Chapter 4). Thus, carbon dioxide and other atmospheric gases may act together to produce a global warming trend.

Monitoring

The principal requirement for improvement of general circulation climatic models is an integrated program of measurements of the distribution in space and time of key physical properties and processes in the ocean and the atmosphere—key properties indicated as essential by the models themselves. These include many kinds of observation that can be made from satellites, such as those of total solar flux, net radiation budget, cloud cover, albedo, extent of snow, sea ice and polar ice sheets, tropospheric and stratospheric aerosols, and atmospheric turbidity. Other properties and processes must be measured *in situ*, including river runoff, soil moisture (to some extent measurable from space), subsurface ocean temperatures, precipitation, sea level, carbon dioxide, and ozone. Many of the necessary measurements are included in the Global Atmospheric Research Program under the joint aegis of the World Meteorological Organization and the International Council of Scientific Unions. Almost as important as the actual measurements are methods of data compilation and processing that will compress the enormous amounts of data that will be obtained into usable information that can be handled in practical models of the ocean-atmosphere system (see Chapter 8).

Overview and Recommendations

The benefits to be gained from establishing a global monitoring system are not only scientific but also practical, for such a system can provide us with early warning of climatic change. Climate cannot be predicted in the sense that weather can (see Chapter 9). Weather is determined by the short-term behavior of the atmosphere and oceans, which is, in turn, determined primarily by initial conditions. The statistical properties that define climate are determined primarily by boundary conditions and the basic parameters of the system; initial conditions are not likely to be as relevant to these properties. The potential climate changes discussed in this report are largely responses to a change in one of those basic parameters: the total quantity of carbon dioxide in the air, waters, and organic matter of the earth. An effective global monitoring system, through the observation of regional and local climatic changes as well as other changes in physical parameters as outlined above under Recommendations, may be expected to provide evidence of such responses before their consequences become unpleasantly obvious.

APPENDIX A: ENERGY CONSUMPTION IN 2075

Hans H. Landsberg and Harry Perry, Resources for the Future, Inc.

INTRODUCTION

The following calculations require a preliminary note. Anyone having engaged in projections for even a decade ahead is aware of the fragility of such estimates. Vulnerability tends to increase with time horizons as well as with widening geographic scope. Thus a 100-year projection must offhand seem foolhardy. It probably is. As the only excuse for having engaged in it the authors offer the insistent demand for it by those asked to speculate on the consequences of energy consumption far enough into the future to enable them to deal with significant cumulative effects.

Undertaking a methodical projection then has the advantage of being subject to inspection and scrutiny but does not bring one any closer to a "forecast." In this context it is desirable to proceed with the least sophistication. Thus the reliance on population and per capita energy consumption in preference to any more complex formula. Both involve arbitrariness and provide fuel for disagreement. As providing orders of magnitude, however, they seem serviceable and keep the discussion on the level of simplicity that matches their merit. It is hoped that no reader or user will be tempted to elevate them to any higher status or divorce them from their sole intended function as tools for the calculations that utilize them.

ASSUMPTIONS CONCERNING POPULATION

World population is expected to stabilize at some time in the future, but the net reproduction rate (NRR) should reach 1.0 at different times for different regions. Table A.1 indicates the five-year range when the NRR is estimated to reach 1.0 for each region of the world and the population that would exist in each region in 2075 as a consequence. The year range selected for the NRR to reach 1.0 was a "most probable" estimate for each region.

Based on these estimates, world population would be about 10.7 billion in 2075.

ASSUMPTION CONCERNING ENERGY PER CAPITA CONSUMPTION

Annual per capita energy consumption now varies widely among regions. For the world as a whole, it is about 52.5×10^6 Btu. For the United States the annual per

Overview and Recommendations

TABLE A.1 World Population in 2075[a]

Region	Year Range NRR Equals 1.0	Population in 2075 (millions)
United States	1975–1980	285
North America (excluding United States)	1975–1980	36
Western Europe	1975–1980	579
Oceania	1985–1990	31
Latin America	2010–2015	989
Japan	1995–2000	139
Other Asia	2019–2024	4901
Africa	2020–2025	1430
Soviet Union	1975–1980	332
Communist East Europe	1990–1995	139
Communist Asia	2000–2005	1835
TOTAL WORLD		10696

[a] Source: T. Frejka (1973).

capita consumption is 306×10^6 Btu, while for Africa it is only 9.2×10^6 Btu, or 3 percent of the level of the United States.

Estimates of future world per capita energy consumption were made by Weinberg and Hammond (1971). They projected that annual world per capita consumption would stabilize at about twice the current U.S. level, or about 600×10^6 Btu. In a more recent article, Puiseux (1975) estimated that an "advanced civilization" could be achieved with a world per capita energy consumption of 185×10^6 Btu to 264×10^6 Btu. He also estimated a world population of 10 billion people.

Table A.2 shows an estimate of the levels of energy consumption that might be reached in different regions by 2075. For the regions that now have large per capita energy consumptions, the values shown in Table A.2 are the maximum that is believed will be used in any future year. Values of per capita consumption below 265×10^6 Btu per year are for regions that would not yet have reached a maximum per capita consumption by 2075 since they currently have such a small base of energy use. North American values already exceed the maximum and have been allowed to increase at a moderate pace.

TABLE A.2 Per Capita Consumption and Total Energy Consumption by Regions, 2075

Region	Per Capita Consumption (Btu $\times 10^6$)	Total Consumption (Btu $\times 10^{15}$)
United States	741	211
North America (excluding United States)	623	23
Western Europe	290	168
Oceania	577	18
Latin America	130	128
Japan	497	69
Other Asia	40	194
Africa	64	92
Soviet Union	760	252
Communist East Europe	580	81
Communist Asia	58	107
TOTAL WORLD		1343

Overview and Recommendations

LEVELS OF ENERGY CONSUMPTION AND HEAT RELEASE

By combining Table A.1 with column 1 of Table A.2, it is possible to determine the heat released annually in 2075 for each of the regions shown on the tables. These data are shown in column 2 of Table A.2. The heat released per km² for each region in 2025 and 2075 can be calculated and is shown and compared with the heat that each region would receive from solar energy in those years. These data are shown in Table A.3.

Table A.4 shows the projected world consumption of fuels for 1973, 2025, and

TABLE A.3 Estimated Energy Densities by Region

	Btu × 10⁹/km²		Man-made Heat Release as Percentage of Solar Energy	
Region	2025	2075	2025	2075
United States	20.93	22.56	0.289	0.311
North America (excluding United States)	1.56	1.87	0.022	0.026
Western Europe	33.60	44.03	0.464	0.608
Oceania	1.76	2.13	0.024	0.029
Latin America	5.54	6.24	0.077	0.087
Japan	182.80	186.29	2.523	2.571
Other Asia	10.65	12.50	0.147	0.172
Africa	2.57	3.03	0.036	0.042
Soviet Union	10.89	11.26	0.150	0.155
Communist Eastern Europe	60.39	63.65	0.834	0.879
Communist Asia	6.28	9.08	0.087	0.126
TOTAL WORLD	8.63	9.89	0.119	0.136

TABLE A.4 World Heat Release[a] and Carbon Dioxide and Particulate Inputs[b]

	1973	2025	2075	2090[c]	2170[d]	Ultimate Total
1. Heat[e]						
Gt of coal equivalent per year	7.6	42	48			7500
Quadrillion (10¹⁵) Btu per year	220	1170	1340			210000
Cumulative percent of total recoverable reserves of fossil fuels	2.5	17	42			100
2. Carbon Dioxide[f]						
Annual production, Gt of carbon	4.8	16		37	13	5000
Annual increase in air (airborne portion), Gt of carbon	3.0	9		21	0	900
Airborne portion/production, percent	56[g]	53		57	0	18
Annual increase of atmospheric carbon relative to 1860, percent	0.5	1.3		3.4	0	—
Total atmospheric carbon, Gt	700	945		1950	2900	1500
3. Particulates[e]						
Emissions, Mt	140	800	2800			

[a] Including all nonrenewable energy forms.
[b] Including coal, oil, and natural gas for carbon dioxide and particulate calculations.
[c] Year of peak carbon dioxide production in Revelle-Munk model (see Chapter 10).
[d] Year of peak atmospheric carbon dioxide in Revelle-Munk model (see Chapter 10).
[e] Calculated with Perry-Landsberg model (see Chapter 1)
[f] Calculated with Revelle-Munk model (see Chapter 10).
[g] Average airborne fraction from 1959 to 1973 inclusive.

2075. Comparing these values with those in Table A.3 using a total inhabited surface area of 136 × 10^6 km^2 indicates that by 2075 all the energy needed could still be provided by fossil fuels, mostly from the production of coal.

REFERENCES

Bacastow, R. B., and C. D. Keeling (1973). Atmospheric carbon dioxide and radiocarbon in the natural carbon cycle: Changes from A.D. 1700 to 2070 as deduced from a geochemical model, in G. M. Woodwell and E. V. Pecan (1973).

Baes, C. F., Jr., H. E. Goeller, J. S. Olson, and R. M. Rotty (1976). *The Global Carbon Dioxide Problem*, ORNL-5194, August. Avail. from NTIS.

Berger, W. H. (1977). *Benthic Processes and Geochemistry of Interstitial Waters of Marine Sediments*, K. A. Fanning and F. T. Manheim, eds., Joint Oceanographic Congress, Edinburgh.

Berger, W. H., and J. V. Garner (1975). On the determination of Pleistocene temperatures from planktonic foraminifera. *J. Foraminifera Res. 5*, 102.

Berger, W. H., and J. S. Killingley (1977). Glacial-Holocene transition in deep-sea carbonates: selective dissolution and the stable isotope signal, *Science 197*, 563.

Bohn, H. L. (1976). Estimate of organic carbon in world soils, *Soil Sci. Soc. Am. J. 40*, 468.

Bolin, B. (1977). Changes of land biota and their importance for the carbon cycle, *Science 196*, 613.

Botkin, D. B., J. F. Janak, and J. R. Wallis (1973). Estimating the effects of carbon fertilization on forest composition by ecosystem simulation, in G. M. Woodwell and E. V. Pecan (1973).

Bowen, H. J. M. (1966). *Trace Elements in Biochemistry*, Academic Press, New York.

Broecker, W. S. (1974). *Chemical Oceanography*, Harcourt Brace Jovanovich, Inc., New York.

Broecker, W. S. (1977). The fate of fossil fuel CO$_2$—a research strategy (to be published). Presented at WMO Scientific Workshop on CO$_2$, Washington, D.C., 1976.

Committee on Impacts of Stratospheric Change (1976). *Halocarbons: Environmental Effects of Chlorofluoromethane Release*. National Academy of Sciences, Washington, D.C.

Dryssen, D., and D. Jagner, eds. (1972). *The Changing Chemistry of the Oceans* (Nobel Symposium 20), Wiley-Interscience, New York.

FAO (1976). Paper presented at the 12th session of the Latin American Forestry Commission, Havana.

FAO-UNESCO (1971). FAO (Food and Agr. Org.), United Nations, *Soil Map of the World, 1:5,000,000*, UNESCO, Paris.

Frejka, T. (1973). *Reference Tables to the Future of Population Growth—Alternative Paths to Equilibrium*, The Population Council.

Garrels, R. M., F. T. Mackenzie, and C. Hunt (1975). *Chemical Cycles and the Global Environment*, W. Kauffman, Los Altos, California.

Goreau, T. J. (1977). *Nature 265*, 525.

GPS No. 16 (1975). *Report of the Study Conference on the Physical Basis of Climate and Climate Modeling*. Wijk, Sweden. WMO, Geneva, Switzerland.

Guinasso, N. L., Jr., and D. R. Schink (1975). Quantitative estimates of biological mixing rates in abyssal sediments, *J. Geophys. Res. 80*, 3032.

Guinasso, N. L. (1976). *EOS 57*, 150.

Hardman, L. L., and W. A. Brun (1971). Effect of atmosphere enriched with carbon dioxide on different developmental stages from growth and yield components of soybeans, *Crop Sci. 11*, 886.

Hughes, T. J. (1974). ISCAP *Bulletin No. 3:* Study of Unstable Ross Sea Glacial Episodes, Institute for Quaternary Studies, Orono, Maine.

Hughes, T. J. (1977). West Antarctic ice streams, *Rev. Geophys. Space Phys. 15*, 1.

Kellogg, W. W. (1977). Global influences of mankind on the climate, in *Climate Change*, J. Gribbin, ed., Cambridge University Press, Cambridge, England, in press.

Manabe, S., and R. Wetherald (1975). The effects of doubling the CO$_2$ concentration on the climate of a general circulation model, *J. Atmos. Sci. 32*, 3.

Matthews, W. H., W. W. Kellogg, and G. D. Robinson, eds. (1971). *Inadvertent Climate Modification, Study of Man's Impact on Climate* (SMIC), MIT Press, Cambridge, Mass.

Michel, R. L., and H. E. Suess (1975). Bomb tritium in the Pacific Ocean, *J. Geophys. Res. 80*, 4139.

Ocean Science Committee (1975). *The Ocean's Role in Climate Prediction*, NRC Ocean Affairs Board, National Academy of Sciences, Washington, D.C.

Östlund, H. G., M. O. Rinkel, and C. G. Rooth (1969). Tritium in the equatorial Atlantic current system, *J. Geophys. Res. 74*, 4535.

Östlund, H. G., H. G. Dorsey, and C. G. Rooth (1974). GEOSECS North Atlantic radiocarbon and tritium results, *Earth Planet. Sci. Lett. 23*, 69.

Persson, R., (1976). *Forest Resources in Africa, Part II, Regional Analysis,* Department of Forest Survey, Royal College of Forestry, Stockholm.

Proceedings of the Dahlem Workshop on Global Chemical Cycles and Their Alteration by Man (1977). To be published.

Proceedings of the ERDA Workshop on Significant Environmental Concerns, Miami, Fla., March 7-11 (1977). To be published.

Proceedings of the SCOPE Workshop on Biogeochemical Cycling of Carbon (1977). To be published.

PSAC (1965). President's Science Advisory Committee. *Restoring the Quality of Our Environment,* The White House, Washington, D.C., p. 127.

Puiseux, L. (1975). World energy consumption and production over the next 50 years, *Pugwash Newsletter 12,* 3 (Jan.)

Revelle, R. R. (1976). The resources available for agriculture, *Sci. Am. 235,* 164.

Sundquist, E., D. K. Richardson, W. S. Broecker, and T. H. Peng (1977). *The Fate of Fossil Fuel CO_2 in the Ocean,* N. R. Anderson and A. Malahoff, eds., Plenum Press, New York.

Sverdrup, H. U., M. W. Johnson, and R. H. Fleming (1942). *The Oceans,* Prentice-Hall, New York.

Swedish Royal College of Forestry (1974). *Forest Resources in Southeast Asia,* Department of Forest Survey, Royal College of Forestry, Stockholm.

U.S. Committee for the Global Atmospheric Research Program (1975). *Understanding Climatic Change: A Program for Action,* National Academy of Sciences, Washington, D.C.

U.S. Domestic Council (1974). *A United States Climate Program,* Environmental Resources Committee, Subcommittee on Climate Change.

Waggoner, P. E. (1969). Environmental manipulation for higher yields, in *Physiological Aspects of Crop Yield,* Am. Soc. Agron. and CSSA, Madison, Wisconsin.

Weinberg, A. M., and R. P. Hammond (1972). Global effects of increased use of energy, in *Peaceful Uses of Atomic Energy,* Proceedings of the Fourth International Conference on the Peaceful Uses of Atomic Energy, jointly sponsored by the United Nations and the International Atomic Energy Agency (IAEA, Vienna), Vol. 1.

Whittaker, R. H., and G. E. Likens (1975). The biosphere and man, in *Primary Productivity of the Biosphere,* H. Lieth and R. H. Whittaker, eds., Springer Verlag, New York.

Wilson, C. L., and W. H. Matthews, eds. (1970). *Man's Impact on the Global Environment, Study of Critical Environmental Problems* (SCEP), MIT Press, Cambridge, Mass.

Woodwell, G. M., and E. V. Pecan (1973). *Carbon and the Biosphere,* Proceedings of the 24th Brookhaven Symposium on Biology, Technical Information Center, Office of Information Services, USAEC.

I

THE CONTEXT

Projected World Energy Consumption

1

HARRY PERRY and HANS H. LANDSBERG
Resources for the Future, Inc.

1.1 INTRODUCTION

Basic to any evaluation of the potential impact of man's use of energy resources on climate are estimates of how much energy will be consumed at various locations, which resources, and how much of each will be used to supply the demand, how much will be converted to other energy forms before use, and what pollutants and how much of each will be emitted as a result of use of the projected mix of fuel forms.

Climatic effects from energy consumption may occur as the result of either heat that is released (over that incident upon the earth from the sun) or the effect that pollutants may have on both the amount of solar energy that reaches the earth and the amount that is radiated from the earth to space.

In the following sections, energy consumption and pollution emissions are estimated for eleven geographic regions and for four different years, the last year being 2025. Section 1.2 presents best estimates of energy resources for the world and for component parts, separately for nonrenewable and renewable sources, paying particular attention to the recoverability factor, i.e., how much of the resource in place can be mobilized for use. Section 1.3 presents estimates of energy demand, based upon projections of population and Gross National Product (GNP). A detailed exposition of the merits precedes the estimates. Section 1.4 gives estimates of the heat, carbon dioxide, and particulates that can be expected to be released by energy use.

1.2 WORLD ENERGY SUPPLIES

NONRENEWABLE WORLD ENERGY RESOURCES

Table 1.1 lists the currently used nonrenewable world energy resources by fuel form and by two categories of resource availability. The "proved recoverable" category is the quantity of each resource whose existence is known with some relatively high degree of precision and that can be produced using currently available extraction technologies at current costs. The "estimated total remaining recoverable" category is that quantity of each fuel that is thought to be included in the earth's crust, will be discovered, and will be recover-

TABLE 1.1 World Developed Nonrenewable Energy Resources, 1972[a]

	Proved Recoverable, Btu × 10^{18} (J × 10^{21})	Estimated Total Remaining Recoverable, Btu × 10^{18} (J × 10^{21})
Coal	24.0 (25.3)	168.0 (177.2)
Crude oil and natural gas liquids	3.5 (3.7)	12.0 (12.7)
Natural gas	2.0 (2.1)	11.0 (11.6)
Oil shale and tar sands	2.0 (2.1)	15.8[c] (16.7)
U$_3$O$_8$[b]	1.3 (1.4)	2.5 (2.6)
	32.8 (34.6)	209.3 (220.8)

[a] Sources: World Energy Conference, *Survey of Energy Resources, 1974*; World Power Conference, *Survey of Energy Resources, 1965*; H. R. Linden and J. D. Parent, "Analysis of World Energy Supplies," *Survey of Energy Resources, 1974*; Senate Committee on Interior and Insular Affairs, *U.S. Energy Resources: A Review As Of 1972*, Part I, Committee Print 93-40 (92-75) (U.S. Govt. Printing Office, Washington, D.C., 1974); *Oil and Gas J.*, Dec. 31, 1973, p. 86.

[b] Assuming 1.5 percent efficiency use of the U$_3$O$_8$ and cost cutoff level of $10 and $15 per pound of U$_3O_8$ for "proved" and "remaining," respectively. For characterization of these highly restrictive assumptions, see text.

[c] Assuming use of oil shale containing 15 or more gallons/ton (51.5 liters per metric ton or more) this number would exceed 50 × 10^{18} Btu (52.8 × 10^{21} J). The 15.8 × 10^{18} was derived by using the lowest value for oil shale and tar sands as reported in the WEC.

able using current or new technology that might be developed and at prices that may prevail in the future. The category includes four types of resource: proved resources; those that have been discovered and cannot now be produced at current prices with current technology but could be with improved technology or higher prices; those that would now be commercially producible but have not yet been found; and those that, if found now, would not be commercially producible with present technology at present prices but would be expected to be recoverable in the future with new technology and at future prices.

The accuracy of information even about "proved and currently recoverable resources" is quite variable since the precision of these estimates is different for each country. As a result, a variety of values for "proved recoverable" resources can be found in published literature for the various fuel forms. The spread in the published values for the "estimated total remaining recoverable" category is usually much wider, since such estimates involve a much higher degree of judgment about what new resources will be found, what new technologies will be developed, and what the future prices of the various energy resources will be.

The basic sources used in preparing Table 1.1 were World Energy Conference (WEC), *Survey of Energy Resources, 1974*; World Power Conference, *Survey of Energy Resources, 1965*; H. R. Linden and J. D. Parent, "Analysis of World Energy Supplies," *Survey of Energy Resources, 1974* (World Energy Conference, 1974); Senate Committee on Interior and Insular Affairs, *U.S. Energy Resources, A Review As Of 1972*, Part I, Committee Print 93-40 (92-75), 1974; and *Oil and Gas Journal*, December 31, 1973, p. 86.

Table 1.1 indicates the "best" estimate* for each category and for each fuel. The "best" values to be used for tar sands and oil shale are very uncertain. If all the estimated oil shale resources containing 51.5 liters or more per metric ton (15 gallons per ton) were included, the values in Table 1 would be much larger than the 15.8 × 10^{18} Btu (16.67 × 10^{21} J) actually shown.

In a recently published report of the National Academy of Sciences,[†] the total remaining recoverable resources of crude oil, natural gas, and coal were estimated at 10 × 10^{18}, 7.2 × 10^{18}, and 264 × 10^{18} Btu (10.55 × 10^{21}, 7.60 × 10^{21}, and 278.52 × 10^{21} J). The crude oil estimates are 16 percent lower and the natural gas estimates are nearly 34 percent lower than those in this paper. However, the coal estimates are about 57 percent higher than those shown in Table 1.1. Considering the great uncertainties involved in these kinds of estimates the agreement between this report and the unpublished data is considered to be very good for oil and gas and good for coal.

The uranium estimates are highly conservative and possibly misleading on two grounds. First, they do not take into account the possible development of the breeder reactor. The efficiency of light-water reactors assumed in the tabulation is about one fortieth of that of the breeder. Thus, in terms of energy, the estimated uranium resources would be 40 times as large if used in a breeder reactor. Second, the convention for considering both reserves and what is called here "remaining recoverable resources" of uranium has been to set up limiting cost levels. These have for a long time been $8 per pound of U$_3O_8$ and have only in the last few years been raised to $15 (always in current dollars), with estimates up to $30 first established in 1975. Nothing comparable is available for most foreign countries. In the context of rapidly rising prices of competing fuels and the declining sensitivity of power costs to the price of uranium (given the rapid escalation of plant cost), uranium above—and perhaps *far* above the $15 cutoff—will qualify as "economic." This would be especially true for the breeder in which the relative importance of fuel costs is very small. For these reasons the figures here shown, which are limited to $15 material, are highly conservative and subject to change. During 1975, purchases in the United States were indeed made at prices as high as $40, though these should not be taken as establishing a new price floor. Lifting the "no breeder" constraint alone would raise the level several tens of times. Raising the $15 constraint to, for example, $30 would at least double the amount. Raising it to $100, which is not an unreasonable proposition, could raise it by a factor of 10 or 20.

Figure 1.1 is a plot of the estimated depletion rates of

*U$_3$O$_8$ values are for ore minable at $33 per kg ($15 per lb) and with an efficiency of use of 1.5 percent in light-water reactors.
†Committee on Mineral Resources and the Environment, *Mineral Resources and the Environment* (National Academy of Sciences, Washington, D.C., 1975).

Projected World Energy Consumption

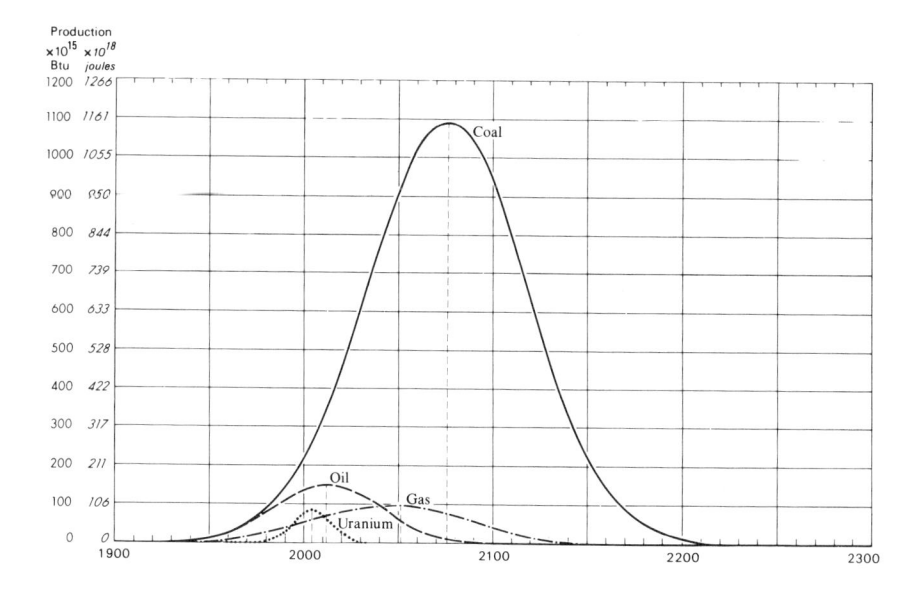

FIGURE 1.1 Idealized depletion curve for the world's (less the United States) coal, oil, gas, and uranium resources, 1900–2300. The general equation for the depletion curves is

$$f(x) = P \times \exp\{-\tfrac{1}{2}[(x-\mu)/\sigma]^2\},$$

where the constants are

coal $P = 1096.5 \times 10^{15}$, $\mu = 2076$, $\sigma = 42$;
oil $P = 146.25 \times 10^{15}$, $\mu = 2012$, $\sigma = 30$;
gas $P = 99.25 \times 10^{15}$, $\mu = 2046$, $\sigma = 40$;
uranium $P = 60 \times 10^{15}$, $\mu = 2005$, $\sigma = 12$.

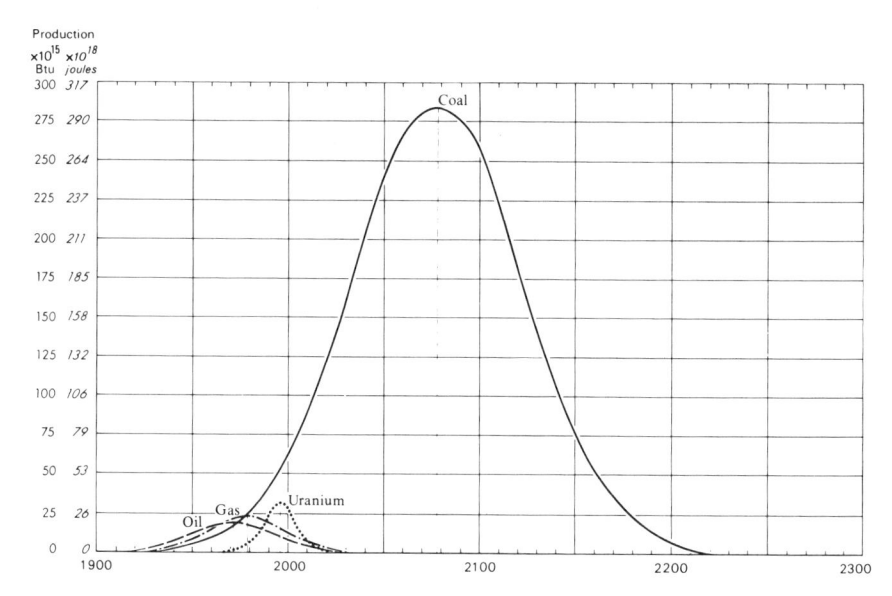

FIGURE 1.2 Idealized depletion curve for U.S. coal, oil, gas, and uranium resources, 1900–2300. The general equation for the depletion curves is

$$f(x) = P \times \exp\{-\tfrac{1}{2}[(x-\mu)/\sigma]^2\},$$

where the constants are

coal $P = 284.5 \times 10^{15}$, $\mu = 2078$, $\sigma = 45$;
oil $P = 19.5 \times 10^{15}$, $\mu = 1972$, $\sigma = 22$;
gas $P = 24 \times 10^{15}$, $\mu = 1979$, $\sigma = 19$;
uranium $P = 31 \times 10^{15}$, $\mu = 1996$, $\sigma = 9$.

world developed nonrenewable energy resources* (excluding the United States) by each type of resource using idealized depletion curves. If the breeder reactor or fusion is commercially available by 2000 to 2010, then *supplies* of energy will present no problem for the world in the period to 2025 and well beyond. The possible shortfall (see below) of world energy supplies that could occur by 2025 if only nonrenewable resources, hydroelectricity, and light-water reactors are used would have to be filled by geothermal resources, wind power, and solar energy. Should these energy resources not be developed by that time in the quantities needed, the depletion rates of the nonrenewable resources would have to be accelerated over that shown in the idealized depletion

*Oil shale and tar sands were not included as part of the nonrenewable resource base in the estimates used in Figures 1.1 and 1.2.

curves, and the bell-shaped curves shown in Figure 1.1 would be compressed toward an earlier total depletion date.

The approximate shape of the curves for each of the resources would actually be determined by the *total* demand for heat units for the world for each of the future years. This, in turn, can be estimated from regional world population and world GNP in those years. (See Section 1.3.) With a knowledge of total demand, total resource availability, and the general shape that the depletion curve for nonrenewable resources takes, a share of the total demand can be allocated to each resource.

Figure 1.1 does not include U.S. resources or their rate of depletion. This is shown in Figure 1.2. As drawn, these figures treat the rest of the world and the United States as acting independently of each other as far as energy resource production and use are concerned. In fact, if there is free access by the United States to world oil and gas resources,

TABLE 1.2 World Developed Nonrenewable Energy Resources, by Region (Recovery Factor Included)

	Coal, Btu × 10^{18} (J × 10^{21})	Oil, Btu × 10^{18} (J × 10^{21})	Gas, Btu × 10^{18} (J × 10^{21})	Oil Shale and Tar Sands, Btu × 10^{18} (J × 10^{21})	U_3O_8, Btu × 10^{18} (J × 10^{21})
North America (excluding United States)	1.2 (1.3)	0.24 (0.25)	0.42 (0.44)	2.0 (2.1)	0.5 (0.5)
Western Europe	5.3 (5.6)	0.08 (0.08)	0.28 (0.30)	0.4 (0.4)	0.5 (0.5)
Oceania	4.2 (4.4)	0.03 (0.03)	0.04 (0.04)	0.9 (0.9)	0.3 (0.3)
Latin America	0.6 (0.6)	1.07 (1.13)	1.07 (1.13)	1.9 (2.0)	Neg. Neg.
Asia (excluding Communist)					
Japan	Neg. Neg.	Neg. Neg.	Neg. Neg.	Neg. Neg.	Neg. Neg.
Other Asia	2.1 (2.2)	3.72 (3.93)	4.11 (4.34)	2.5 (2.6)	Neg. Neg.
Africa	1.0 (1.1)	1.16 (1.22)	1.16 (1.22)	3.3 (3.5)	0.5 (0.5)
U.S.S.R. and Communist Eastern Europe					
U.S.S.R.	61.0 (63.3)	3.54 (3.74)	2.86 (3.02)	0.5 (0.5)	— —
Communist E. Europe	0.6 (0.6)	0.01 (0.01)	Neg. (Neg.)	Neg. (Neg.)	— —
Communist Asia	60.0 (63.3)	1.15 (1.21)	0.06 (0.06)	2.5 (2.6)	— —
World (excluding United States)	136.0 $(143.5)^b$	11.0 $(11.61)^b$	10.0 (10.55)	14.0 $(14.8)^b$	1.8 $(1.9)^b$
United States	32.1 (33.9)	1.1 (1.16)	1.1 (1.16)	1.8 (1.9)	0.7 (0.7)
TOTAL WORLD	168.1 $(177.4)^b$	12.1 $(12.77)^b$	11.1 $(11.71)^b$	15.8 $(16.7)^b$	2.5 $(2.6)^b$

^{a} At $15 per pound of U_3O_8.
^{b} Converted figures (in parentheses) may not add because of round-off error.

then depletion of these two resources in the rest of the world would occur more rapidly.

Curves similar to those reproduced in Figure 1.1 (for the rest of the world in total) and in Figure 1.2 (for the United States) were prepared for each fuel for each region based on the data in Table 1.2 in order to estimate production by individual fuel and geographic area in future years.*

---

*A particular set of uncertainties is associated with the estimates of nuclear energy: (a) The stipulated efficiency of use of 1.5 percent is equivalent to eliminating any variety of breeder reactor; (b) the upper limit of $15/lb of U_3O_8 puts a lid on estimates that was significantly exceeded even in 1975; and (c) the rapidly escalating capital cost of nuclear reactors is likely to permit much higher cost of U_3O_8 to be tolerated. Any substitution of nuclear for fossil fuel-generated electricity would ease the pressure on climate to the extent that emissions other than heat are concerned. The case we have built, therefore, may be thought of as biased on the "worst approach" side.

RENEWABLE ENERGY RESOURCES

There are at least five potentially important renewable energy resources: hydroelectric, geothermal, tidal, solar, and fusion energy. Of these energy forms, hydroelectric is by far the most widely used, solar and geothermal are used to a small extent, tidal applications are minuscule and almost certain to remain so, and the feasibility of fusion technology has yet to be proven. Table 1.3 shows the developed and undeveloped hydroelectric energy resources of the United States and the rest of the world in megawatts of potential capacity. Figure 1.3 shows the estimated amount of worldwide (less the United States) hydroelectric energy that could be available at various future times using what are believed to be reasonable estimates for the rate at which new hydroelectric capacity could be installed.

Figure 1.4 gives a similar projection for the United States. Even if all the potential hydroelectric sites were developed and operated at the assumed conditions, only about 1 per-

Projected World Energy Consumption 39

TABLE 1.3 Potential World Water-Power Capacity[a]

Region	Potential, 10^3 MW (10^9 Btu/h)	Percentage of Total, %	Development, 10^3 MW (10^9 Btu/h)	Percentage Developed, %
North America	313 (1068)	11	59 (201)	19
South America	577 (1969)	20	5 (17)	
Western Europe	158 (539)	6	47 (160)	30
Africa	780 (2662)	27	2 (7)	
Middle East	21 (72)	1	—	
Southeast Asia	455 (1553)	16	2 (7)	
Far East	42 (143)	1	19 (65)	
Australasia	45 (154)	2	2 (7)	
U.S.S.R., China and Satellites	466 (1590)	16	16 (55)	3
TOTAL	2857 (9751)[b]	100	152 (519)	

[a]*Source: Energy Resources*, A Report to the Committee on Natural Resources of the National Academy of Sciences–National Research Council (NAS-NRC, Washington, D.C., 1962), p. 99, Table 3.
[b]Converted figures (in parentheses) may not add because of round-off error.

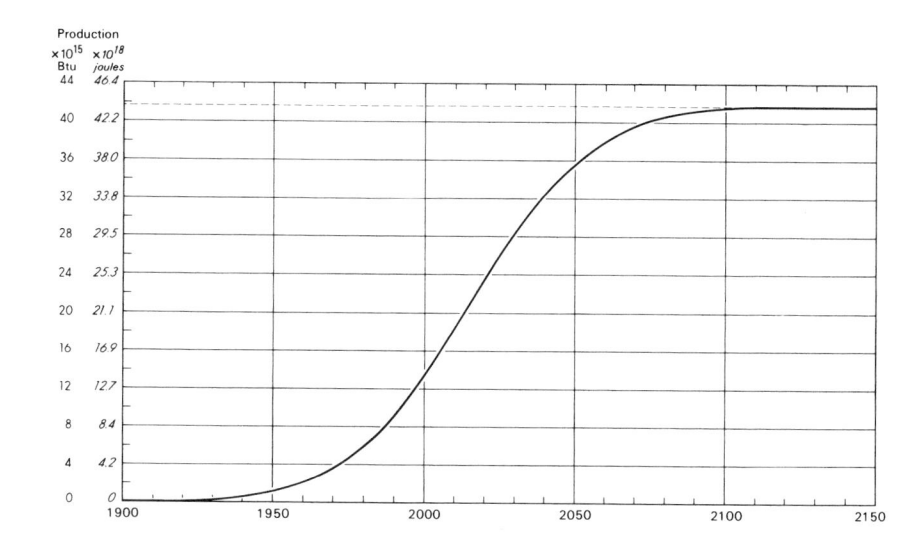

FIGURE 1.3 World (excluding the United States) hydroelectric production potential, 1900–2150. The equation for the production potential is

$$f(x) = P \int_{-\infty}^{x} \frac{1}{\sqrt{2\pi}\sigma} \exp\left\{-\frac{1}{2}[(x'-\mu)/\sigma]^2\right\} dx',$$

where $P = 41.7 \times 10^{15}$, $\mu = 2015$, and $\sigma = 33$.

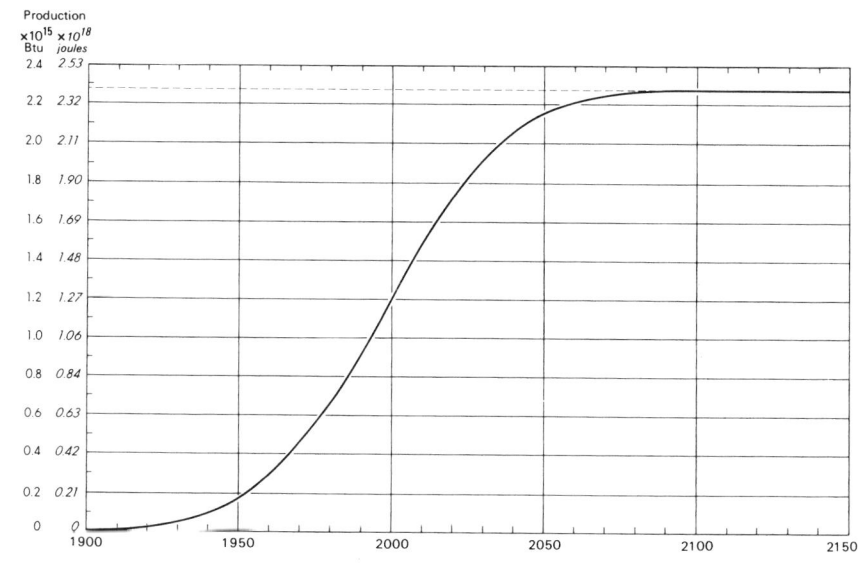

FIGURE 1.4 U.S. hydroelectric production potential, 1900–2150. The equation for the production potential is

$$f(x) = P \int_{-\infty}^{x} \frac{1}{\sqrt{2\pi}\sigma} \exp\left\{-\frac{1}{2}[(x'-\mu)/\sigma]^2\right\} dx',$$

where $P = 2.275 \times 10^{15}$, $\mu = 1998$, and $\sigma = 33$.

cent of the energy consumed in the United States and about 3 percent of that of the world would be provided from this renewable resource by the year 2000.

In the analysis that follows, the use of geothermal and solar energy has been assumed to remain small even to the year 2025 in order to test whether there would be sufficient nonrenewable resources to supply demand and to estimate the *maximum* amount of pollutants that could be released. In fact, small amounts of geothermal energy are being and will be used in the future, and by 2025 solar energy could be making a major contribution to energy supply.

FUTURE AVAILABILITY OF NONRENEWABLE AND HYDROELECTRIC FUEL SUPPLIES BY REGION

From the idealized depletion curves for *each* of the nonrenewable fuels, estimates of fuel production were made for *each region* in 1980, 1985, 2000, and 2025. Estimates of hydroelectric production for the different areas of the world for these same years were also made.

The sum of the production rates for each fuel form for all the geographic areas (for 1980, 1985, 2000, and 2025) is not the same as the estimates that are obtained when *all* the energy resources for *each fuel form* for the world (less the United States) are *first aggregated* (Figure 1.1) and then depletion estimates made for each fuel. Although not shown in the accompanying tables for 1980 and 1985, the total estimate made on an aggregated resource basis is less than results from summing the individual depletion estimates for each of the geographic areas. However, although the total is smaller when using the aggregate world values for some individual fuels, the aggregate estimate is larger for uranium and coal and smaller for oil and gas. For the years 2000 and 2025, the production rates that are estimated using the aggregate method of determining resource production (shown in Table 1.4 as "World excluding

TABLE 1.4 Energy Resource Production by Regions, 2025

Regions	Coal, Btu × 10¹⁵ (J × 10¹⁸)	Oil, Btu × 10¹⁵ (J × 10¹⁸)	Gas, Btu × 10¹⁵ (J × 10¹⁸)	Hydro-electricity, Btu × 10¹⁵ (J × 10¹⁸)	Uranium, Btu × 10¹⁵ (J × 10¹⁸)	Total, Btu × 10¹⁵ (J × 10¹⁸)
North America excluding United States	1.9 (2.0)	0.9 (0.9)	0.2 (0.2)	2.4 (2.5)	8.0 (8.4)	13.4 (14.1)
Western Europe	34.5 (36.4)	0.2 (0.2)	0.0 (0.0)	2.4 (2.5)	0.5 (0.5)	37.6 (39.7)
Oceania	9.2 (9.7)	0.0 (0.0)	0.9 (0.9)	0.6 (0.6)	3.1 (3.3)	13.8 (14.6)
Latin America	2.2 (2.3)	6.5 (6.9)	16.5 (17.4)	5.8 (6.1)	—	31.0 (32.7)
Japan	—	—	—	—	—	—
Other Asia	22.8 (24.1)	21.3 (22.5)	51.0 (53.8)	6.7 (7.1)	—	101.8 (107.4)
Africa	16.6 (17.5)	9.4 (9.9)	27.1 (28.6)	6.0 (6.3)	4.5 (4.7)	63.6 (67.1)
U.S.S.R.	93.1 (98.2)	45.9 (48.4)	25.1 (26.5)	5.9 (6.2)	[b]	170.0 (179.4)
Communist East Europe	5.4 (5.7)	0.8 (0.8)	—	[a]	[b]	6.2 (6.5)
Communist Asia	60.6 (63.9)	20.9 (22.0)	—	[a]	[b]	81.5 (86.0)
World (excluding United States)	246.3 (259.8)	105.9 (111.7)[c]	120.8 (127.4)	29.8 (31.4)[c]	16.1 (17.0)[c]	518.9 (547.4)[c]
United States	141.8 (149.8)	1.1 (1.2)	1.3 (1.4)	1.8 (1.9)	0.1 (0.1)	146.1 (154.1)
Total Nonrenewable	388.1 (409.4)[c]	107.0 (112.9)[c]	122.1 (128.8)	31.6 (33.3)[c]	16.2 (17.1)[c]	665.0 (701.6)
TOTAL RENEWABLE						508.0
TOTAL ALL ENERGY						1,173.0 (1,237.6)

[a] Totals included with U.S.S.R.
[b] Figures unavailable.
[c] Converted figures may not add because of round-off error.

Projected World Energy Consumption

United States") are *greater* than those obtained by surveying the estimated production of the different regions, largely because of differences that arise from the regional and aggregate estimates of coal production in those years. This difference occurs because estimates of the regional production rates for coal were made on the basis of optimizing coal use for each geographic area's need, while the aggregate estimates optimized coal use for the total world energy requirements.

1.3 ENERGY DEMAND

INTRODUCTION

Worldwide energy demand has grown at an exponential rate during the entire period in which reasonably reliable world energy statistics have been collected. Energy consumed in selected years in the different regions and the percentage shares of these regions are shown in Table 1.5. World energy consumption over the period 1925 to 1968 had an average growth rate of about 3.5 percent per year, but the rate has been increasing. Between 1925 and 1938, the rate was less than 2 percent per year; between 1960 and 1965 it was 5½ percent per year.

Growth rates, however, have varied widely among the different regions, as shown by the shifting percentages of the total used by the various regions in the values given in the bottom half of Table 1.5. From 1925 to 1972, the U.S. share of the total decreased from 48.3 percent to 32.8 percent; that of Western Europe from 34.8 percent in 1925 to only 19.6 percent in 1968. On the other hand, the U.S.S.R. increased its share from 1.7 to 15.7 percent of the total in this period. The less developed areas too increased their percentage share, indicating a growth rate in energy demand that exceeded that of North America and Western Europe. However, these regions still consume only a small portion of the world's energy, since their growth in energy use has been from a small (starting) base.

TABLE 1.5 Energy Consumption, by Major World Regions, Selected Years, 1925-1971[a,b]

Region	1925	1938	1950	1955	1960	1965	1967	1968	1971
Million Metric Tons									
North America	748.9	706.9	1276.3	1461.3	1659.5	2040.2	2230	2359	2529
of which: United States	717.7	669.4	1201.0	1370.3	1550.1	1881.6	2055	2173	2327
Western Europe	517.0	619.2	583.9	748.2	849.5	1117.2	1168	1242	1388
Oceania	15.6	18.1	29.3	37.3	45.8	61.2	67	72	80
U.S.S.R. and Communist E. Europe	80.5	244.0	464.1	691.2	934.5	1255.8	1376	1433	1591
U.S.S.R.	25.3	176.3	303.3	461.0	640.6	880.6	989	1025	1112
Communist E. Europe[c]	55.1	67.7	160.8	230.2	293.9	375.1	387	408	479
Communist Asia[c]	23.7	27.3	43.1	98.3	235.3	323.0	255	332	477
Latin America	24.7	38.7	66.2	105.4	153.5	199.5	224	245	274
Asia[d]	60.3	112.4	105.8	157.7	247.7	385.1	471	522	635
Japan	30.5	62.4	45.8	66.5	111.0	188.6	249	280	342
Other Asia	29.8	50.0	60.0	91.2	136.7	196.5	222	242	293
Africa	13.9	23.4	42.0	58.8	70.2	92.6	97	102	121
WORLD	1484.5	1790.1	2610.9	3358.2	4196.1	5474.6	5888	6306	7095
Percentage Distribution, %									
North America	50.4	39.5	48.9	43.5	39.5	37.3	37.9	37.4	35.6
of which: United States	48.3	37.4	46.0	40.8	36.9	34.4	34.9	34.5	32.8
Western Europe	34.8	34.6	22.4	22.3	20.2	20.4	19.8	19.7	19.6
Oceania	1.1	1.0	1.1	1.1	1.1	1.1	1.1	1.1	1.1
U.S.S.R. and Communist E. Europe	5.4	13.6	17.8	20.6	22.3	22.9	23.4	22.7	22.4
U.S.S.R.	1.7	9.8	11.6	13.7	15.3	16.1	16.8	16.3	15.7
Communist E. Europe[c]	3.7	3.8	6.2	6.9	7.0	6.9	6.6	6.5	6.8
Communist Asia[c]	1.6	1.5	1.7	2.9	5.6	5.9	4.3	5.3	6.7
Latin America	1.7	2.2	2.5	3.1	3.7	3.6	3.8	3.9	3.9
Asia[d]	4.1	6.3	4.1	4.7	5.9	7.0	8.0	8.3	8.9
Japan	2.1	3.5	1.8	2.0	2.6	3.4	4.2	4.4	4.8
Other Asia	2.0	2.8	2.3	2.7	3.3	3.6	3.8	3.8	4.1
Africa	0.9	1.3	1.6	1.8	1.7	1.7	1.6	1.6	1.7
WORLD	100.0	100.0	100.0	100.0	100.0	100.0	100.0	100.0	100.0

[a] All figures based on coal equivalents.
[b] Sources: Data for 1925-1968 from J. Darmstadter, with P. D. Teitelbaum and J. G. Polach, *Energy in the World Economy: A Statistical Review of Trends in Output, Trade, and Consumption Since 1925* (Johns Hopkins U. Press, Baltimore, Md., 1972), p. 10; data for 1971 from United Nations, *World Energy Supplies*, Series J, No. 16 (UN, New York, 1973).
[c] It should be borne in mind that throughout this study pre-World War II data for "Communist Eastern Europe" refer to the present countries of that area with the exception of prewar East Germany, which appears (as part of Germany) within Western Europe. Prewar "Communist Asia" refers to Mainland China, while the postwar figures for that region also include North Korea, Mongolia, and (beginning in 1955) North Vietnam.
[d] Throughout this study, unless otherwise specified, "Asia" means non-Communist Asia.

ENERGY CONSUMPTION AND GROSS NATIONAL PRODUCT

A relatively good (but far from perfect) correlation has been shown to exist between energy consumption and GNP. A typical plot of this relationship is shown in Figure 1.5, in which energy consumption per capita for the 49 countries is plotted against the GNP per capita. The plotted points, using 1965 data, fall within a narrow band of the line representing the best fit of all the data with a correlation coefficient of 0.87. If the hydroelectric data used in this plot are corrected to take into account the fossil fuel inputs that would be required to produce this energy (increasing the energy consumption from this resource by a factor of about 3), the correlation coefficient is increased to 0.89.

The reasons that per capita energy consumption and GNP are highly correlated are in general quite obvious, but they are not well enough understood when it comes to changes in the relationship over time or differences between countries. For example, it is hard to explain persuasively why in the United States a steady (but not constant) decline in energy consumption per unit of GNP was sharply reversed in the latter part of the 1960's. It has also not been possible to disaggregate energy use by consuming sectors in sufficient detail to explain the marked differences in energy use per unit of GNP between the United States and Western Europe or the changes in energy use per unit of GNP that occur over time in nearly all countries. For example, it is not understood why the United States used 2.75 kg of coal equivalent (2.75×10^{-3} metric ton of coal equivalent) per dollar of GNP in 1965, while France used only 1.57 kg (1.57×10^{-3} metric ton); or why the amount of energy used per unit of GNP produced in Spain rose from 0.67 during 1960–1965 to 1.48 in 1965–1971.

In spite of the difficulty in explaining in detail the empirical relationship that has been found to exist, its persistence over time and the high correlation coefficient between the two variables suggest that it is useful for projecting world

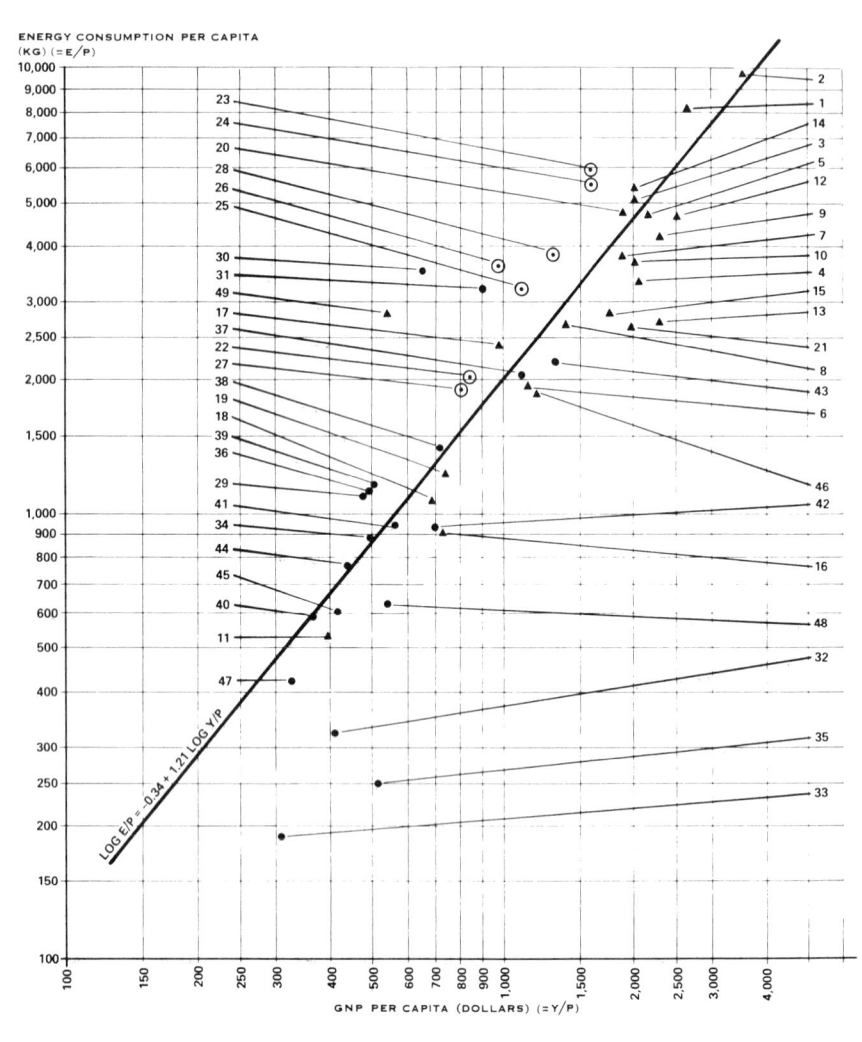

FIGURE 1.5 GNP per capita and energy consumption per capita: 49 selected countries, 1965 (see Table 1.6). Source: J. Darmstadter, with P. D. Teitelbaum and J. G. Polach, *Energy in the World Economy: A Statistical Review of Trends in Output, Trade, and Consumption Since 1925* (Johns Hopkins U. Press, Baltimore, Md., 1972), pp. 66–67.

Country Legend:
- ▲ North America, Western Europe, Oceania, South Africa, Japan
- ⊙ East European Communist countries and U.S.S.R.
- ● Latin America, Other Africa and Non-Communist Asia

Projected World Energy Consumption 43

TABLE 1.6 GNP per Capita and Energy Consumption per Capita (see Figure 1.5)

		(GNP per Capita) ($74) GNP per Capita ($65)		Energy Consumption			
				Per Capita		Per $1 of GNP (In 1974 $)	
Country	Number on Chart	Dollars	Rank	kg of Coal Equiv.	Rank	kg of Coal Equiv.	Rank
North America							
Canada	1	2,658 (4,079)	2	8,077	2	3.04 (1.98)	7
United States	2	3,515 (5,394)	1	9,671	1	2.75 (1.79)	10
Western Europe							
Belgium-Luxembourg	3	1,991 (3,055)	10	5,152	6	2.59 (1.69)	12
France	4	2,104 (3,228)	7	3,309	16	1.57 (1.02)	37
West Germany	5	2,195 (3,368)	6	4,625	8	2.11 (1.38)	20
Italy	6	1,254 (1,924)	20	1,940	28	1.55 (1.01)	38
Netherlands	7	1,839 (2,822)	13	3,749	12	2.04 (1.33)	22
Austria	8	1,365 (2,095)	17	2,584	23	1.89 (1.23)	24
Denmark	9	2,333 (3,580)	4	4,149	10	1.78 (1.16)	27
Norway	10	2,015 (3,092)	8	3,621	13	1.80 (1.17)	26
Portugal	11	396 (608)	46	520	45	1.31 (0.85)	42
Sweden	12	2,495 (3,828)	3	4,604	9	1.85 (1.21)	25
Switzerland	13	2,331 (3,577)	5	2,699	21	1.16 (0.76)	45
United Kingdom	14	1,992 (3,057)	9	5,307	5	2.66 (1.73)	11
Finland	15	1,750 (2,685)	14	2,825	19	1.61 (1.05)	33
Greece	16	772 (1,185)	29	904	39	1.17 (0.76)	44
Ireland	17	981 (1,505)	24	2,359	24	2.41 (1.57)	16
Spain	18	686 (1,053)	33	1,080	36	1.57 (1.02)	35
Yugoslavia	19	743 (1,140)	30	1,217	32	1.65 (1.08)	32
Oceania							
Australia	20	1,910 (2,931)	12	4,697	7	2.46 (1.60)	13
New Zealand	21	1,970 (3,023)	11	2,603	22	1.32 (0.86)	40
U.S.S.R. and Communist E. Europe							
Bulgaria	22	829 (1,272)	27	2,011	27	2.43 (1.58)	15
Czechoslovakia	23	1,561 (2,395)	16	5,870	3	3.76 (2.45)	3
East Germany	24	1,562 (2,397)	15	5,534	4	3.54 (2.31)	6
Hungary	25	1,094 (1,679)	22	3,188	18	2.91 (1.90)	8
Poland	26	980 (1,504)	25	3,552	14	3.62 (2.36)	5
Romania	27	778 (1,194)	28	1,916	30	2.46 (1.60)	14
U.S.S.R.	28	1,340 (2,056)	18	3,819	11	2.85 (1.86)	9
Latin America							
Mexico	29	475 (729)	42	1,104	35	2.33 (1.52)	17
Trinidad and Tobago	30	646 (991)	34	3,505	15	5.43 (3.54)	1
Venezuela	31	882 (1,353)	26	3,246	17	3.68 (2.40)	4
Costa Rica	32	414 (638)	45	317	47	0.77 (0.50)	48
Guatemala	33	318 (488)	49	188	49	0.59 (0.38)	47
Jamaica	34	492 (755)	41	873	40	1.77 (1.15)	28
Nicaragua	35	527 (809)	38	247	48	0.70 (0.46)	49
Panama	36	495 (760)	40	1,112	34	2.25 (1.47)	19
Puerto Rico	37	1,094 (1,679)	23	2,126	26	1.94 (1.26)	23
Argentina	38	718 (1,102)	31	1,471	31	2.05 (1.34)	21
Chile	39	497 (763)	39	1,119	33	2.25 (1.47)	18
Peru	40	367 (563)	47	577	44	1.57 (1.03)	36
Uruguay	41	573 (879)	35	958	37	1.67 (1.09)	31
Asia							
Cyprus	42	702 (1,077)	32	927	38	1.32 (0.86)	41
Israel	43	1,325 (2,033)	19	2,248	25	1.70 (1.11)	30
Lebanon	44	438 (672)	43	770	41	1.76 (1.15)	29
Hong Kong	45	421 (646)	44	605	43	1.44 (0.94)	39
Japan	46	1,222 (1,875)	21	1,926	29	1.58 (1.03)	34
Malaysia and Singapore[a]	47	332 (509)	48	424	46	1.28 (0.83)	43
Africa							
Libya	48	542 (832)	36	613	42	1.13 (0.74)	46
South Africa	49	535 (821)	37	2,761	20	5.16 (3.36)	2

[a]Including Sabah and Sarawak.

energy use in the future. Therefore, energy consumption estimates for the world and for selected regions have been based on population and GNP projections for the world and individual areas.

HISTORICAL CORRELATION OF ENERGY CONSUMPTION AND GNP

Energy consumption per capita and GNP per capita were correlated for each of the regions listed in Table 1.5 by using data for each region for selected years. A typical plot is shown in Figure 1.6, where the energy consumption per capita and GNP per capita for Japan for selected years from 1925 to 1972 are plotted. The slope of the line indicates how much energy is used for the particular region (or country) per unit of GNP produced. Slopes less than ½ indicate that energy consumption growth rates are lower than GNP growth rates. This is true for the highly industrialized areas such as the United States, Western Europe, and Oceania. The balance of North America, Oceania, Japan, Eastern Europe, and Africa have slopes approximately equal to ½, while Latin America, Other Asia, U.S.S.R., and Communist Asia have slopes much steeper than ½. Except for the U.S.S.R., which has a high energy consumption per unit of GNP that probably results from a planned economy dedicated to a rapid energy-intensive industrialization, the other areas with a high slope are the less developed areas with very low GNP. (Not nearly enough work has been done to rationalize this situation, and it is just possible that the most important element is the substitution of statistically recorded commercial for nonrecorded, noncommercial energy.)

Within each of the areas (or countries) there was a high

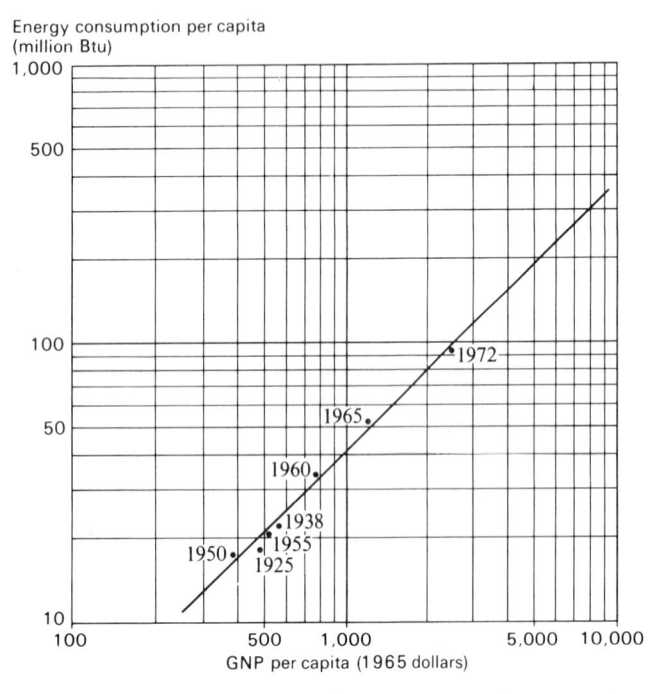

FIGURE 1.6 GNP per capita and energy consumption per capita: Japan. $\log (E/P) = -1.3178 + 0.9777 \log (Y/P)$.

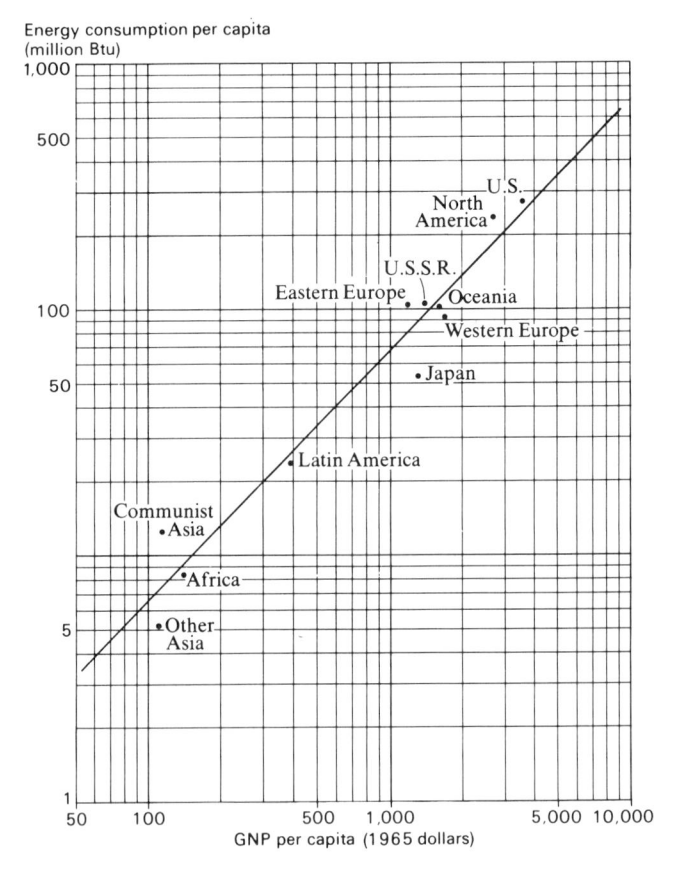

FIGURE 1.7 GNP per capita and energy consumption per capita: 1965. $\log (E/P) = -1.2023 + 1.0118 \log (Y/P)$.

correlation coefficient for the data for the years that were used. Unfortunately, energy consumption and GNP information over long periods are not always available. Data for the United States, other North America, U.S.S.R., and Japan were obtainable for the period 1925–1972, while data for most of the other areas covered the period 1950–1972.

CROSS-SECTIONAL CORRELATION OF ENERGY CONSUMPTION AND GNP

Energy consumption per capita and GNP per capita were correlated for related years by plotting these data for each of the regions (where available) for given years. A typical plot is shown in Figure 1.7, reflecting the data for 1965 for the 11 regions for which individual energy consumption data have been tabulated. Similar tabulations were made for 1925 (four data points), 1938 (four data points), 1950, 1955, 1960, and 1965 (eleven data points), and 1972 (seven data points).*

It was observed that the slope of the line that best fitted the data became less steep over time, indicating that for the world as a whole energy use per unit of GNP has been de-

*In these plots the energy inputs for hydroelectricity were taken as 3412 Btu per kWh. If average energy inputs for generating this amount of hydroelectricity were used, the correlations obtained would be improved slightly.

TABLE 1.7 Energy Consumption by Regions for the Years 1980, 1985, 2000, and 2025[a]

Region	1980 Btu × 10^{15} (J × 10^{18})	1985 Btu × 10^{15} (J × 10^{18})	2000 Btu × 10^{15} (J × 10^{18})	2025 Btu × 10^{15} (J × 10^{18})
United States[b]	73 (77)	84 (89)	117 (123)	196 (207)
North America (less U.S.)[b]	7 (7)	8 (8)	12 (13)	19 (20)
Western Europe[b]	48 (51)	54 (57)	81 (85)	123 (130)
Oceania[b]	3 (3)	4 (4)	7 (7)	15 (16)
Latin America[c]	12 (13)	16 (17)	39 (41)	114 (120)
Japan[b]	15 (16)	19 (20)	31 (33)	68 (72)
Other Asia[c]	17 (18)	22 (23)	55 (58)	165 (174)
Africa[c]	5 (5)	7 (7)	19 (20)	78 (82)
U.S.S.R.[b]	43 (45)	56 (59)	97 (102)	244 (257)
Communist Eastern Europe[b]	17 (18)	20 (21)	41 (43)	77 (81)
Communist Asia[c]	18 (19)	22 (23)	38 (40)	74 (78)
TOTAL WORLD	258 (272)	312 (329)	537 (567)	1173 (1238)

[a] Converted figures (in parentheses) may not add because of round-off error.
[b] Developed countries: low population, low growth.
[c] Developing countries: high population, low growth.

clining, although consumption of energy has been rising at an accelerated pace.

This appears to be the result of two factors. The geographic areas with the lowest GNP have the highest energy growth rate per unit of GNP so that the observed Btu per capita versus GNP per capita relationships at the lower end of the curve move up the ordinate more rapidly over time than those of other geographic areas. At the same time, the geographic areas with the highest GNP have the lowest energy growth rate per unit of GNP, and this also contributes to the flattening of the slope of the line in plots similar to those shown in Figure 1.7.

ESTIMATES OF FUTURE ENERGY CONSUMPTION

Estimates of the growth in GNP and population for the United States and the rest of the world were made by individual regions and were aggregated to obtain estimates for the entire world.* From these, estimates of energy

*Estimates of population and GNP for the individual regions of the world that are used in this study are based on projections made be Ronald Ridker at Resources for the Future for a report now being prepared for the National Institutes of Health.

consumption for the United States and for the other regions were made by two different methods. In the first, the historical correlation (Figure 1.6) was used to establish Btu consumption per capita in future years for each individual region. In the second, the cross-sectoral correlations (Figure 1.7) were used to estimate Btu consumption per capita in future years. The values estimated by these two methods were compared and a "best" value selected. This value was determined by taking into consideration the level of GNP in different regions at different future times, the rate at which energy consumption in the region had been growing and the likelihood that it would continue, future changes that might be expected in the structure of the economy of the region, and the probable impact on consumption of higher energy prices as low-cost nonrenewable resources are exhausted.

With the best average value of Btu consumption per capita, the total consumption in selected years was calculated using the estimated population in each of the regions for that year. These consumption estimates for individual regions are shown in Table 1.7, in which a high population-low economic growth situation was postulated for developing countries and a low population-low economic growth situation was selected for developed countries.

TABLE 1.8 Estimated Average Sulfur Content of Oil and Coal, by Geographic Region[a]

	Percentage Sulfur	
	Oil, %	Coal, %
United States	0.65	1.54
North America less United States	0.83	0.72
Western Europe	1.12	0.94
Oceania	0.02	0.70
Latin America	2.22	2.99
Japan	—	1.90
Other Asia	2.05	0.89
Africa	0.30	0.71
U.S.S.R.	1.02	2.00
Communist Eastern Europe	0.14	3.19
Communist Asia	0.50	1.35

[a]Sources: U.S. Department of the Interior, Bureau of Mines Information Circular 642/1968, *Qualitative and Quantitative Aspects of Crude Oil Composition* (U.S.—oil); U.S. Department of the Interior, Federal Energy Administration, Project Independence Blueprint Final, Task Force Report, *Coal*, November 1974 (U.S.—coal); U.S. Department of the Interior, Bureau of Mines Information Circular 7059/1967, *Sulfur Content of Crude Oil of the Free World* (rest of world—oil); World Energy Conference, *Survey of Energy Resources*, 1974, Appendix 2 (rest of world—coal).

1.4 HEAT AND AIR POLLUTION EMISSIONS FROM ENERGY USE

Estimated energy consumption, which is equivalent to heat releases from energy use, is shown in Table 1.7 for each region for four different years. In 2025, it is estimated that 1173×10^{15} Btu (1237×10^{18} J) will be released, which would still be less than 0.1 percent of the solar energy reaching the earth. Even in Western Europe, where heat releases from energy use are relatively large and concentrated in a small area, the heat from energy use would still be less than 0.1 percent of that of solar energy in 2025.

Air pollution emissions from the use of energy have been calculated for each region for the year 2025, when the emission rate would be largest, using the following assumptions:

1. Energy resources produced in a region (Table 1.4) will be used to supply regional demand to the extent that production has been estimated to be able to meet demand.
2. Where regional demand exceeds regional production, emission estimates were made in two ways: (a) assuming that a new renewable, nonpolluting energy resource would

TABLE 1.9 Carbon Dioxide Emissions in 2025 (Nonpolluting Fuels Supplying Regional Shortfalls)[a]

Region	Coal, tons × 10⁹ (metric tons × 10⁹)	Oil, tons × 10⁹ (metric tons × 10⁹)	Gas, tons × 10⁹ (metric tons × 10⁹)	Total, tons × 10⁹ (metric tons × 10⁹)
United States	14.7 (13.3)	0.1 (0.1)	0.1 (0.1)	14.9 (13.5)
North America less United States	0.2 (0.2)	0.1 (0.1)	Neg. (Neg.)	0.3 (0.3)
Western Europe	3.6 (3.3)	Neg. (Neg.)	0 (0)	3.6 (3.3)
Oceania	0.9 (0.8)	0 (0)	0.1 (0.1)	1.0 (0.9)
Latin America	0.2 (0.2)	0.4 (0.4)	0.8 (0.7)	1.4 (1.3)
Japan	— (—)	— (—)	— (—)	— (—)
Other Asia	2.3 (2.1)	1.4 (1.3)	2.5 (2.3)	6.2 (5.7)
Africa	1.7 (1.5)	0.7 (0.5)	1.3 (1.2)	3.7 (3.2)
U.S.S.R.	9.6 (8.7)	3.3 (3.0)	1.2 (1.1)	14.1 (12.8)
Communist Eastern Europe	0.5 (0.4)	0.1 (0.1)	— (—)	0.6 (0.5)
Communist Asia	6.3 (5.7)	1.4 (1.3)	— (—)	7.7 (7.0)
				53.5 (48.5)

[a]Converted figures (in parentheses) may not add because of round-off error. These quantities are given in terms of weight of carbon dioxide, as opposed to the usual practice of giving the weight of the carbon alone. Conversion may be effected by multiplying these figures by 12/44.

TABLE 1.10 Carbon Dioxide Emissions in 2025 (Regional Shortfalls Supplied by Coal)[a]

Region	Coal, tons × 10^9 (metric tons × 10^9)	Oil, tons × 10^9 (metric tons × 10^9)	Gas, tons × 10^9 (metric tons × 10^9)	Total, tons × 10^9 (metric tons × 10^9)
United States	19.8 (18.0)	0.1 (0.1)	0.1 (0.1)	20.0 (18.2)
North America less United States	0.8 (0.8)	0.1 (0.1)	Neg. (Neg.)	0.9 (0.9)
Western Europe	12.4 (11.2)	Neg. (Neg.)	0 (0)	12.4 (11.2)
Oceania	1.0 (0.9)	0 (0)	0.1 (0.1)	1.1 (1.0)
Latin America	8.8 (8.0)	0.4 (0.4)	0.8 (0.7)	10.0 (9.1)
Japan	7.0 (6.3)	— (—)	— (—)	7.0 (6.3)
Other Asia	8.9 (8.1)	1.4 (1.3)	2.5 (2.3)	12.8 (11.7)
Africa	3.3 (3.0)	0.7 (0.5)	1.3 (1.2)	5.3 (4.7)
U.S.S.R.	17.3 (15.7)	3.3 (3.0)	1.2 (1.1)	21.8 (19.8)
Communist Eastern Europe	8.0 (7.2)	0.1 (0.1)	— (—)	8.1 (7.3)
Communist Asia	6.3 (5.7)	1.4 (1.3)	— (—)	7.7 (7.0)
				107.1 (97.2)

[a] Converted figures (in parentheses) may not add because of round-off error. These quantities are given in terms of the weight of carbon dioxide as opposed to the usual practice of giving the weight of the carbon alone. Conversion may be effected by multiplying these figures by 12/44.

be available to meet the deficiency of nonrenewable resources and (b) that the deficiency would be met by coal, the most polluting of the nonrenewable resources and the fuel in greatest supply.

Emissions of carbon dioxide have been estimated on the following bases:

1. For coal, 207 pounds of CO_2 would be emitted per million Btu burned (89 metric tons per 10^{12} J).
2. For oil, 166 pounds of CO_2 would be emitted per million Btu burned (71 metric tons per 10^{12} J).
3. For gas, 118 pounds of CO_2 would be emitted per million Btu burned (51 metric tons per 10^{12} J).

Emissions of particulates have been estimated on the following bases:

1. For coal, an average ash content of 15 percent and an ash collection efficiency of 70 percent (mechanical collectors) was assumed. This would yield emissions of about 4 lb of particulates per million Btu (1.7 metric tons per 10^{12} J).
2. For oil, the ash content is assumed to be 0.5 percent and no collection equipment is used. This yields emissions of 0.3 lb per million Btu (0.1 metric ton per 10^{12} J).

3. For gas, no particulates are assumed to be emitted.

Uncontrolled emissions of sulfur oxides have been estimated using the average sulfur content for oil and coal for the various regions, as shown in Table 1.8.

Based on these assumptions, world carbon dioxide emissions in 2025 would be 53.5×10^9 tons (48.5×10^9 metric tons) when shortfalls of nonrenewable resources are supplied by nonpolluting fuels (Table 1.9), and 107.1×10^9 tons (97.2×10^9 metric tons) when the shortfall is completely supplied by coal (Table 1.10). This is about 3 to 6½ times the amount emitted worldwide in 1972.

World particulate emissions in 2025 would be—for the two assumptions about energy supply—791×10^6 and 1818×10^6 tons (717×10^6 metric tons and 1648.9×10^6 metric tons), as shown in Tables 1.11 and 1.12, respectively, and are about 5 to 11½ times the emissions rate in 1972.

Uncontrolled world sulfur oxide emissions in 2025 for the two assumptions regarding energy supply would be 548.5×10^6 and 1361.7×10^6 tons (497.5×10^6 and 1235.1×10^6 metric tons), as shown in Tables 1.13 and 1.14, respectively, and about 2⅓ and 6 times the emission rates estimated in 1972. If sulfur oxide emissions are controlled at the current levels for new source performance standards in the United States (1.2 lb of SO_2 per million

TABLE 1.11 Particulate Emissions in 2025 (Nonpolluting Fuels Supplying Regional Shortfalls)[a]

Region	Coal, tons × 10^6 (metric tons × 10^6)	Oil, tons × 10^6 (metric tons × 10^6)	Total, tons × 10^6 (metric tons × 10^6)
United States	284 (258)	Neg. (Neg.)	284 (258)
North America less United States	4 (4)	Neg. (Neg.)	4 (4)
Western Europe	69 (63)	Neg. (Neg.)	69 (63)
Oceania	18 (16)	Neg. (Neg.)	18 (16)
Latin America	4 (4)	1 (1)	5 (5)
Japan	— (—)	— (—)	— (—)
Other Asia	46 (42)	3 (3)	49 (44)
Africa	33 (30)	1 (1)	34 (31)
U.S.S.R.	186 (169)	7 (6)	193 (175)
Communist Eastern Europe	11 (10)	Neg. (Neg.)	11 (10)
Communist Asia	121 (110)	3 (3)	124 (112)
			791 (717)

[a] Converted figures (in parentheses) may not add because of round-off error.

TABLE 1.12 Particulate Emissions in 2025 (Regional Shortfalls Supplied by Coal)[a]

Region	Coal, tons × 10^6 (metric tons × 10^6)	Oil, tons × 10^6 (metric tons × 10^6)	Total, tons × 10^6 (metric tons × 10^6)
United States	383 (347)	Neg. (Neg.)	383 (347)
North America less United States	15 (14)	Neg. (Neg.)	15 (14)
Western Europe	240 (218)	Neg. (Neg.)	240 (218)
Oceania	18 (16)	Neg. (Neg.)	18 (16)
Latin America	170 (154)	1 (1)	171 (155)
Japan	136 (123)	— (—)	136 (123)
Other Asia	172 (156)	3 (3)	175 (159)
Africa	62 (56)	1 (1)	63 (57)
U.S.S.R.	334 (303)	7 (6)	341 (309)
Communist Eastern Europe	152 (138)	Neg. (Neg.)	152 (138)
Communist Asia	121 (110)	3 (3)	124 (112)
			1818 (1649)

[a] Converted figures (in parentheses) may not add because of round-off error.

TABLE 1.13 Sulfur Oxide Emissions in 2025 (Nonpolluting Fuels Supplying Regional Shortfalls)[a]

Region	Coal, tons × 10^6 (metric tons × 10^6)	Oil, tons × 10^6 (metric tons × 10^6)	Total, tons × 10^6 (metric tons × 10^6)
United States	181.2 (164.3)	0.4 (0.4)	181.6 (164.7)
North America less United States	1.1 (1.0)	0.4 (0.4)	1.5 (1.4)
Western Europe	26.9 (24.4)	0.1 (0.1)	27.0 (24.5)
Oceania	5.3 (4.8)	0.0 (0.0)	5.3 (4.8)
Latin America	5.5 (5.0)	8.1 (7.3)	13.6 (12.3)
Japan	— (—)	— (—)	— (—)
Other Asia	16.8 (15.2)	24.5 (22.2)	41.3 (37.5)
Africa	9.8 (8.9)	1.6 (1.5)	11.4 (10.3)
U.S.S.R.	154.5 (140.1)	26.2 (23.8)	180.7 (163.9)
Communist Eastern Europe	14.3 (13.0)	0.1 (0.1)	14.4 (13.1)
Communist Asia	67.9 (61.6)	3.8 (3.4)	71.7 (65.0)
			548.5 (497.5)

[a] Converted figures (in parentheses) may not add because of round-off error.

TABLE 1.14 Sulfur Oxide Emissions in 2025 (Regional Shortfalls Supplied by Coal)[a]

Region	Coal, tons × 10^6 (metric tons × 10^6)	Oil, tons × 10^6 (metric tons × 10^6)	Total, tons × 10^6 (metric tons × 10^6)
United States	245.0 (222.2)	0.4 (0.4)	245.4 (222.6)
North America less United States	4.5 (4.1)	0.4 (0.4)	4.9 (4.4)
Western Europe	93.5 (84.8)	0.1 (0.1)	93.6 (84.9)
Oceania	6.0 (5.4)	0.0 (0.0)	6.0 (5.4)
Latin America	211.4 (191.7)	8.1 (7.3)	219.5 (199.1)
Japan	107.2 (97.2)	— (—)	107.2 (97.2)
Other Asia	63.5 (57.6)	24.5 (22.2)	88.0 (79.8)
Africa	18.3 (16.6)	1.6 (1.5)	19.9 (18.0)
U.S.S.R.	277.4 (251.6)	26.2 (23.8)	303.6 (275.4)
Communist Eastern Europe	201.8 (183.0)	0.1 (0.1)	201.9 (183.1)
Communist Asia	67.9 (61.6)	3.8 (3.4)	71.7 (65.0)
			1361.7 (1235.1)

[a] Converted figures (in parentheses) may not add because of round-off error.

Btu or 0.5 metric ton of SO_2 per 10^{12} J), emissions will be 386.3×10^6 and 840.5×10^6 (350.4×10^6 and 762.3×10^6 metric tons), respectively, for the two different assumptions about energy supply.

Regional emissions of heat will be concentrated in the densely populated regions of the United States, U.S.S.R., Western Europe, and Japan. Carbon dioxide and particulate emissions will originate in these same areas but will be dispersed relatively quickly. The rate at which these emissions (heat, carbon dioxide, and particulates) will be dispersed and removed from the atmosphere, and the effect on climate are discussed in the following papers.

2
The Changing Climate

J. MURRAY MITCHELL, JR.
National Oceanic and Atmospheric Administration

2.1 WEATHER STATISTICS AND CLIMATE DYNAMICS

The term "weather" refers to the total panoply of atmospheric conditions as they vary from moment to moment and from place to place on the earth.

The term "climate" always connotes "average weather" but properly includes consideration of weather extremes, joint frequency distributions, and many other measures of weather variability in both time and space. The classical view of climate leaves it at that: a purely statistical concept in which climate is nothing more than the sum of its parts (i.e., the collective of weather as experienced at a point or over a designated area of the earth, over a period of years). Such a view of climate, useful for many practical purposes, is better referred to as *climatography*.

Another, more enlightened view of climate differs from the classical view in that it recognizes climate as a basic physical entity and weather as the momentary, transient behavior of the atmosphere striving to satisfy the requirements dictated by the climate for horizontal and vertical transfer of mass, momentum, and energy. In this view, weather is the means to a climatic end: the maintenance of a global-scale atmospheric heat engine striving to transport solar heat from where it is in superabundance (principally the tropics) to where it is in deficit (principally the polar regions). This view of climate, far more illuminating in the context of understanding (rather than merely describing) overall atmospheric behavior, introduces the notion of *climate dynamics*.

We can briefly characterize climate and its origins in the following way.

The climate of any locality on the earth is fundamentally determined by a hierarchy of environmental influences, which range in scale from the general astronomical setting of our planet down to minute details of the local environment within and immediately surrounding that locality. The atmosphere itself plays a very important intermediary role in the way that these diverse influences combine at the point to help to establish its climate and also contributes an important influence of its own to local climate.

This hierarchy of climatic influences involves a wide variety of environmental factors. Some of these factors affect the general characteristics and distribution of climate over very large regions of the earth's surface (the so-called *macroclimate*), while others affect climate with regard to its more detailed characteristics and distribution within smaller and smaller regions (*mesoclimate* and *microclimate*).

Generally speaking, the size of the region over which the climate (or some characteristic of climate) is systematically influenced by a given factor—whether natural or man-made—bears a rather close relationship to the geographical dimensions of the factor itself. For example, the distribution of solar radiation over the whole earth, dictated by the seasonally varying geometry of the earth's orbit and axis of spin relative to the sun, affects climate (together with atmospheric circulation) on the scale of the whole earth. On the other hand, the climatic influence of a small lake, for example, extends over an area not much larger than the lake itself. Thus we can conclude that, in general, global-scale qualities of climate are governed by global-scale environmental factors and not by local-scale factors, except under special circumstances for which local-scale factors are operating in unison over large regions of the earth. When considering the climatic impacts of power-plant cooling towers, for example, questions of the geographical scale of such impacts arise that have to be kept in mind.

2.2 THE RECORD OF PAST CLIMATES

For more than a century it has been known from a variety of geological evidence that the earth's climate was not always what it is today. It has been only in the last few years, however, that highly significant developments in the new and rapidly developing science of quantitative paleoclimatology have begun to provide us with insights into many details of the earth's climatic history, especially that of the last million years. These insights lend valuable perspective to the climate of the present.

Briefly stated, the climate of the earth is now known beyond any doubt to have been in a more or less continual state of flux. Changeability is evidently a characteristic of climate on all resolvable time scales of variation, from that of aeons down to those of millennia and centuries. The lesson of history seems to be that climatic variability must be recognized and dealt with as a fundamental quality of climate and that it would be potentially perilous for modern civilization to assume that the climate of future decades and centuries will be free of similar variability.

With this lesson in mind, some salient features of the earth's climatic history are worth outlining here.

Geological evidence leaves little doubt that, during the past billion years or so, the prevalent condition of global climate was one of relative warmth—as much as 10°C warmer than now—and almost totally free of polar ice (see Figure 2.1). This warm condition was, however, punctuated by relatively short ice-age intervals of the order of 10 million years' duration and separated by a few hundreds of millions of years. Begin-

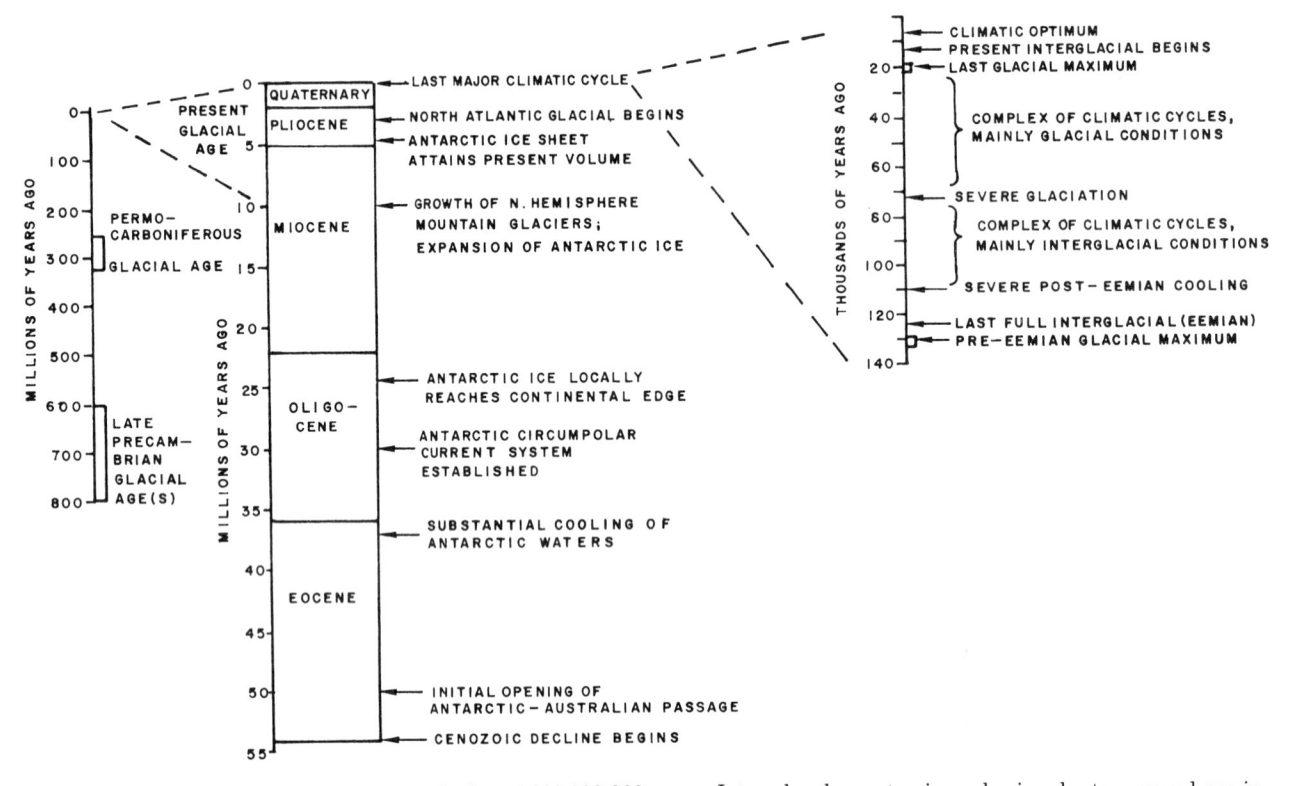

FIGURE 2.1 History of glacial ages over the last 1,000,000,000 years. Intervals when extensive polar ice sheets occurred are indicated as glacial ages on the left. An outline of significant events in the Cenozoic climate decline is given in the middle, and the significant climate events during the last major glacial-interglacial cycle are given at right. Source: U.S. Committee for the Global Atmospheric Research Program, *Understanding Climatic Change: A Program for Action*, National Academy of Sciences, Washington, D.C., 1975.

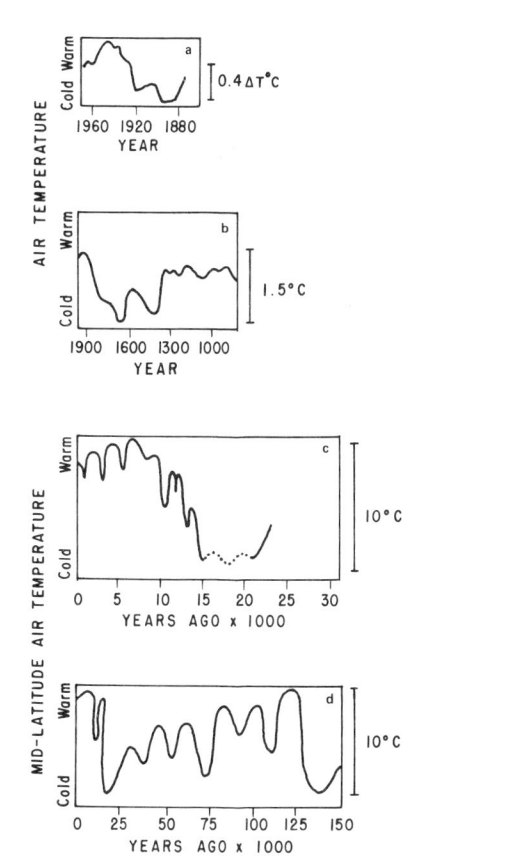

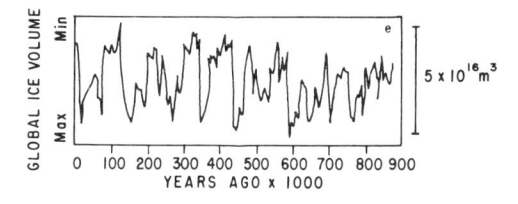

FIGURE 2.2 General trends in global-scale climate for the past million years: (a) changes in the five-year average surface temperatures over the region 0–80° N; (b) winter severity index for eastern Europe; (c) generalized northern hemisphere air-temperature trends, based on fluctuations in Alpine glaciers, changes in tree lines, marginal fluctuations in continental glaciers, and shifts in vegetation patterns recorded in pollen spectra; (d) generalized northern hemisphere air-temperature trends based on midlatitude sea-surface temperature, pollen records, and worldwide sea-level records; (e) fluctuations in global ice-volume recorded as changes in isotopic composition of fossil plankton in a deep-sea core. Source: GARP Publications Series No. 16, *The Physical Basis of Climate and Climate Modeling*, Report of the International Study Conference in Stockholm, 29 July–10 August 1974, World Meteorological Organization, International Council of Scientific Unions, Geneva, 1975.

ning roughly 50 million years ago, something happened to bring about a gradual deterioration of climate. This deterioration culminated, about 2 million years ago, in the arrival of a new mode of climate, characterized by a long sequence of perhaps as many as 20 major glacial–interglacial oscillations, which presumably continues to grip the world today.

The history of glacial–interglacial events is revealed in some detail for the past one million years by oxygen-18 analysis of deep-sea sediments and by several other kinds of paleoclimatic indicators. In that period of time, and most particularly during the last half-million years, the principal variations of ice volume and other conditions of climate appear to have been of a quasi-periodic nature, with a characteristic wavelength close to 100,000 years (see Figure 2.2). Lesser ice volume changes seem also to have occurred, with characteristic periods of about 20,000 and 40,000 years. The world has been in a relatively warm extreme-interglacial phase during the past 10,000 years, similar to other relatively brief warm phases that have occurred earlier at roughly 100,000-year intervals. In these interglacial phases, ice has been confined primarily to the Antarctic and Greenland ice sheets. In the intervening glacial phases, continental ice sheets have developed primarily in high northern latitudes, which covered a maximum of about 9 percent of the earth's surface and reached a volume of about 75×10^6 km^3—some three times present-day figures. The variations of ice conditions have been accompanied by midlatitude temperature changes of the order of 6 to 10 °C and by sea-level changes of the order of 100 m.

Since recovery of the earth from the last major glacial stage about 10,000 years ago, global climate has been found from a variety of paleoclimatic indications to have varied within narrower limits (see Figure 2.3). An intercontinental survey of mountain glacier moraines and tree lines has revealed three periods of glacier expansion (each of about 1000 years' duration), alternating with three periods of glacial contraction (each of 1000 to 2000 years' duration) in the past 8000 years. There is evidence, derived primarily from pollen analyses of lake and bog sediments, that significant shifts of the earth's vegetation zones accompanied these glacial variations, in what is sometimes referred to as the "neoglacial cycle." General temperature levels are believed to have varied by about 1 or 2 °C during the course of this cycle.

Narrowing our attention to the past millennium, we can begin to call into play various additional climatic indices such as human chronicles, tree-ring sections, and annually laminated ice cores from Greenland and Antarctica, to add further detail to our reconstruction of climatic history (see Figure 2.4). The period from about A.D. 1430 to A.D. 1850 was one of comparatively cold climate and expanded mountain glaciation. A member of the "neoglacial cycle," noted above, this cold period is commonly referred to as the "Little Ice Age." Earlier centuries were milder, although probably not everywhere as warm as the climate of today. From long instrumental records and human chronicles, typical fluctuations of 30-year averages of climatic variables over the past several centuries have been as follows: Major circulation features, such as the centers of subpolar lows and subtropical highs, and the position of the Intertropical Convergence Zone, have varied by 2 or more degrees of latitude. The positions of midlatitude troughs and ridges have shifted east or west

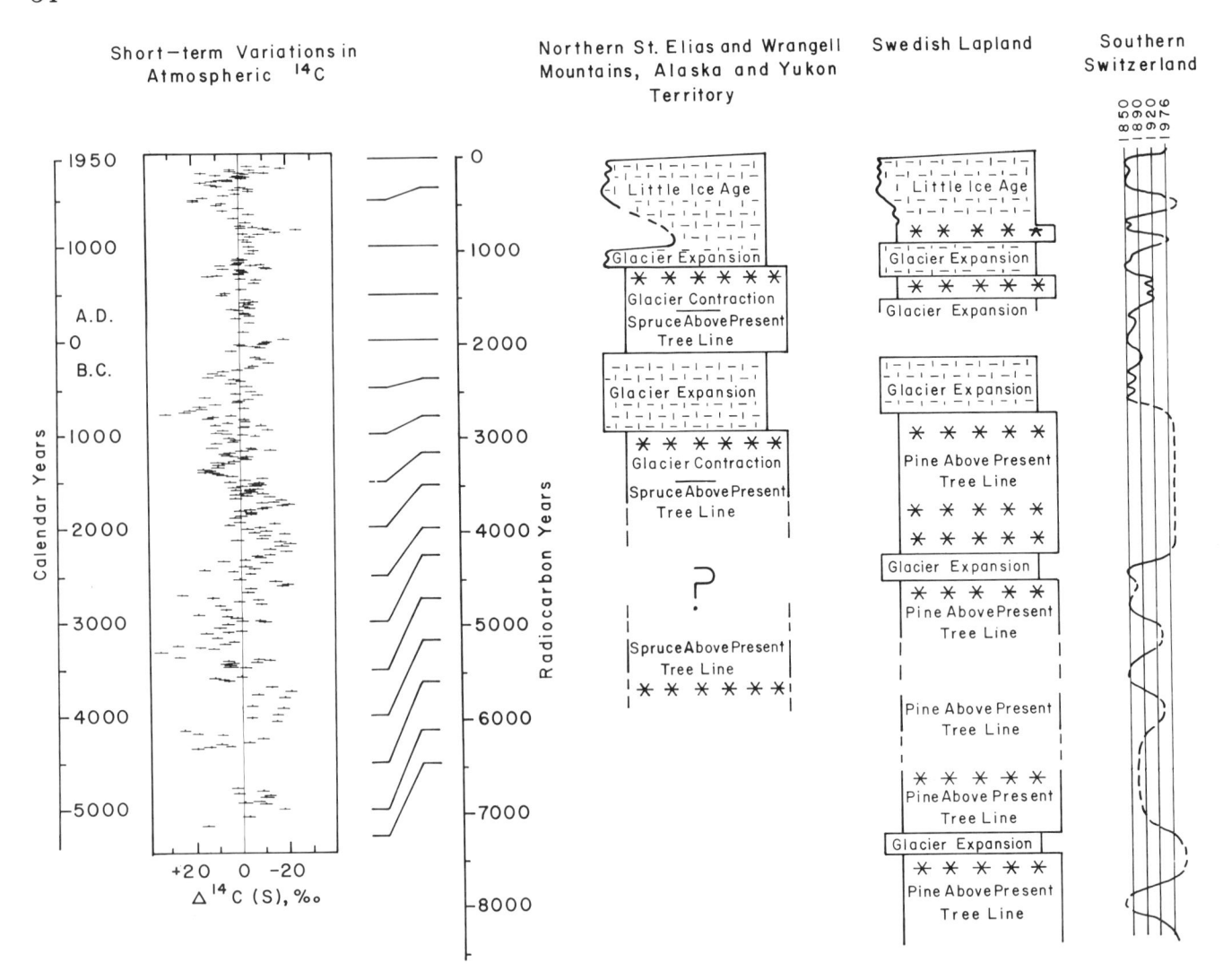

FIGURE 2.3 Holocene short-term atmospheric ^{14}C variations, glacial events, and tree-line fluctuations. Short-term ^{14}C variations are from Suess (1970; Figures 1 and 2). Glacial and tree-line events in St. Elias and Wrangell Mountains are from Denton and Karlén (1977); those in Lapland are from Karlén (1973, 1976; Karlén and Denton, 1976). Glacial variations in southern Switzerland are from Röthlisberger (1976).

References:

Denton, G. H., and W. Karlén (1977). Holocene glacial and tree-line variations in the White River Valley and Skolai Pass, Alaska and Yukon Territory, *Quaternary Res. 7*, 63.

Karlén, W. (1973). Holocene glacier and climatic variations Kebnekaise Mountains, Swedish Lapland, *Geografiska Ann. 55*, 29.

Karlén, W. (1976). Lacustrine sediments and tree-limit variations as indicators of Holocene climatic changes in Lapland, northern Sweden, *Geografiska Ann. 58*, 1.

Karlén, W., and G. H. Denton (1976). Holocene glacial variations in Sarek National Park, northern Sweden, *Boreas 5*, 25.

Röthlisberger, F. (1976). Gletscher- und Klimaschwankungen im Raum Zermatt, Ferpècle und Anolla, *Die Alpen 52 (3)*, 59.

by 10 to 20° longitude. Regional variations of temperature have been of the order 1-2 °C, and those of precipitation, of the order of 10-20 percent. Much larger variations have been determined to occur over periods shorter than 30 years. If repeated in the near future, variations of these magnitudes would probably have highly significant impacts on water resources and agricultural productivity.

In the first half of the twentieth century, the world was enjoying a full recovery, at least temporarily, from the Little Ice Age (see Figure 2.5). There are indications that, at that time, the circumpolar westerlies contracted toward the poles, and that, in the northern hemisphere at least, the amplitude of the planetary waves underwent a decrease. Also at that time, a general warming of the earth occurred, which was most pronounced in the Atlantic sector of the sub-Arctic. A rapid worldwide retreat of mountain glaciers and a poleward extension of the ranges of many flora and fauna took place then.

There is considerable evidence that, between the 1940's and about 1970, the climatic changes of the earlier part of this century had tended to undergo a reversal. Temperatures had mostly fallen, especially in the Arctic and the Atlantic sub-Arctic, where sea ice has been increasing. The circulation of the northern hemisphere appears to have shifted in a manner suggestive of an increasing amplitude of the planetary waves and of greater extremes of weather conditions in many areas of the world. (The situation in the southern hemisphere has not been so well documented.) These events have culminated, at times in the last several years, in the emergence of anomalous conditions in the monsoon belt of the tropics and in widespread drought in the Sahel zone of Africa and in northwest India (see Figure 2.6). To what extent these calamitous recent events are related to each other as manifestations of a globally coherent fluctuation of climate is not clear. In any event, they dramatize the fact that climatic variability, whether globally coherent or not, is to be expected no less on time scales of months and years than on time scales of centuries and millennia. An evident faltering of these tendencies of climate, in just the last five or ten years, attests also to the ephemeral nature of all climatic "trends." Such is the nature of climate and climatic variations.

2.3 ORIGINS OF CLIMATE VARIABILITY

Mindful as we are of the remarkable variability of climate in the remote past, the lesser variability of present-day climate catches us by surprise only in the sense that we lack a satisfactory explanation for it, and we do not know how to

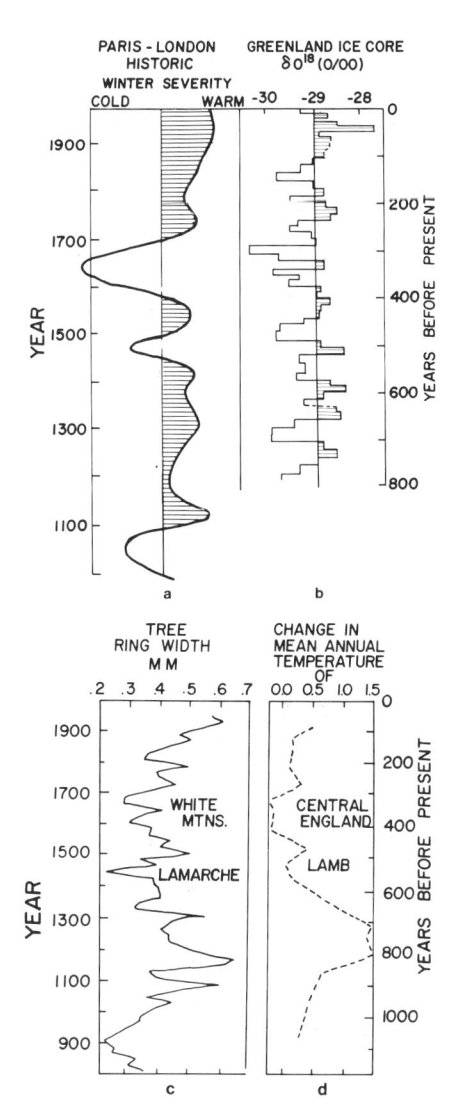

FIGURE 2.4 Climatic records of the past 1000 years. (a) The 50-year moving average of a relative index of winter severity compiled for each decade from documentary records in the region of Paris and London (Lamb, 1969). (b) A record of $\delta^{18}O$ values preserved in the ice core taken from Camp Century, Greenland (Dansgaard et al., 1971). (c) Records of 20-year mean tree growth at the upper treeline of bristlecone pines, White Mountains, California (LaMarche, 1974). At these sites tree growth is limited by temperature with low growth reflecting low temperature. (d) The 50-year means of observed and estimated annual temperatures over central England (Lamb, 1966).

References:
Lamb, H. H. (1969). Climatic fluctuations, in *World Survey of Climatology, 2, General Climatology*, H. Flohn, ed., Elsevier, New York, pp. 173-249.
Dansgaard, W. S., S. J. Johnsen, H. B. Clausen, and C. C. Langway, Jr. (1971). Climatic record revealed by the Camp Century ice core, in *The Late Cenozoic Glacial Ages*, K. Turekian, ed., Yale U.P., New Haven, Conn., pp. 37-56.
LaMarche, V. C., Jr. (1974). Paleoclimatic inferences from long tree-ring records, *Science 183*, 1042.
Lamb, H. H. (1966). Climate in the 1960s, *Geog. J. 132*, 183.

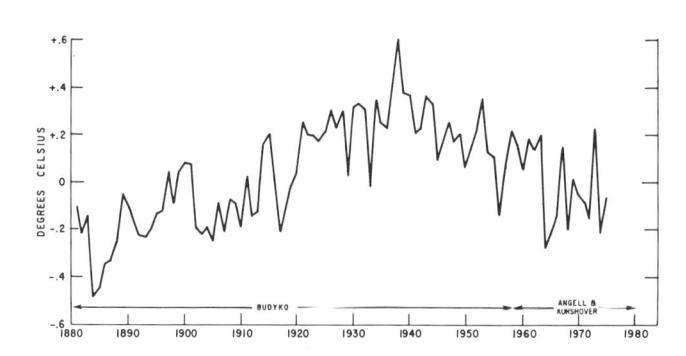

FIGURE 2.5 Recorded changes of annual mean temperature of the northern hemisphere as given by Budyko (1969) and as updated after 1959 by H. Asakura of the Japan Meteorological Agency (unpublished results).

Reference:
Budyko, M. I. (1969). The effect of solar radiation variations on the climate of the earth, *Tellus 21*, 611.

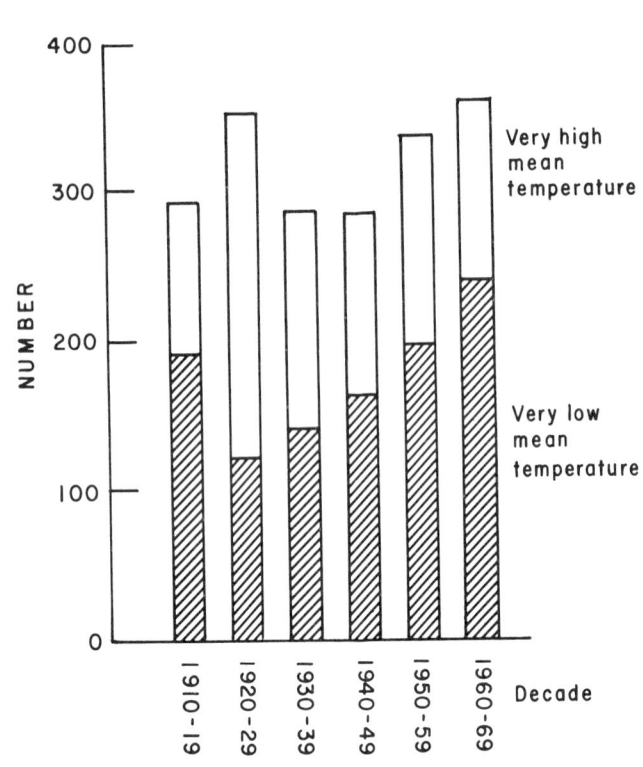

FIGURE 2.6 Total number of very low and very high monthly mean surface temperatures observed in each decade since 1910, at 65 stations distributed over the world. Length of each column indicates total number of temperatures in the decade that lie outside the range expected 90 percent of the time in all 60 years of record at each station. The hatched part of each column indicates total number of very low temperatures and the unhatched part the number of very high temperatures. Source: Japan Meteorological Agency (1974). *Report of a Study on Recent Unusual Weather and Climatic Trend in the World and the Outlook for the Future.* (Summarized edition, in English.) Mimeographed, 13 pp.

predict it in advance. Today, we find ourselves uneasy because through our use of energy we may be significantly disturbing the natural climate system. This uneasiness is justified. For surely we cannot hope to grasp all the ramifications of the forces we set in motion when we alter our atmospheric environment through the massive use of energy, if we fail to understand the source of natural climatic variations that will be taking place at the time. By the same token, if we fail to understand the source of natural climatic variations, how can we know when and whether our use of energy is responsible in any way for new developments in the climate that might be disadvantageous to society?

If we agree to the proposition that an understanding of natural climatic variability is a prerequisite to the intelligent assessment of the climatic impacts of energy use, we are faced with a problem: our understanding of natural climatic variability is not yet at hand, nor is it likely to be at hand for many years. In order to achieve it, an enormously difficult task lies before us in developing credible mathematical models of the climate system (see Figure 2.7). We cannot hope for success without a national and international commitment to devote the considerable resources of various kinds that are required to tackle so complex and so challenging a scientific problem.

At this stage, some assessment can nevertheless be made as to the *nature* of the problem to be solved, if we are to understand climatic variability and to predict it.

First, we can point to the fact that those parts of the climate system that we already understand—at least in principle—constitute a highly complex fluid-dynamical system, involving the coexistence of water in all three of its phases, in which there are many highly nonlinear interactions. While these nonlinear interactions are the source of many difficulties faced in developing successful climate models, they are also to be recognized as the source of the very climatic variability that we seek to understand.

Second, our knowledge of the climate system is thus far confined primarily to the atmospheric part of the system. While the atmosphere is perhaps the most complex part of the system, the atmosphere alone cannot account for the bulk of climatic variation, except on very short time scales. To take into account the longer time scales, it is necessary to consider the interaction of the atmosphere with other parts of the fluid envelope of the earth, which impart correspondingly longer relaxation phenomena to the climate system. To begin with, numerical simulation schemes (models) of the atmosphere-ocean-sea ice system are required, following the lead of some relatively crude models now under development. It is hoped that such models, although they admittedly do not contain all the elements of the complete climate system, will ultimately succeed in capturing the essence of climatic variability on time scales of months, years, and decades. It is possible that much of the variability on still longer time scales (centuries and perhaps even millennia) can also be captured in joint atmosphere-ocean-sea ice models that include deep-ocean processes.

Ultimately a capability to model climatic variability on the time scales of glacial-interglacial events would be highly desirable, although the relevance of such models to the problems of climate in the time frame of direct concern to us here may seem more remote. In any case, the modeling of climatic variability on such long time scales would require incorporation of ice-sheet dynamics into the climate system.

Present understanding of the causes of climate variability is meager at best. Most causal hypotheses fall into one of two schools of thought.

One school of thought holds that climate varies mainly as a result of environmental changes outside the climate system. A familiar example is the notion that climate responds in a more or less deterministic manner to variations of total solar irradiance, presumed but never proven to accompany the 11-year or longer sunspot cycles. It cannot be ruled out that such external influences on climate are real, but even if they are real there is much evidence to indicate that they can account only for a small fraction of the total variability of climate, the bulk of which requires a different explanation.

The other school of thought holds that most, if not all, of climatic variability originates from purely internal stochastic

The Changing Climate 57

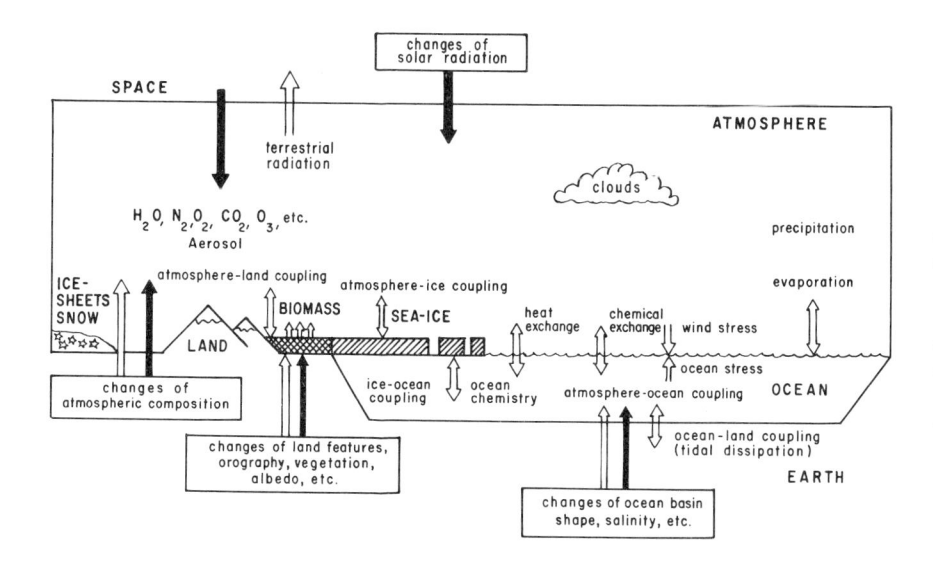

FIGURE 2.7 Schematic illustration of the components of the coupled atmosphere-ocean-ice-land surface-biomass climatic system. The full arrows (→) are examples of external processes, and the open arrows (⇒) are examples of internal processes in climatic change. Source: GARP Publications Series No. 16, *The Physical Basis of Climate and Climate Modeling*, Report of the International Study Conference in Stockholm, 20 July–10 August 1974, World Meteorological Organization, International Council of Scientific Unions, Geneva, 1975.

variation of the climate system itself. Indeed, present and planned climate modeling efforts generally subscribe implicitly or explicitly to this view. This is as it should be, for it is only through experiments with climate models capable of dealing with internal stochastic processes that one can hope to gain a realistic measure of the capacity of the climate system to vary through internal dynamics as well as external forcing.

In this connection, it has to be emphasized that in a highly interactive system, which involves as wide a range of reaction times as that of the real climate system of the earth, many opportunities exist for the internal stimulation of slowly varying modes of system behavior. In the joint atmosphere-ocean part of the climate system, for example, oceanic thermohaline reaction times are likely to be one or more orders of magnitude slower than the slowest (thermal) reaction time of the atmosphere. Under these circumstances, the stimulation of what are essentially random variations of the ocean part of the system would, to the extent that such variations are transmitted to the atmospheric part of the system, give rise to what are perceived as nonrandom variations in the atmosphere. There is ample reason to believe that interannual variations of climate to a considerable extent owe their existence to ocean-atmosphere interactions of this "random" form. More generally, what may appear to be a nonrandom form of variability in one part of the climate system may be recognizable as having its origin in a strictly random (albeit dynamically constrained) form of variability in another part of the system. If such internal sources of climatic variability are not recognized for what they are, the variability might easily be mistaken for an effect of variable external forcing.

It is conceivable that all climatic variations, even including the ice ages, owe themselves to "random" processes in those parts of the climate system that possess extremely slow reaction mechanisms. The plausibility of this view of climatic variation, together with its implications for climate predictability, remains to be explored.

2.4 A LOOK TO THE FUTURE

Our very superficial understanding of climate dynamics translates to a very modest capability for prediction of climate. One of the principal motivations for climate-dynamics research is to gain enough insight into climate system behavior to enable the skillful prediction of climate. Yet, it has to be recognized that our understanding is not yet adequate to assure us that climate *can* be predicted with practically useful skill for an extended period of time into the future.

In the face of the many urgent societal needs for information about future climates, it seems appropriate to inquire further into the present situation with regard to climate prediction.

On the one hand, it can be observed that the pioneers in the field of modern climate-dynamics research have been reluctant to engage in any kind of speculation about future climatic developments. This is so despite the fact that those pioneers are probably in the best position of all to speculate intelligently about the future. This reticence is not difficult to understand. It is born of a healthy respect for the limitations of present generations of climate models, which necessarily omit too much of the physics that they will eventually have to include before real-time predictions are likely to bear much relationship to reality.

On the other hand, one can point to the fact that in the last 20 years or so, a number of deterministic long-range projections of climate have found their way into the scientific literature. The authors of such projections have based them on various kinds of simplistic reasoning, sometimes involving "blind" statistical extrapolation of what are essentially unverified periodicities in past climate and other times involving certain external physical processes (e.g., solar activity) that they have supposed to have an overriding influence on the course of global climate. It cannot be flatly asserted that all of these projections are without merit. Unfortunately, however, where merit exists it will be likely to remain obscure until climate-dynamics research finally progresses to a point where it can confirm that merit. By then, the modeling framework available to illuminate climatic dynamics would make possible a much more comprehensive assessment of reality.

If society is to embrace any deterministic climate projection as a basis for sound future planning, it should be clear that two criteria have to be met. First, the scientific basis of the projection must be rational. Second, the projection must be accompanied by a suitably objective and unbiased measure of its own reliability. No deterministic projections of future climate now available can be said to meet both of these criteria.

Making reliable projections of future climate of any kind will likely remain an elusive goal until our knowledge of climate dynamics is adequate to provide the answers to two sets of fundamental questions.

The first set of fundamental questions to be answered concerns the natural climate system: (1) Were man's various impacts on environment to be held in the future at their present-day levels, how would the earth's climate evolve from its present state in the decades and centuries ahead? (2) Is the answer to (1) knowable? Or is it the case that the natural evolution of climate is a probabilistic process in time, such that, after a relatively short interval, the state of global climate will be essentially unrelated to its present state? (3) To the extent that the answer to (1) is knowable, to what further extent is it within man's power to find that answer?

The second set of fundamental questions to be answered concerns man himself in relation to his potential capacity for interfering in the workings of the natural climate system: (4) On the basis of each of a variety of credible alternative scenarios as to future increases and proliferation of his impacts on environment, can man assess the consequences of those changing impacts on the natural evolution of climate in the decades and centuries ahead? (5) Can man establish which, if any, alternative scenarios would lead to "unacceptable" climatic consequences and are therefore to be avoided? (6) If the answer to (5) is yes, will man come into possession of such knowledge in time to avert possible calamity?

The very essential matter of climate *predictability* is addressed by question (2). Should climate turn out to be inherently unpredictable after a short interval of future time, then the capacity of science ever to provide answers to the other questions posed here will be limited, perhaps severely so.

Up to this point we have mainly been referring to *deterministic* projections of climate, by which we mean statements as to the "expected" climate in a specific future interval of time or in each of a series of future intervals. Also to be considered are *probabilistic* projections of climate, in which the conditions to be projected for specified future intervals of time are not the "expected" climate as such but rather the probability density distribution of the full range of outcomes of climate that the statistics of *past* climates indicate to be possible. Such probabilistic projections are normally assessed through statistical inference rather than physical inference and are based essentially on past experience rather than physical understanding with regard to climatic variability. Projections of this kind are always meaningful, in the sense that even if no organized time-series structure whatever is evident in past climatic behavior, future projections can still be made. In that case, the projections would reduce to unconditional statements to the effect that the future probability of any given climatic condition (at *any* time in the future) is identical to that of the same climatic condition in the past.

The record of past climates deserves our very close scrutiny, to learn the full extent of organization in the time-series structure of climatic variations that could be used to frame improved probabilistic projections of future climate. To date, surprisingly little has been done along these lines.

Viewing past climates in the perspective of events of the past million years, we can at least venture the conclusion that, in the long run, natural forces will be more likely to relax our global climate toward a glacial condition than to maintain climate indefinitely in its present, unusually warm interglacial condition. When the return to a glacial condition may begin and at what rate global climate will be altered when it does begin are by no means obvious from the paleoclimatic record.

Viewing past climates in the somewhat shorter perspective of the postglacial period, the sequence of "neoglacial" events appears to be a dominant feature. It would be tempting to suggest that the last neoglacial event, the "Little Ice Age" of the fifteenth to nineteenth centuries, is behind us and that the next comparable event will not arrive for another thousand years or more. On the other hand, we should keep in mind evidence that the Little Ice Age was appreciably shorter than earlier neoglacial events. This invites an alternative view of the comparative warmth of the twentieth century, as being not a recovery from the Little Ice Age but perhaps only an interruption of it.

Beyond such vague statements as these, disturbingly little can be said about the probable course of natural climate in the decades and centuries ahead. Moreover, a number of man's activities are being ingested by the climate system, adding further uncertainty to an already uncertain future of climate. The need for better understanding of the climate system, in all its aspects, is a clear and present one.

II

EFFLUENTS OF ENERGY PRODUCTION

Effluents of Energy Production: Particulates

3

GEORGE D. ROBINSON
Center for Environment and Man, Inc.

3.1 PARTICLES IN THE ATMOSPHERE

Particles are a normal constituent of the atmosphere. If we exclude the larger water droplets in clouds or fog that have formed in slightly supersaturated air, particles can properly be called a trace constituent, with a global average mass mixing ratio near the surface of about 10^{-8}, or 10 μg^{-3}, in the units conventionally used in monitoring. We are concerned here with the proportion of this load that can be attributed to man's activities and with any effects that a change of this proportion might have on climate.

3.2 SOURCES OF ATMOSPHERIC PARTICLES

Particles may be characterized as wind-raised dust; wind-raised sea salt; direct products of combustion, soot, ash, condensed organic materials, etc.; indirect products of combustion, i.e., particles formed by chemical reactions in the atmosphere from the gaseous products of combustion—sulfates, organic nitrates, sulfuric and nitric acid; volcanic particles; and particles formed in the atmosphere from such products of plant and animal life and decay as terpenes, H_2S, and NH_3. Table 3.1 summarizes two attempts to estimate the annual production of particles by these various means. The material is several years old, but more recent work has not reduced the uncertainties indicated, which are certainly not overstated: the "nitrate" entry is particularly suspect. Table 3.1 also attempts to separate "natural" and "anthropogenic" sources; there is very little basis on which to make this separation in some categories, particularly "forest fires" and "soil dust."

Figure 3.1 shows the range of particle radius with which we are concerned and indicates the general nature of the radius-number distributions that are observed.

3.3 SINKS OF ATMOSPHERIC PARTICLES

In the steady state, production and loss of atmospheric particles balance and a "mean residence time" may be defined as the ratio of loading to production rate. The estimate of mean loading given above was obtained from an assumed production of 1.8×10^{15} g yr^{-1} and a residence time of

10 days, which is the average residence time of a water molecule in the atmosphere. The major removal mechanisms are fallout under gravity without condensation; processes in which water condenses on a particle that later falls out in rain—conventionally termed rainout and the most effective sink; and processes in which the particle is captured by falling rain or snow and carried to the ground—conventionally termed washout. Some particles such as nitrates and certain organic materials may in some circumstances volatilize or decompose to gaseous products. The shape of the large radius end of the size-distribution curve is strongly influenced by dry fallout. The shape of the small-radius end depends to a great extent on coagulation processes, which do not change the mass mixing ratio but greatly affect optical properties.

It will be clear from the nature of the removal processes, particularly those involving precipitation, that the residence time of a particle can be greatly influenced by the location and time of its production. A particle may exist in the atmosphere for a time measured in minutes or in years; this is true of both natural and man-made particles. Volcanic particles, for example, are often injected high in the atmosphere, away from the immediate influence of precipitation processes

TABLE 3.1 Global Summary of Source Strengths for Atmospheric Particulate Matter

	Strength (Tg/yr)			
Source	Natural		Anthropogenic	
Primary particle production				
Fly ash from coal	—		36	
Iron and steel industry emissions	—		9	
Nonfossil fuels (wood, mill wastes)	—		8	
Petroleum combustion	—		2	(10-90)
Incineration	—		4	
Agricultural emission	—		10	
Cement manufacture	—		7	
Miscellaneous	—		16	
Sea salt	1000			
Soil dust	(428-1100)	200		(?)
Volcanic particles	4			
Forest fires	(3-150)	3	—	(?)
SUBTOTAL		1207	92	
Gas-to-particle conversion				
Sulfate from H$_2$S	(130-200)	204	—	
Sulfate from SO$_2$	—		147	(130-200)
Nitrate from NO$_x$	(60-430)	432	30	(30-35)
Ammonium from NH$_3$	(80-270)	269	—	
Organic aerosol from terpenes, hydrocarbons, etc.	(75-200)	200	27	(15-90)
SUBTOTAL		1105	204	
TOTAL	(773-2200)	2312	296	(185-415)

MODIFICATION OF THE TROPOSPHERE

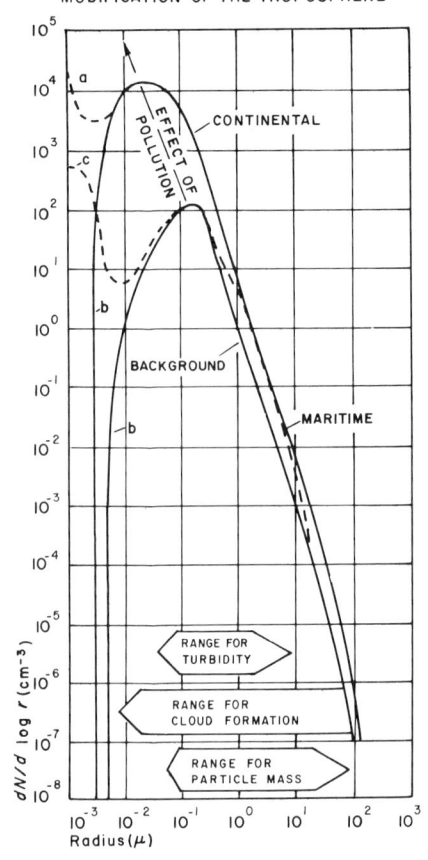

FIGURE 3.1 Typical comprehensive size distribution for the principal tropospheric regimes and the size ranges important for turbidity, cloud formation, and mass concentration of particles. Curves a and b refer to possible variation of the size distribution with and without continuous production of very small particles. The arrow indicates the effect of pollution on the location of the maximum of the size distribution. After *Inadvertent Climate Modification, Report of the Study of Man's Impact on Climate*. MIT Press, Cambridge, Mass., 1971.

so that although they are a minor entry in Table 3.1, they have received considerable attention as a potential agent of climatic change. The same considerations apply to particles injected into the stratosphere by aircraft.

3.4 ANTHROPOGENIC PARTICLES

The major sources of particles associated with man are industrial production and processing of materials such as metals and cement; combustion associated with industrial, commercial, and domestic needs including transportation; and agriculture. We look first at the industrial/domestic sector.

Increasing populations, and increase in specific energy consumption, potentially increase particle production by processing and combustion, and increasingly within the last few decades steps have been taken to control emissions. Figure 3.2 illustrates the degree of success of control measures within the United States.

Urban measurements of SO_2 concentration may be as relevant to considerations of atmospheric particle load as are the urban particulate load measurements themselves. Figure 3.3 illustrates the effect of fuel controls on urban SO_2 concentrations. Draconian control measures have been effective, urban SO_2 concentrations declining in proportion to the controlled proportion of S in the fuel burned. Figure 3.4 shows the trend in SO_2 concentration at a selection of stations. We emphasize the measurements of SO_2 because of its fate in the atmosphere. Its mean global life as a gas is estimated to be two to three days. In the winter of 1968-1969, an estimate for the State of Connecticut was 2-3 hours (Hilst, 1970). Recent work in the St. Louis, Missouri, area suggests a few hours, with some indication that it disappears more quickly from the general urban plume produced by many small sources than from the highly concentrated stack plumes of major power plants. Possible sinks are absorption on ground surfaces and conversion to sulfuric acid with production of acid particles, some of which are in turn converted to $(NH_4)_2SO_4$, NH_4HSO_4, or $CaSO_4$. Such particles are universally present in the atmosphere. The production rate of Table 3.1 seems to imply that about 75 percent of SO emission is converted to sulfate particles before leaving the atmosphere, but some recent measurements in highly polluted atmospheres find only about 10 percent conversion. Altshuler (1973) shows that at nonurban sites in eastern United States, concentrations of gaseous SO_2 and particulate SO_4^{-2} are about equal and of order 10 μg^{-3}. Concentrations of SO_4^{-2} of up to 30 μg^{-3} have been measured at rural sites in Sweden.* Investigation of details of the conversion is just beginning, but there seems little doubt that it occurs with a time constant, to some extent controlled by the concentrations of other pollutants, of hours to days and at efficiencies of not less than 10 percent in regions of high emission and perhaps more than 50 percent global average.

We briefly mention particle production by agricultural practices. The first major contribution to the global load of particles is by wind-raised dust. The American dust-bowl phenomenon of the 1930's has not recurred, in part because the accompanying climatic conditions have not recurred in their full severity, in part because of the success of positive countermeasures. There may now be a similar problem in China, and there is certainly a major windborne dust problem in overgrazed arid regions in Africa and India. The second contribution from agriculture is to the particle load due to combustion, particularly by the "slash and burn" type agriculture in less arid tropical and subtropical regions of Africa and Southeast Asia. The particle load from these agricultural practices is not negligible in the context of local climatic effects, but the associated immediate primary resource depletion is such as to call for remedial action on a short time scale without regard to possible longer-term climatic effects. In this discussion of climate and energy, we will therefore not further consider the sources associated with agricultural malpractice.

3.5 ATMOSPHERIC PARTICLES AND WEATHER AND CLIMATE

The properties of atmospheric particles that affect the processes of weather and climate are those concerned with condensation of water on the particle, which, for convenience, we will term "nucleation properties," and with interaction of the particle with solar and terrestrial radiation, which we will term "optical properties." Since condensation or evaporation of water on or from the particle change its size and composition, the nucleation properties are not independent of the optical properties.

NUCLEATION PROPERTIES

Many of the constituents of atmospheric particles are hygroscopic or deliquescent. Figure 3.5(a) shows how the size of such particles might be expected to vary with the surround-

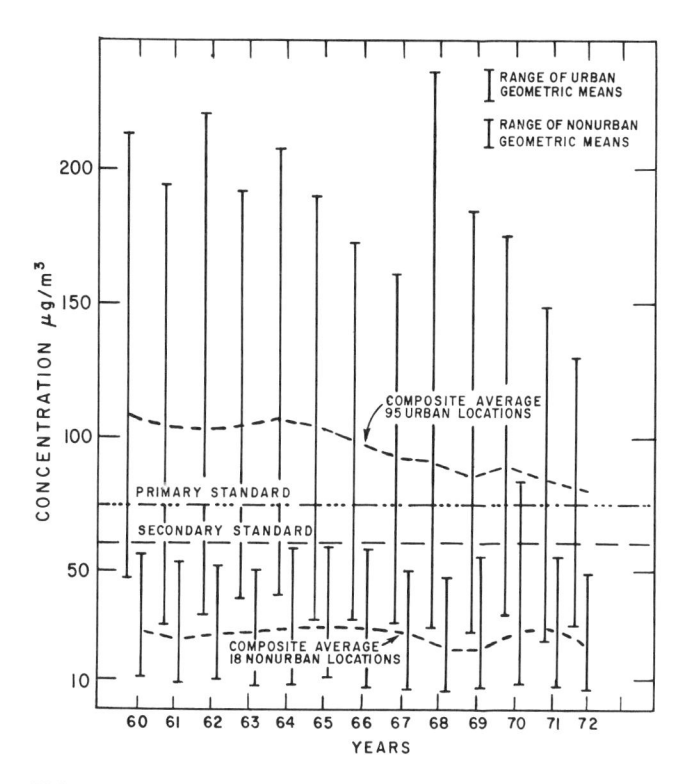

FIGURE 3.2 Composite levels of total suspended particulate at urban and nonurban NASN stations. From "Monitoring and Air Quality Trends Report 1972," U.S. Environmental Protection Agency, EPA-450/1-73-004.

*The measurements were made by L. Granat and S. Larssen and examined and communicated to me by R. J. Charlson, A. H. Vanderpol, and A. P. Waggoner of the University of Washington, Seattle, Washington.

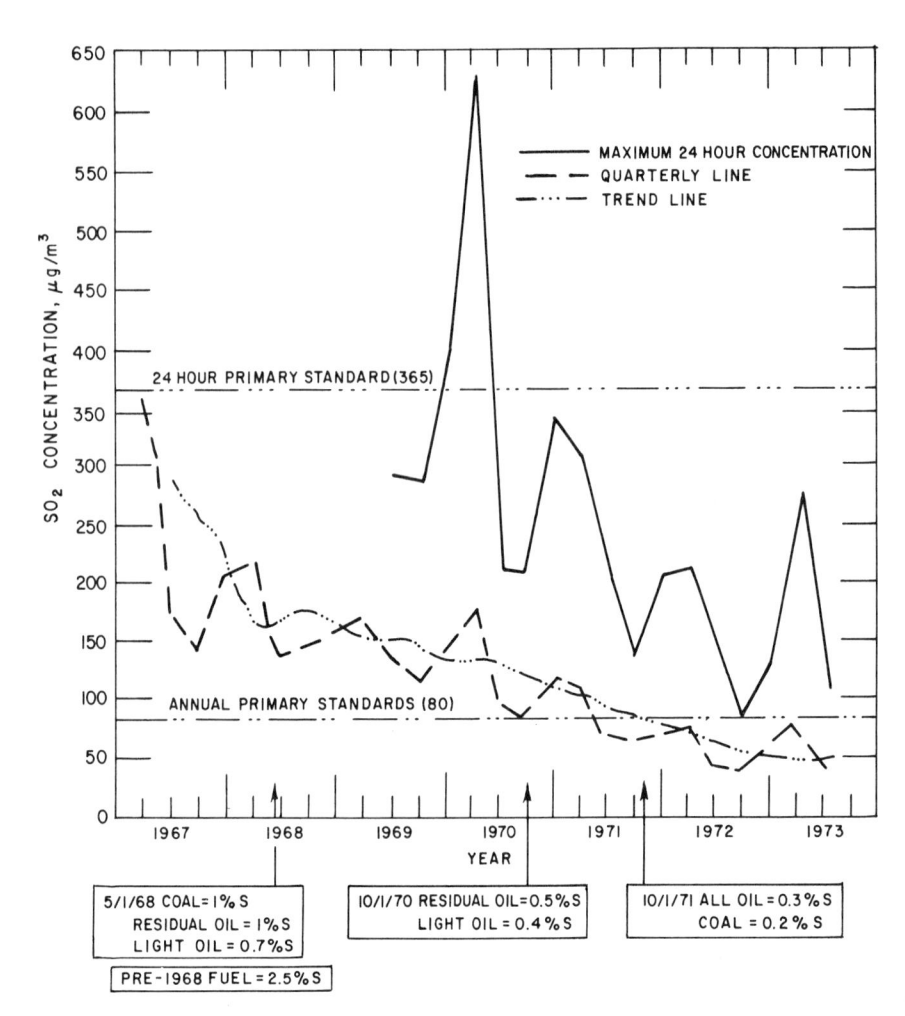

FIGURE 3.3 Comparison on SO₂ trends at Bayonne, New Jersey, with regulations governing percent sulfur content in fuel. From "Monitoring and Air Quality Trends Report 1972," U.S. Environmental Protection Agency, EPA-450/1-73-004.

ing relative humidity. Charlson et al. (1974) have confirmed this type of behavior for particles in the atmosphere. Figure 3.5(b) illustrates some of their observations, taken near St. Louis, Missouri. The size actually attained by hygroscopic particles, and by all particles in a supersaturated atmosphere, is controlled by the rate of supply of water to the particle—the equilibrium radius may not be reached at any time as relative humidity changes at humidities below 100 percent, and at higher humidities only the larger particles may be effective in forming cloud droplets. Effectively, all particles in the atmosphere are condensation nuclei, but in practice only a proportion of them—"cloud condensation nuclei"—may be active in cloud formation on any one occasion. The number and size distribution of cloud or fog droplets forming in cooling air depend on the nature and size distribution of the particles present—clouds and fogs forming in polluted air will have properties different from those forming in unpolluted air. Subsequent precipitation from and dissolution of the clouds may be affected, and bulk optical properties will be different.

Certain rare atmospheric particles have the property of initiating the freezing process in supercooled droplets at higher temperatures than that at which they freeze spontaneously (about −40 °C). Certain pollutant particles might

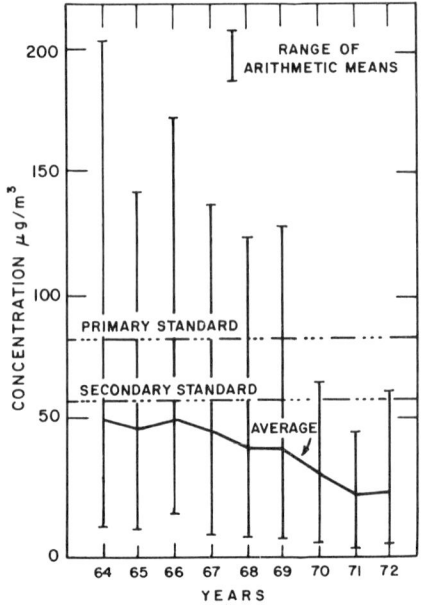

FIGURE 3.4 Composite levels of sulfur dioxide at 32 NASN stations. From "Monitoring and Air Quality Trends Report 1972," U.S. Environmental Protection Agency, EPA-450/1-73-004.

Effluents of Energy Production: Particulates

have this property. Silicates in wind-raised dust are perhaps the most common "natural" freezing nuclei. Some material processing, e.g., steel manufacture, might produce such nuclei artificially. Artificial introduction of particularly effective freezing nuclei is the basis of "rainmaking" operations, but there is no evidence at present of consistent substantial production of similarly effective nuclei by industrial and domestic activity, although localized effects have been reported.

Direct sampling from aircraft and radar observations have demonstrated changes in the rain-producing processes and particle numbers and size distributions in summer clouds in the St. Louis area. There is also some indication that the areal distribution of precipitation is affected by the presence of the city, although in the latter case it is not yet possible to distinguish clearly between added particles and added heat as causative agents.

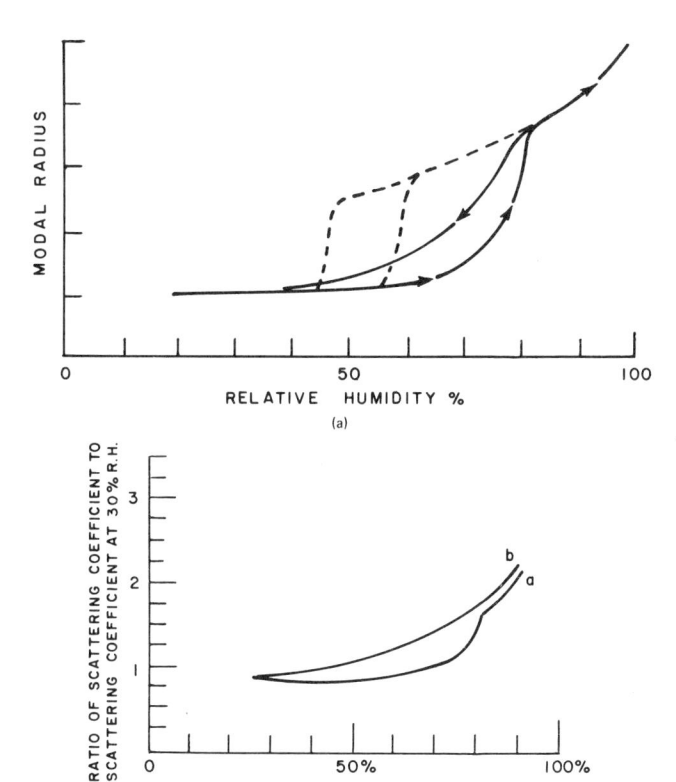

FIGURE 3.5 (a) Relation of size and relative humidity for a typical sample of urban pollutant particles containing hygroscopic and deliquescent material, e.g., H_2SO_4 and $(NH_4)_2SO_4$. Note the hysteresis and the indeterminacy of the relation for decreasing relative humidity. (b) Measured relation between relative humidity and the scattering coefficient of air samples. Tyson, Missouri. (Data from D. S. Covert, University of Washington.) Curve a, 2330h, September 24, 1973 (deliquescent particle behavior). Curve b, 1223h, September 23, 1973 (hygroscopic particle behavior).

OPTICAL PROPERTIES OF SINGLE PARTICLES

The interaction of a spherical homogeneous particle with electromagnetic radiation can be computed to any required degree of accuracy in terms of the ratio of its radius to the wavelength of the radiation and the "complex refractive index," which characterizes refractive and absorptive properties of the material. The properties of meteorological interest are the optical cross section of the particle, the proportion of radiation incident on this cross section that is absorbed, and the polar diagram of the intensity of scattered radiation. The absorbed fraction is usually expressed as $(1 - \omega_0)$, where ω_0 is the "albedo for single scatter." The complex refractive index must be known for all wavelengths of solar and terrestrial radiation carrying significant energy. Figures 3.6 and 3.7 are examples, respectively computed and observed, of optical cross sections and polar scattering diagrams of particles found in the atmosphere.

OPTICAL PROPERTIES OF ASSEMBLIES OF PARTICLES

Many of the problems of the optical effects of suspensions of particles in the atmosphere can be discussed with sufficient accuracy by single scattering approximations—the supposition that a photon traversing the depth of the atmosphere is very unlikely to encounter more than one particle. This is particularly true if we are interested only in the total energy carried by the various streams of radiation. In this approximation, Figure 3.8 represents, schematically, the radiative effect of a layer of particles. The layer is illuminated from above by the direct solar beam and by a diffuse flux of scattered radiation. Some of the incident solar radiation has also been absorbed. The layer itself absorbs some radiation and scatters some upward and downward. Of the radiation transmitted downward, direct and diffuse, some is absorbed and some scattered upward, both from the atmosphere and the ground. Some of this is trans-

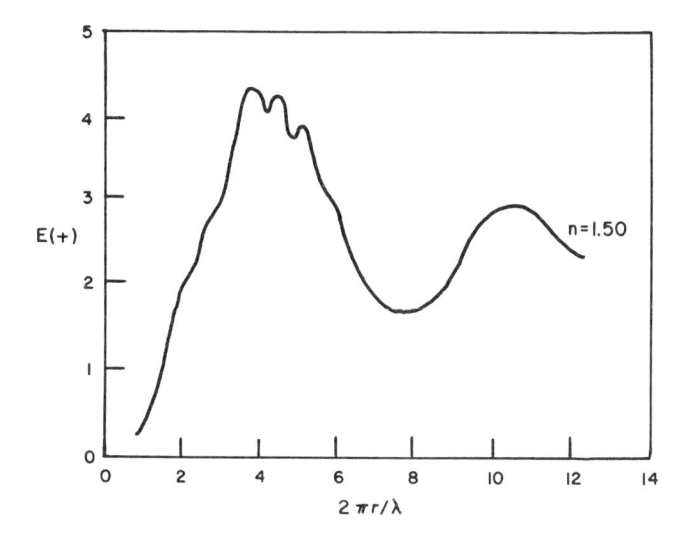

FIGURE 3.6 Ratio of optical and geometric cross section for particles with a refractive index of 1.5.

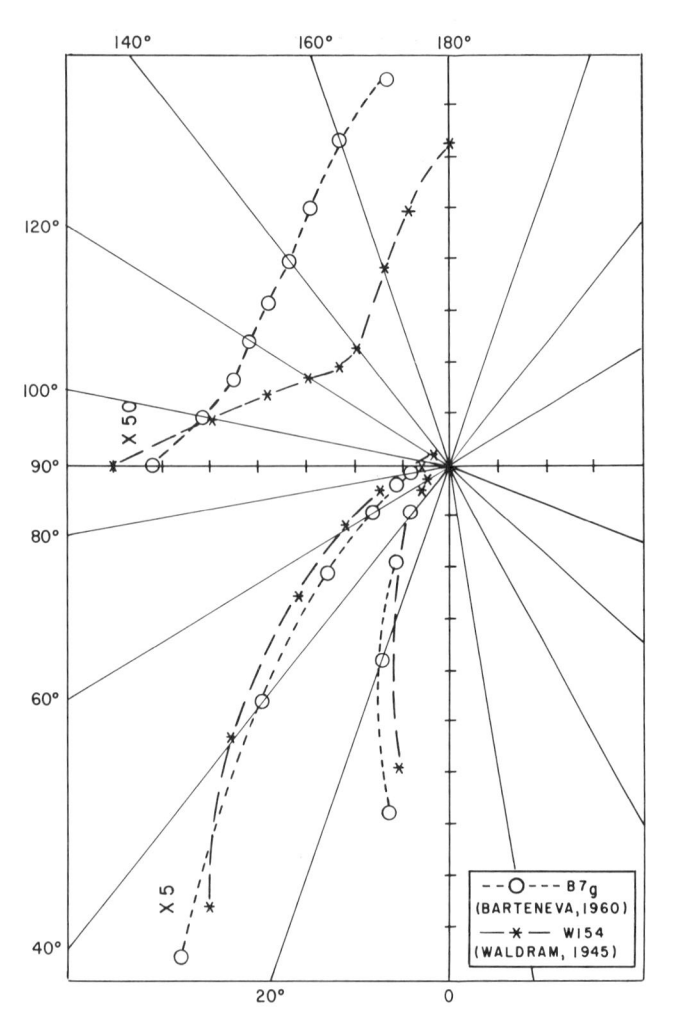

FIGURE 3.7 Phase function for atmospheric scattering.

mitted, with or without scattering by the particle layer, and some is rescattered downward. The process continues as illustrated. The proportion scattered downward or upward varies with the direction of the incident radiation, as shown in Figure 3.9, so that the proportion of the radiation in the direct solar beam that is scattered upward and lost to space is greater for low than for high solar elevation. In particular, the proportion of backscatter from a diffuse flux is greater

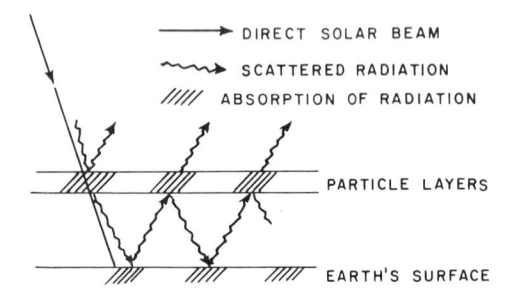

FIGURE 3.8 Effect of a particle layer on albedo (schematic).

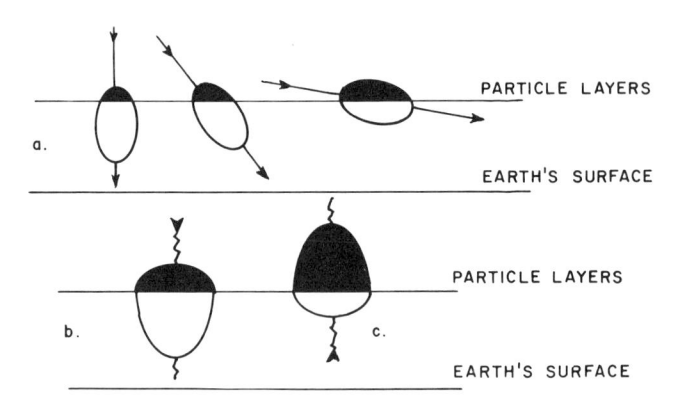

FIGURE 3.9 Upward and downward scattering of solar radiation by a particle (schematic).

than that from a normally incident beam and less than that from a beam at grazing incidence. For high solar incidence and a sufficiently high underlying albedo, the possibility thus arises that a layer of particles might redirect to earth more energy from the upward diffuse flux than it redirects to space from the predominantly direct beam incident from above. A particle that itself absorbs no radiation and scatters some to space may in this way actually increase the radiation absorbed by the planet. If the particle itself absorbs, the likelihood of an increase in planetary absorption is even greater. For a detailed numerical solution of this problem we require, for each wavelength of radiation, a specification of the number density, the size distribution, the distribution with height, and the single scatter albedo of the particles (or the complex refractive index of the material of the particles). If there are present particles of different materials or if the refractive index varies with size or if there are nonspherical particles, we are faced with difficulties that have not yet been resolved. We must also know the absorptive and scattering properties of the overlying atmosphere and the albedo and angular variation of scattering of the underlying atmosphere-surface system. Although complex, the problem can be solved to a useful degree of approximation in a horizontally homogeneous atmosphere. Several authors have attacked the problem, and we will examine some examples of their work in Section 3.7. In general, they show that the interaction of atmospheric particles and solar radiation affects climate and weather processes in two ways: it may change (either increase or decrease) the energy absorbed by the planet, and it may change the pattern of heating and cooling in the surface-atmosphere system, initially perturbing the static stability of the atmosphere. Change in the planetary albedo means change in the equivalent radiative temperature of the planet. The difficulty of inferring conditions on a planet's surface from its equivalent radiative temperature is well illustrated by considering the planet Venus, which is irradiated at about twice the intensity of earth and has about half its absorptivity. Its equivalent radiative temperature is thus not greatly different from that of the earth, but its surface temperature is believed to be about 700 K.

Computation of the transmission and reflection of radiation from clouds is not different in principle from that for any other collection of particles. The single scatter approximation is, of course, inappropriate, and the assumption of spherical particles must be made so that there are uncertainties where ice clouds are concerned. The optical properties of pure water are reasonably well known, and the albedo of a cloud containing a given amount of pure water can be shown to increase as the number of cloud particles increases. Clouds formed in polluted air containing a large number of cloud condensation nuclei would, therefore, be expected to have a higher albedo than those formed from cleaner air, other factors being unchanged. The simple picture might, however, be changed if absorbing material were present in the polluting nuclei.

The same equations of radiative transfer apply to terrestrial as to solar radiation, but in practice very different computations are required because of the nature of the source and absence of a direct beam. The effects of particles (excluding, of course, cloud droplets) are mainly determined by their small size relative to the wavelengths of concern. For the radiation carrying maximum energy flux, the parameter $2\pi a/\lambda$ is lower by a factor of 20 in the terrestrial radiation than in the solar radiation; and for a given refractive index, the optical cross section and single scatter albedo are much smaller at terrestrial than at solar wavelengths. The major difficulty is to know the complex refractive index at terrestrial wavelengths.

In summary, we have in principle the means for computing the effects of homogeneous spherical particles on energy transfer in a stratified horizontally homogeneous atmosphere. We need to know, and to predict, the number, size, and material of all the particles and the refractive index of this material over a wide range of wavelengths. We need to know where the particles are and will be in the atmosphere. We need to know the reflectivity of the earth's surface over a wide range of wavelengths.

In principle, we can compute the required radiative properties of clouds; but in practice, because some limit must be set to computation (and because the atmosphere is usually not homogeneously stratified), it is desirable to know the bulk reflectivity and absorptivity of clouds and regard them as bounding surfaces.

3.6 SOME OBSERVATIONS OF THE RADIATIVE EFFECTS OF PARTICLES

At present, we cannot clearly identify any effect of changes in atmospheric particle content on climate. There are some suggestive correlations of volcanic activity and surface temperature, but they are far from establishing a causal relation.

There are numerous phenomena that can be described as effects of particles on weather. The most obvious concerns visibility—the distance at which a black object large enough to be comfortably resolved can be seen against the horizon. Contrast determines visibility, and contrast is reduced by scattering of light into the line of sight to the black object. For an average limiting contrast, the visual range is

$$V = 4/\sigma,$$

where σ is the scattering coefficient of the atmosphere. For a particle-free atmosphere at sea level, σ is about 1.5×10^{-2} km^{-1}, so V is about 270 km. Such ranges have very occasionally been reported in the Antarctic. If there are 10 μg^{-3} of sulfate particles of density 1.67, all of radius 0.3 μm at sea level, V is about 65 km. (This is the maximum optical effect of the loading that we have suggested as a global average.) For the current U.S. urban average of about 80 μg^{-3}, the minimum visibility would be 8 km if the particles were of this material and size. These numbers, although not unrealistic, are computed for the size distribution that produces the maximum optical effect. Recent observations in Sweden (see footnote in Section 3.4) have shown a tenfold variation of scattering coefficient for a given *mass* loading of $SO_4^=$, the maximum effect corresponding to that computed here. The size distribution of the particles is all important.

Figure 3.10 (from Munn, 1973) shows the number of hours with visibility 10 km or less in an industrial region (Windsor, Ontario) and two locations remote from industry (Mont Joli, Quebec, and Gander, Newfoundland). Windsor reflects the urban improvement shown in Figure 3.2, but the frequency of haze at the remote stations is increasing.

Numerous observations of the direct and diffuse components of solar radiation in cities show the effect of particles in the atmosphere. Robinson (1962) examined records from the suburbs of London and Vienna in a period around 1950 and found on average both scattering and absorption in

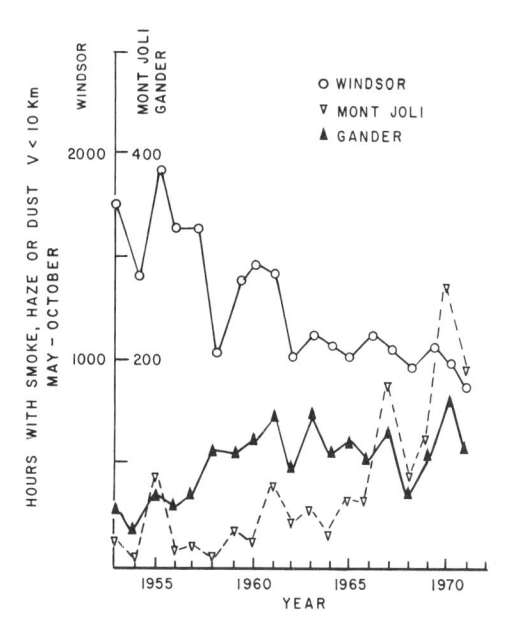

FIGURE 3.10 Secular trend in reports of smoke, haze, or dust at Windsor, Mont Joli, and Gander Airports. (May–October) Source: Munn (1973).

cloudless atmospheres to be about 10 percent in excess of the expectation for clean air. A few observations from aircraft suggested that not more than one fifth of the excess scattering was directed upward. These observations, made before the institution of clean air acts, suggest a modal single scatter albedo of about 0.5 for the particles concerned.

Direct measurements of the optical properties of "industrial haze" in southern England were reported by Waldram (1945), his method of observation being visual photometry of searchlight beams. He made the first direct measurements of absorption of visible radiation by these particles. Figure 3.7 illustrates one of the polar scatter diagrams that he measured, on an occasion when the effective ω_0 was about 0.55. [Figure 3.7 also contains an example of a polar scatter diagram from the extensive observations of Barteneva (1960).]

Charlson and his collaborators (e.g., Lin et al., 1973; Charlson et al., 1974) have recently perfected portable equipment that allows investigation of both optical and nucleating properties of particles and have used it in urban and rural areas. Their investigation is in its early stages, but for urban particles their preliminary measurements indicate single scatter albedos for visible light, mainly in the range 0.8 to 0.4 with the mode near 0.7. They have confirmed that many particles are composed mainly of sulfuric acid and the ammonium sulfates—the nature of the absorbing constituents is not clear. Dzubay and Stevens (1973) collecting particles with radii less than about 1 μm in urban St. Louis write, "... the small [particles] ... make a black deposit ... consistent with the notion that ... the small particles consist of combustion and secondary aerosols. At least 75 percent of the sulphur—is contained in the small particles—significant amounts of the sulphur may be bound to light cations ... H^+, NH_4^+." This suggests absorbing material originating in combustion either carried on sulfate particles or associated with the same size range. The combustion concerned is in a city subject to EPA regulation.

Although Robinson (1947) claimed to have observed it in the London atmosphere, there are yet no incontrovertible observations of modification of the terrestrial radiation field by particles. The effect sought is small, measurement is difficult, and alternative explanations of anomalies are available. There is no reason to question the results of the computations that indicate modification.

Observations in the St. Louis, Missouri, area have shown modification of the nature of convective clouds by the particles in the urban atmosphere, but there has been no direct confirmation of the expected increase in the albedo of clouds forming in polluted air; and because of the wide range of observed cloud albedo, a very extensive statistical investigation would be required. Satellite cloud observations should contain appropriate material. On the other hand, several observers (e.g., Robinson, 1958) have reported absorption of solar radiation in cloud considerably in excess of that which would be expected from pure water clouds—an effect that can be explained by the inclusion of absorbing particles of the type observed in urban polluted air.

3.7 COMPUTATION OF THE RADIATIVE EFFECTS OF POLLUTANT PARTICLES

SURFACE TEMPERATURE

We consider, as examples of the radiative effect of realistic pollutant particle loadings, two theoretical studies, one of particles in the lower atmosphere and one of stratospheric pollution.

Atwater (1970b) has studied the effect on both solar and terrestrial radiation of a low-lying layer of particles simulating urban pollution. He used a size distribution corresponding to that measured in a city atmosphere by Peterson et al. (1969) and made computations for a range of complex refractive index. We select from Atwater's tables the case with refractive index for solar radiation (λ = 500 nm), m = 1.5 − 0.03i and for terrestrial radiation (λ ~10 μm), u = 1.5 − 0.2i. The real part is appropriate to hydrated sulfuric acid. The imaginary part for solar radiation is within the range measured later by Lin et al. (1973) in urban aerosol. The imaginary part for terrestrial radiation is appropriate to aqueous solutions. The scattering and absorption coefficients computed by Atwater are

for solar radiation: $\sigma_{\text{scattering}} = 0.3$ km^{-1}
$\sigma_{\text{absorption}} = 0.06$ km^{-1}

i.e., an "effective single scatter albedo" for the layer of 0.83, and

for terrestrial radiation: $\sigma_{\text{scattering}} = 5 \times 10^{-5}$ km^{-1}
$\sigma_{\text{absorption}} = 0.01$ km^{-1}

The results for solar radiation correspond to a visibility of 12 km. For a 1000-m layer and the global average solar power at the surface, which is around 150 W m^{-2}, the mean heating rate due to absorption of solar radiation is about 0.4 K per day. The perturbation of terrestrial radiation depends not only on local conditions in the polluted layer but on the temperature throughout the atmosphere and particularly on cloud conditions. Atwater made computations for a standard case without cloud, with $\sigma_{\text{absorption}}$ of 0.01 km^{-1}, and he found the mean cooling within the layer of particles to be about 0.5 K per day, approximately equal in magnitude to the solar heating. Atwater's work suggests that in midlatitude cities temperature changes due to perturbation of solar and terrestrial radiation averaged over the year almost exactly balance. The balance is, however, the result of a redistribution of heating and cooling. Daytime heating is reduced at the surface and increased within the particle layer. Nighttime cooling is reduced at the surface and increased within the particle layer. At the surface, there is a moderation of the diurnal temperature cycle and a net decrease in solar radiation. The potential effect on crop growth, particularly early and late in the growing season, deserves more careful examination.

Effluents of Energy Production: Particulates

PLANETARY ALBEDO

Atwater (1970a) also computed the effect on planetary albedo of a layer of particles in the lower atmosphere. He found the relation, which we discussed qualitatively above, between planetary albedo change, surface albedo, absorption by the particle layer, and backscattering by the particle layer. Figure 3.11 illustrates some of his conclusions. In this diagram a particle layer is characterized by absorption and backscattering coefficients, and for each surface albedo a line divides those layers that increase planetary albedo from those that decrease it. Some observations of the properties of particle layers by Waldram and Robinson are plotted on this diagram: actual atmospheric particle layers appear to be close to the neutral line for the average land surface albedo of 0.15. All the observed layers would decrease planetary albedo over snow, all would increase it over quiet ocean waters with albedo around 0.05.

B. M. Herman of the University of Arizona (private communication) has studied the influence of a layer of particles in the lower stratosphere. His computations are for a size distribution similar to that of naturally occurring stratospheric particles as observed by Friend and for visible solar radiation ($\lambda = 500$ nm). He allows for a tropospheric particle loading with optical depth in the zenith of 0.1, which is consistent with our estimate of a mean mass mixing ratio of about 10^{-8} for particles with refractive index (real part) 1.5 and radius 0.5 μm. He integrates for solar position over the day for each month of the year in each $10°$ square of the northern hemisphere, using realistic estimates of the underlying albedo, neglecting optical effects above the perturbing layer. He performs the computations for several values of single scatter albedo. Figures 3.12 and 3.13 show the nature

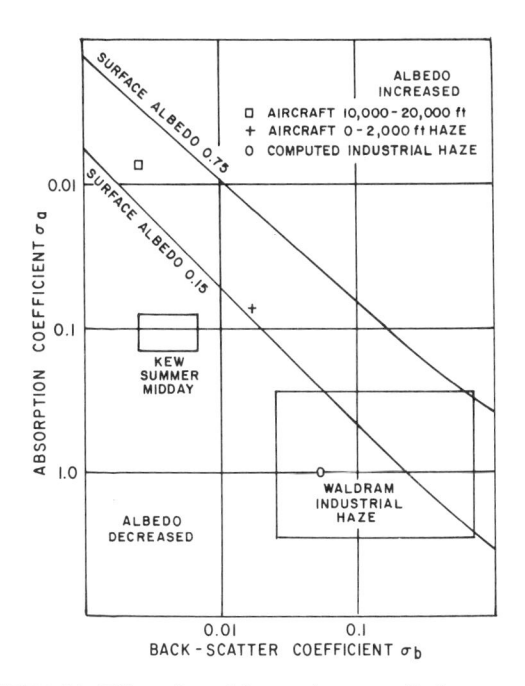

FIGURE 3.11 Effect of particles on planetary albedo.

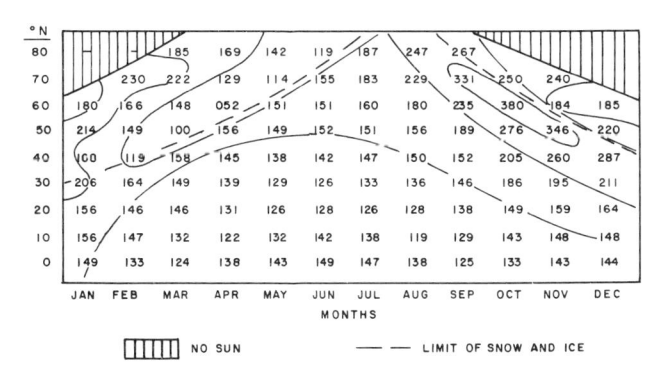

FIGURE 3.12 Albedo *increase* by latitude and month following addition of a 10-km layer of 0.08 μg m^{-3} of particles, $\omega_0 = 1$ (no absorption). Units are 10^5 × fractional albedo change.

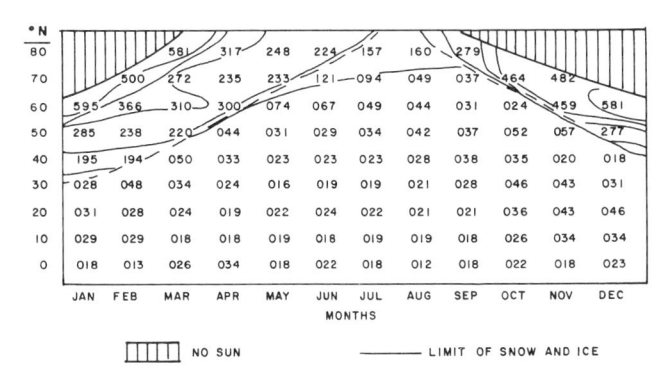

FIGURE 3.13 Albedo *decrease* by latitude and month following addition of a 10-km layer of 0.08 μg m^{-3} of particles, $\omega_0 = 0.9$. Units are 10^5 × fractional albedo change.

of his results for no absorption and for a single scatter albedo of 0.9. In the first case, the particle layer increases the planetary albedo at all locations; in the second case there is a decreased albedo everywhere. The albedo perturbation scales linearly with the mass loading of particles up to loadings more than 20 times that used in preparing Figures 3.12 and 3.13.

To extend Herman's calculations to a realistic situation, it is necessary to know the exact nature of the size distribution of the particles, their complex refractive index over the whole solar spectrum, and the albedo of the underlying surface-atmosphere-cloud system, again over the whole solar spectrum. This we do not know, except that there must be considerable differences between values in the solar infrared and those observed in the visible radiation.

3.8 SUMMARY AND PROSPECTS FOR THE FUTURE

We have seen that the burning of fossil fuel containing sulfur leads to the formation of sulfuric acid particles in the atmosphere. They have a radius around 0.5 μm and are efficient scatterers of solar radiation. If they are pure sulfuric acid particles, they would be expected to absorb about 1 percent of the energy of the solar radiation falling on them, mainly in the infrared. Measurements, however, indicate that in the atmosphere they are often, perhaps always, associated with other material, which causes absorption of more than 10 percent of the intercepted solar radiation. These conditions exist in and near cities in which regulation of particulate emissions has been enforced. Visibility is a sensitive indicator of atmospheric particle loading, and we have noted that while emission control has improved visibility in and near industrial cities, there is a trend toward reduced visibility in locations about 1000 km downwind of regions of major industrial activity. There is no evidence of recent increases in more remote areas (e.g., the southern Pacific Ocean).

Particles of the type in question would be expected to change the optical properties of clouds—their absorption and reflection of solar radiation. There are a few observations of unexpectedly high absorption, but increase of cloud albedo has not yet been firmly established. There is some evidence of a modification of the processes leading to precipitation in clouds formed in air containing combustion-produced particles.

Known mechanisms of removal lead us to expect an average lifetime in the atmosphere of a few days, but for an individual particle this lifetime would be expected to have a very wide range. There should be a close analogy with the life in the atmosphere of water molecules. For these the mean life is 10 days. Many have only a few hours' stay, but some of the small proportion that enter the stratosphere may stay there for several years.

Computation of the energetic consequences of the radiative properties of particles has concentrated on two areas: perturbation of the planetary albedo and perturbation of local heating rates. If observed properties of particles are used in the computations (rather than the optical constants of pure sulfuric acid solutions), we find that the sign of the perturbation of planetary albedo is uncertain: the magnitude of the perturbation must be small. This result is independent of particle loading, at least up to loadings considerably greater than those currently observed, and it holds whether the particles are in the troposphere or the stratosphere. We can greatly increase emission of the kind of particle now produced by combustion in industrial communities without greatly changing the integrated radiative properties of our planet's disk as seen from space, unless, as may be the case, the increased particle loading changes the albedo of cloud. (We should remember that a change of optical properties of the constituent droplets of a cloud does not necessarily mean a change of its albedo.) Current air-quality and emission standards for particles and SO_2 are set by reference to effects on human health, not to effects on weather and climate. There is little doubt that an increase in particles and sulfurous emissions to a magnitude that might have global climatic consequences would be intolerable from the point of view of community health. Unless there is considerable improvement in the physical and economic efficiency of techniques for removal of particles and particle-producing gases, health rather than climatic considerations may be the ultimate constraint on the burning of low-grade fossil fuels.

Computation also shows that although the amount of solar radiation reaching the ground is decreased, the net perturbation of local heating rate by industrial pollution in and near a midlatitude city is small when averaged over the day and year. This aspect of the energetic effects of particle content cannot be satisfactorily studied in isolation. A simple analysis demonstrates this.

Consider the combustion of 1 ton (1000 kg) of coal containing 2.5 percent S. Current practice concerning particulate emission control is observed, but SO_2 in not removed. Twenty-five kilograms of the S is burned. Assume that 50 percent of the resulting SO_2 is converted to $H_2SO_4 \cdot 4H_2O$ particles of radius 0.5 μm. There are 50 kg of these particles. We assume a density of 2×10^3 kg m^{-3} for the material of the particles. The geometrical cross section of this 50 kg of particles is

$$50 \times \frac{3}{4\pi (1.25 \times 10^{-19} \times 2 \times 10^3)} \times \pi (2.5 \times 10^{-13}) \, \text{m}^2.$$

The optical cross section is at least twice this, i.e.,

$$\sim 7.5 \times 10^4 \, \text{m}^2.$$

We assume that the particles have a life in the atmosphere of one day. The global mean (night and day through the year and over the planet) of solar power reaching the surface for 50 percent cloudiness is about 150 W m^{-2}. To compute the solar energy absorbed by the particles, we make two alternative assumptions: (a) They are pure $H_2SO_4 \cdot 4H_2O$ with a single scatter albedo for the whole solar radiation of 0.97. (b) They are contaminated by or associated with other material, and their effective single scatter albedo is 0.9. (This is rather cleaner than the current particle load near St. Louis.) The absorbed solar energy is, for Case (a), $\sim 10^4$ kWh; for Case (b) $\sim 3 \times 10^4$ kWh.

The calorific value of a ton of coal, when its products have mixed with the atmosphere, is $\sim 8 \times 10^3$ kWh. Burning 2.5 percent sulfur coal with no control of SO_2 emission, even with complete removal of all solid emission, leads on average to a net loss of energy at the earth's surface. This is probably compensated by adjustments in the field of terrestrial radiation, but our exercise suggests that climatic effects of particle production should not be considered independently of the effects of heat production, which are the subject of other chapters in this volume, nor of the effects of CO_2 production on the terrestrial radiation field.

3.9 ACKNOWLEDGMENTS

I wish to thank M. A. Atwater of the Center for the Environment and Man, Inc., Hartford, Connecticut; R. J. Charlson and A. P. Waggoner of the Department of Civil Engineering and the Institute for Environmental Studies, University of Washington, Seattle, Washington; and B. M. Herman of the Institute of Atmospheric Sciences, University of Arizona, Tucson, Arizona, each of whom allowed me to use results of his unpublished work during preparation of this paper.

REFERENCES

Altshuler, A. P. (1973). Atmospheric sulphur dioxide and sulphate. Distribution of concentration at urban and non-urban sites in the United States, *Environ. Sci. Technol. 7*, 709.

Atwater, M. A. (1970a). Planetary albedo changed due to aerosols, *Science 170*, 64.

Atwater, M. A. (1970b). Investigation of the radiation balance for polluted layers of the urban environment, Ph.D. Thesis, New York U.

Barteneva, O. D. (1960). Scattering functions of light in the atmospheric boundary layer, *Izv. Geophys. Sci. 12*, 1852 (English transl. p. 1237).

Charlson, R. J., A. H. Vanderpol, D. S. Covert, A. P. Waggoner, and N. C. Ahlquist (1974). $H_2SO_4/(NH_4)_2SO_4$ background aerosol: Optical detection in St. Louis region, *Atmos. Environ. 8*, 1257.

Dzubay, T. G., and R. K. Stevens (1973). Applications of x-ray fluorescence to particulate measurements, preprint from National Environmental Research Center, U.S. EPA.

Hilst, G. R. (1970). Sensitivities of air quality prediction to input errors and uncertainties, in *Proceedings of Symposium on Multiple-Source Urban Diffusion Models*, U.S. EPA, APCO Publ. No. AP-86.

Lin, C. I., M. Baker, and R. J. Charlson (1973). Absorption coefficient of atmospheric aerosol: A method for measurement, *Appl. Opt. 12*, 1356.

Munn, R. E. (1973). Secular increases in summer haziness in the Atlantic provinces, *Atmosphere 11*, 156.

Petersen, C. M., H. J. Paulus, and G. A. Foley (1969). The number-size distribution of atmospheric particles during temperature inversions, *J. Air Pollut. Control Assoc. 19*, 795.

Robinson, G. D. (1947). Notes on the measurement and estimation of atmospheric radiation, *Quart. J. R. Meteorol. Soc. 73*, 127.

Robinson, G. D. (1958). Some observations from aircraft of surface albedo and albedo and absorption of clouds, *Arch. Meteorol. Geophys. Bioklimatol. B9*, 28.

Robinson, G. D. (1962). Absorption of solar radiation by atmospheric aerosol, as revealed by measurements at the ground, *Arch. Meteorol. Geophys. Bioklimatol. B12*, 19.

Waldram, J. M. (1945). Measurement of the photometric properties of the upper atmosphere, *Quart. J. R. Meteorol. Soc. 71*, 319.

Impact of Industrial Gases on Climate

4

CHARLES D. KEELING and ROBERT B. BACASTOW
*Scripps Institution of Oceanography
University of California, San Diego*

4.1 INTRODUCTION

Over 97 percent of the energy demand of the industrial world is met today by burning conventional fossil fuels. Even if conversion to other more costly forms of energy production is pursued vigorously, the annual consumption of fossil fuel is likely to double by A.D. 2000. If the world follows the widely respected economic policy of preferring at any one time the cheapest available fuel, the 4 percent per year growth in consumption of fossil fuel that prevailed for a quarter of a century before 1974 could resume and persist into the next century. A peak annual rate 10 or even 20 times today's rate may occur before depletion of fuel reserves forces a decline in consumption.

Most of the by-products of fossil-fuel combustion are presently injected directly into the atmosphere. While still airborne, they interfere with natural radiative processes, and, if their removal from the air does not keep pace with their input, their accumulation in the air may result in climatic change. Both particles and gases are involved. The latter will now be discussed with major emphasis on the principal fuel by-product, carbon dioxide. The discussion will emphasize global-scale impacts. Particulate matter is dealt with in an accompanying paper by Robinson (see Chapter 3).

Additional amounts of gases are produced industrially for purposes not related to energy production. These gases, to the extent that they may influence climate globally, will also be considered.

The combustion process and industrial gases, in addition to interfering directly with atmospheric radiation, may reduce the abundance of other atmospheric gases, notably ozone and to a minor degree, oxygen. A change in the concentration of the latter, because of its transparency, would not influence atmospheric radiation significantly, but changes in ozone concentration might influence the radiative balance in the stratosphere where ozone is relatively abundant, and indirectly even near the earth's surface. Thus the impact of industry on ozone will be addressed.

Finally, changing concentrations of industrial gases in air may have more subtle indirect influences on climate. For example, additions to the air of carbon dioxide from industry might promote plant growth, which in turn might alter the earth's albedo and hydrological budget. Although such possibilities deserve investigation, they are too remote to justify further discussion here.

4.2 CLIMATIC IMPACT OF CARBON DIOXIDE

Carbon dioxide, CO_2, is an important natural factor, as are water vapor and ozone, in controlling the temperature of the atmosphere. This gas is nearly transparent to visible light but is a strong absorber of infrared radiation, especially at wavelengths between 12 and 18 μm, where a considerable proportion of the outgoing radiation from the earth's surface is transmitted to outer space. An increase in atmospheric CO_2 to levels appreciably above the preindustrial concentration of about 290 ppm (mole fraction in parts per million) might act, much like adding glass to a greenhouse, to increase the temperature of the lower atmosphere.

As discussed below, CO_2 will be produced by man in large quantities relative to the amount now present in the atmosphere. Atmospheric CO_2 concentrations five to ten times the preindustrial level may be attained during the twenty-second century. If high levels are once reached, they will probably decrease only slowly and thus remain well above the preindustrial level for at least a thousand years.

The magnitude of the temperature rise attending an increase in atmospheric CO_2 has been estimated with mathematical models of increasing plausibility and complexity. Early studies, summarized by Manabe and Wetherald (1975), estimated the global average temperature at the earth's surface without regard for changes in the turbulent energy exchanged with the overlying air column and need not be examined here. Manabe and Wetherald (1967) overcame this limitation with a one-dimensional convective adjustment, or "radiative convective," model, which allowed for temperature adjustments in the air column to preserve a reasonable vertical gradient, or lapse rate. They also considered the influence on radiation of varying amounts of water vapor but assumed cloud cover to be unaltered by the change in CO_2 concentration. Their model has been widely quoted and has been used to compare the temperature effect of other industrial gases to that of CO_2, as discussed below.

Recently, Manabe and Wetherald (1975) have developed a three-dimensional model that takes into account both vertical and horizontal atmospheric motions. The model employs an idealized continent and ocean with snow cover, rainfall, and vertical lapse rate of temperature treated as dependent variables. Heat transport by ocean currents is neglected. The distributions of cloudiness at low, middle, and high levels are preset at average values that remain invariant because no general circulation model has yet been devised that predicts clouds reliably. On the basis of what the authors considered to be the best available radiative transfer scheme, the model predicts an average rise of 2.5 °C in air temperature in the lower atmosphere for a doubling of atmospheric CO_2 from 300 to 600 ppm. This rise is about 25 percent higher than predicted by their one-dimensional model. Also, precipitation, not predicted in the one-dimensional model, is increased by 7 percent. The temperature rise is most evident in polar latitudes (see Figure 9.2 of Chapter 9) as a result of decreased snow cover and suppressed vertical air motion. This positive feedback explains in large part why a higher temperature rise is predicted than by their one-dimensional model.

Although the climatic effects of CO_2 levels higher than 600 ppm have not been investigated for the three-dimensional model, the one-dimensional convective adjustment model predicts nearly equal additional rises in average air temperature for each successive doubling of CO_2 level (Augustsson and Ramanathan, 1977). For example, an eightfold increase in CO_2 might raise average air temperatures by 7 °C. Such a global average rise would be comparable with the extreme changes in global temperature believed to have occurred during geologic history.

The Manabe-Wetherald three-dimensional model, in spite of its complexity compared with one-dimensional models, still falls short of an accurate portrayal of atmospheric processes. The neglect of feedback mechanisms for clouds and ocean circulation leaves open the possibility that the temperature rise might be considerably larger or smaller than the model predicts, even possibly of opposite sign, as discussed by Smagorinsky in Chapter 9. Nevertheless, the prediction of a substantial temperature rise cannot be arbitrarily dismissed. As pointed out by Schneider (1975), there is no strong evidence that the present models are more likely to overestimate the rise than to underestimate it.

Our limited knowledge of cloud feedback illustrates this point. As Manabe and Wetherald (1967) and others have shown, increasing either low or middle stratoform clouds, *per se*, would produce lower surface temperatures. It is not clear, however, that these forms of cloudiness necessarily increase with increasing atmospheric CO_2. The Manabe-Wetherald three-dimensional model predicts higher relative humidity in the low troposphere and, probably, as Smagorinsky suggests in Chapter 9, more low clouds. But in the middle troposphere, the model predicts *lower* relative humidity, associated with the stronger hydrological cycle that accompanies added heating from CO_2. If this lower humidity leads to substantially less middle clouds, the cooling effect of more low clouds could be partially or wholly canceled.

With respect to the role of the oceans, progress in modeling is hampered by the long adjustment times involved in the intermediate water circulation. This water, which lies just below the ocean surface at high latitudes and circulates to depths as great as 1000 m at low latitudes, probably contributes significantly to the atmospheric heat budget and should be included in models of climatic change.

Also, all the present modeling attempts have involved so-called equilibrium models, which compare the climate for two or more distinct CO_2 levels each maintained indefinitely. As discussed below, the CO_2 concentration of the atmosphere is likely to vary considerably over the next several hundred years. This time is too short for the atmosphere and ocean to attain equilibrium either with respect to climate or chemical processes.

Near the time of most rapid CO_2 buildup, the associated heating effect may produce unparalleled perturbations in the wind-driven and thermal haline circulations of the oceans. Until climatic models can correctly allow for the slow response of subsurface ocean waters as a climatic feedback, it is likely to prove difficult to predict reliably the regional changes in climate that are of greatest interest to mankind. We may for some time be forced to infer the climatic impact of CO_2 almost exclusively on the basis of predicted global temperature changes.

4.3 CLIMATIC IMPACT OF OTHER INDUSTRIAL GASES

Other industrial gases besides CO_2 have strong infrared absorption bands and, while airborne, will contribute to the atmospheric greenhouse effect. The effectiveness of a specific gas in altering the earth's radiation balance and equilibrium temperature depends on the location of its absorption bands; gases with strong bands in the relatively transparent region of the atmospheric spectra from 8 to 12 μm may impact the radiative balance at much lower concentrations than atmospheric CO_2. Because the spectrum of each gas is different from all others, each has an almost independent, and therefore additive, thermal effect.

At the extreme of sensitivity are the chlorofluorocarbons, or Freons. These gases do not occur naturally in the atmosphere but have been introduced in recent years, especially as propellants in spray cans. Their absorption bands span about half of the atmospheric infrared window region from 8 to 12 μm. Ramanathan (1976) has calculated that if both CF_2Cl_2 and $CFCl_3$ were to attain concentrations of 0.002 ppm, perhaps possible by A.D. 2025 if present rates of injection are maintained, a surface temperature increase of 0.9 °C would occur.

Of intermediate sensitivity is nitrous oxide, N_2O, which occurs naturally and has a present abundance of 0.28 ppm. Yung et al. (1976) have calculated a surface temperature increase of 0.5 °C if its concentration were to double. They believe that such an increase is possible by A.D. 2025. The increase might be produced as a result of accelerated use of nitrogen fertilizers (McElroy et al., 1976) and as a by-product of the combustion of fossil fuel (Weiss and Craig, 1976).

Industrial gases with lesser but not necessarily negligible impacts are CH_4, NH_3, NHO_3, C_2H_2, SO_2, and CH_2Cl_2. Wang et al. (1976) have calculated that a doubling in concentration of these gases would produce surface temperature increases from 0.02 to 0.28 °C, depending on the gas. The combined effect of doubling the two most sensitive ones, CH_4 and NH_3, they predict to be 0.4 °C.

Our knowledge of the rates of injection and removal of these gases from the atmosphere and of recent changes in their atmospheric concentrations is too meager to place much confidence in predictions of future concentration increases. Even for the chlorofluorocarbons, which have recently been studied intensely (Committee on Impacts of Stratospheric Change, 1976) the uncertainties are larger than for CO_2. Additional uncertainties attend the calculation of thermal effects, but since nearly the same one-dimensional radiative convective model was employed for all these gases as was used for CO_2, the thermal effect of each gas compared with CO_2 should involve relatively little uncertainty. Since all these gases tend to warm the lower atmosphere, the combined global impact of all industrial gases is likely to be considerably greater than for CO_2 alone.

This prediction is, however, complicated by the tendency of industrial gases in the stratosphere to react photochemically with each other and with other natural constituents. Several industrial gases persist in the air long enough to mix appreciably into the stratosphere where they are decomposed by the sun's ultraviolet radiation. Some of the reactive by-products, notably of the chlorofluorocarbons and N_2O, attack ozone, O_3, reducing the total amount and somewhat shifting the distribution toward lower altitudes. If concentrations of the gases that attack O_3 rise to the levels suggested earlier, the concentration of O_3 might fall by as much as 25 percent, although a reduction of about 7 percent is more likely (Committee on Impacts of Stratospheric Change, 1976). Because O_3 also contributes strongly to the atmospheric radiation balance, a reduction of 25 percent might produce a surface-temperature decrease of the order of 0.5 °C (Wang et al., 1976). This prediction, however, is much more uncertain than the predictions of thermal effects for the industrial gases, because the calculations are strongly dependent on the vertical distribution of O_3, which varies both spatially and temporally under natural conditions. To account for changes in distribution produced by reaction with industrial gases will require that calculations consider radiative, dynamical, and chemical processes simultaneously. This has not yet been done.

Because O_3 acts as a shield to incoming ultraviolet solar radiation, a reduction of the order of 25 percent would be likely to harm plant and animal life including man. Since the production of the industrial gases that influence O_3 probably can be curtailed without extreme hardship to mankind (Committee on Impacts of Stratospheric Change, 1976) measures will perhaps be taken to prevent a man-made reduction of O_3 large enough to produce a significant greenhouse effect, either from O_3 or from the gases that attack O_3.

If curtailment does not occur, the greenhouse effect of O_3 may cancel out part of the previously predicted effect for the industrial gases.

Still more complicated gaseous interactions may affect the abundance of the industrial gases and, in turn, the naturally occurring constituents of the stratosphere. For example, increases in the production of carbon monoxide, CO, a by-product of combustion, might reduce the amount of OH radical in both the stratosphere and the troposphere. This reduction would, in turn, probably suppress the atmospheric sinks for several other gases. The abundances of CH_4 and the chlorofluorocarbons throughout the air column and of O_3 and H_2O in the stratosphere might all increase (Sze, 1977) with attending increases in the overall greenhouse effect. Increase in stratospheric H_2O could be especially significant. Water vapor in the stratosphere occurs naturally at concentrations far below saturation. A doubling of concentration perhaps possible by this mechanism might cause an increase in surface temperature of 1.0 °C (Wang et al., 1976).

Our knowledge of the climatic effects of the other industrial gases is evolving rapidly, and this paper cannot hope to give definitive answers. The reader may consult Wang et al. (1976) and a comprehensive review article by Bach (1976) for further details. In comparing these other gases with CO_2, it should be realized, however, that none of them has the extremely long removal time that large quantities of industrial CO_2 will exhibit. Stopping production of any one of them would bring near restoration to pre-emission conditions for that gas within, at most, half a century or so.

4.4 RECENT CHANGES IN ATMOSPHERIC CO_2 ABUNDANCE

The remainder of this paper will be devoted to estimating probable future concentrations of atmospheric carbon dioxide. As a basis for such estimates we shall first discuss observed past changes.

The concentration of CO_2 in the atmosphere has been monitored since 1957 at two remote stations: Mauna Loa Observatory, Hawaii, and the South Pole as a joint effort of the U.S. Government (National Oceanic and Atmospheric Administration) and the Scripps Institution of Oceanography (Machta, 1973; Keeling et al., 1976a, 1976b). Except for an interruption in both records in 1964 (resulting from funding problems), average concentrations, precise to about 0.2 ppm, have been obtained for almost every month since late 1958.

Plots of the data (Figures 4.1 and 4.2) reveal seasonal variations that do not, however, obscure evidence of a steadily rising background level, which we will call the "secular increase." Seasonally adjusted data (Figures 4.3 and 4.4), reveal this secular increase more clearly. A comparison of the two records (Figure 4.5) indicates that variations in the secular increase have differed in the separate hemispheres. In particular, the South Pole record shows an approximately four-year periodicity not readily apparent at Hawaii. There is evidence (Bacastow, 1976; Newell and Weare, 1977) that this periodicity is related to the Southern Oscillation in barometric pressure, to sea-surface temperature changes, and to the El Niño phenomenon (Flohn, 1975). The record thus may reflect large-scale changes in uptake and release of CO_2 by the oceans in response to variable upwelling of subsurface waters or variable rates of exchange at the air-sea boundary.

Shorter atmospheric CO_2 records from other stations (Kelley, 1969; Lowe, 1974) and aircraft data (Bolin and Bischof, 1970; Bischof, 1973; Pearman and Garratt, 1973) are as yet too short to establish variations in the secular trend, but all these records tend to substantiate the average rates of increase found for Hawaii and the South Pole.

The annual production of CO_2 by industry is almost double the annual rise in atmospheric CO_2 abundance. The ratio of rise to production, which we will call the "airborne fraction," has varied considerably from year to year as can be seen by comparing the annual averages of input with the increase

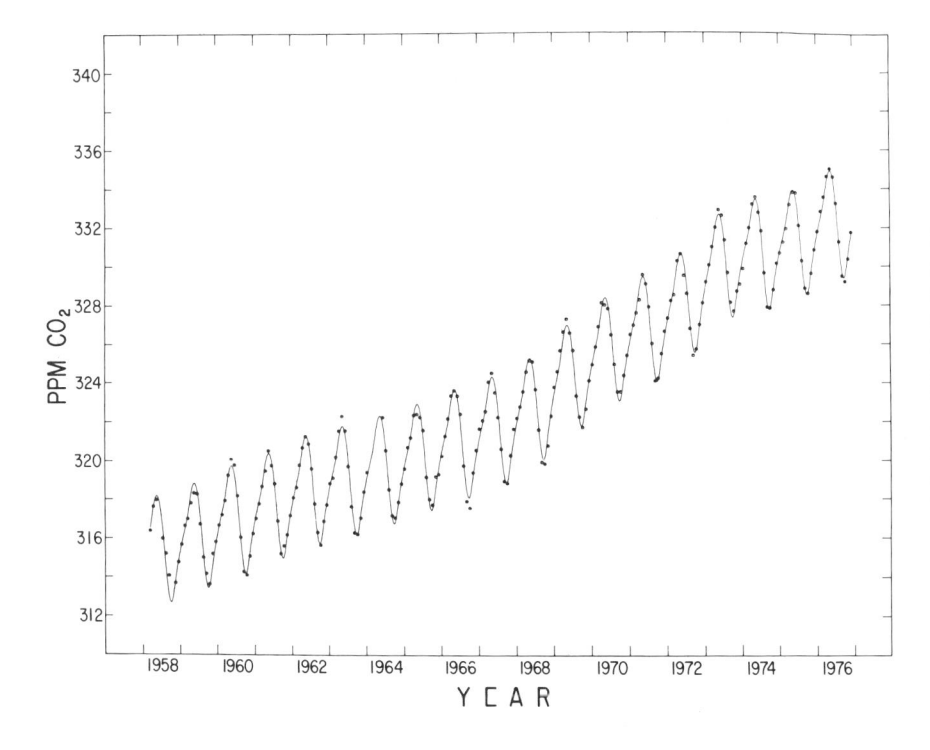

FIGURE 4.1 Trend in the concentration of atmospheric CO_2 at Mauna Loa Observatory, Hawaii, at 19.5° N, 155.6° W (Keeling et al., 1976b and to be published). The dots indicate the observed monthly average concentration based on continuous measurements. The oscillating curve is a fit of the average annual variation superimposed on a spline function representation of the seasonally adjusted secular trend. The spline function was derived by the method of Reinsch (1967).

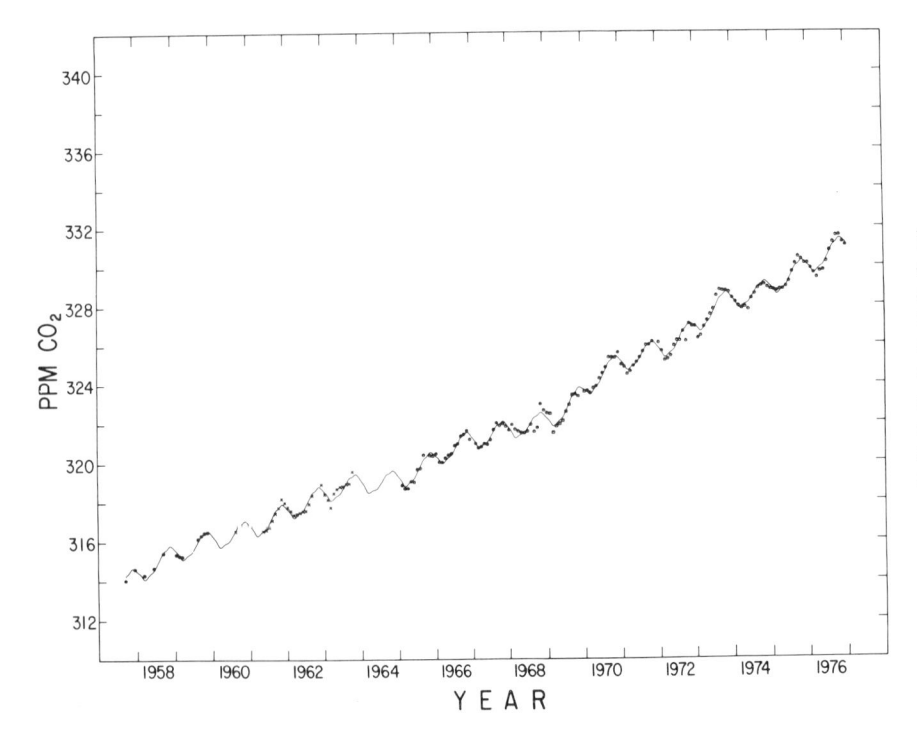

FIGURE 4.2 Trend in the concentration of atmospheric CO_2 at the South Pole (Keeling et al., 1976a and to be published). The dots indicate monthly averages based on flask analyses, except for 1961 through 1963 for which monthly averages based on continuous measurements are plotted. The oscillating curve is a fit of the average annual variation superimposed on a spline function representation of the seasonally adjusted trend.

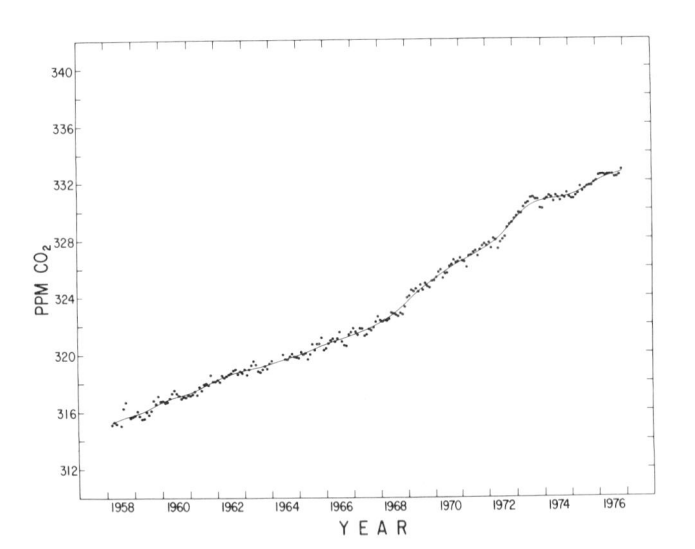

FIGURE 4.3 Seasonally adjusted concentration of atmospheric CO_2 at Mauna Loa Observatory, Hawaii. Dots indicate seasonally adjusted monthly averages. The smooth curve indicates the spline function used to derive the curve in Figure 4.1.

based on the records for Hawaii and the South Pole combined (see Figure 4.6 and Table 4.1). This variability clearly reflects the approximately four-year periodicity already noted.

The airborne fraction averaged over 15 years of record, 1951 through 1973, is computed to be 56 percent. This estimate may, however, be in error by as much as 10 percent because of variability in the atmospheric data and possible

TABLE 4.1 Atmospheric CO_2 Budget

Year	Observed CO_2 Increase[a] (ppm)	Fossil Fuel Input (ppm)[b]	Airborne Fraction (%)
1955		0.98	
1956		1.03	
1957		1.08	
1958	(0.89)	1.10	(81)
1959	0.80	1.15	70
1960	0.55	1.19	46
1961	0.90	1.24	73
1962	0.76	1.30	58
1963	0.50	1.38	36
1964	0.40	1.46	27
1965	0.90	1.52	59
1966	0.76	1.60	48
1967	0.68	1.63	42
1968	0.98	1.73	57
1969	1.51	1.83	83
1970	1.31	1.97	66
1971	0.85	2.04	42
1972	1.47	2.12	69
1973	1.41	2.26	62
1974	(0.45)	2.26	
1975		2.26	
1959–1973 (inclusive)	13.78	24.42	56.43

[a] Average of the seasonally adjusted records of Mauna Loa Observatory and the South Pole as plotted in Figures 4.3 and 4.4. 1958 is estimated; 1974 is preliminary.
[b] As fraction of atmosphere $\times 10^6$. Annual production of CO_2 in grams of carbon per year (see Table 4.2 for sources of data) was converted to ppm by multiplying by the factor 290 ppm/$(N_{uo} + N_{lo})$, consistent with the numerical values of model parameters listed by Bacastow and Keeling (1973, p. 133).

Impact of Industrial Gases on Climate

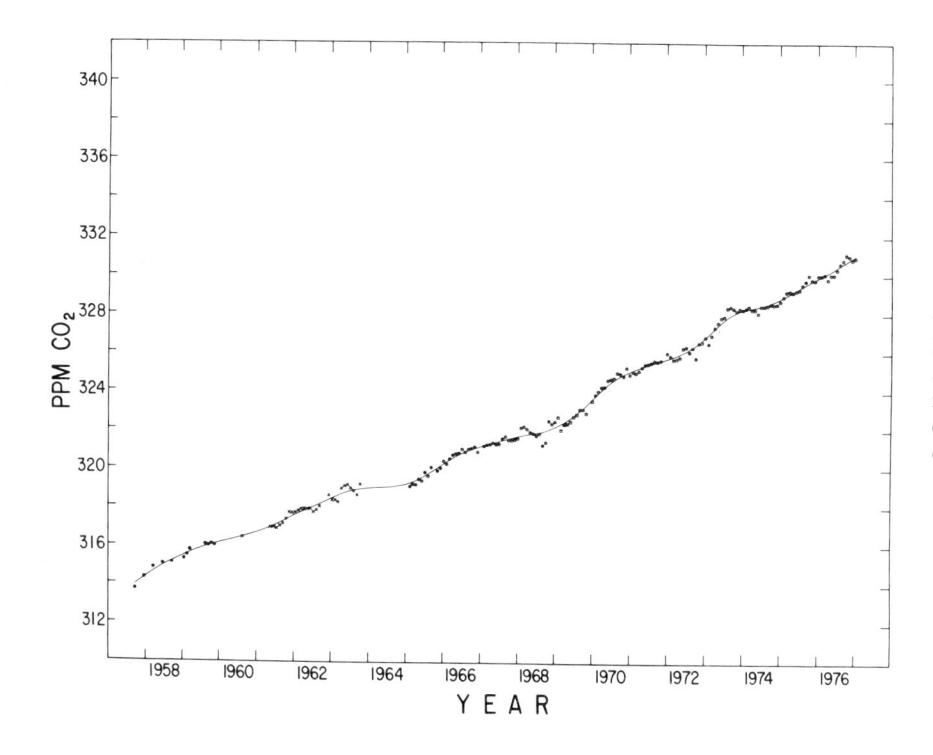

FIGURE 4.4 Seasonally adjusted concentration of atmospheric CO_2 at the South Pole. Dots indicate seasonally adjusted monthly averages based on flask samples or continuous measurements as in Figure 4.2. The smooth curve indicates the spline function used to derive the curve in Figure 4.2.

systematic errors in computing the amount of CO_2 produced from fossil fuel (Keeling, 1973).

That portion of the industrial CO_2 that is not accounted for by an increase in atmospheric CO_2 we infer to have been taken up by ocean water and possibly by land plants. A small additional amount may have been consumed by accelerated chemical erosion of carbonate or silicate rocks on land. The apportioning of industrial CO_2 between these carbon pools and sinks cannot easily be verified, however. No accelerated CO_2 uptake is directly evident over land or in seawater. Indeed, with respect to uptake over land, no way is known to detect the small changes in storage of organic carbon that may have occurred recently as a result of an increase in atmospheric CO_2 or for other reasons. A net out-

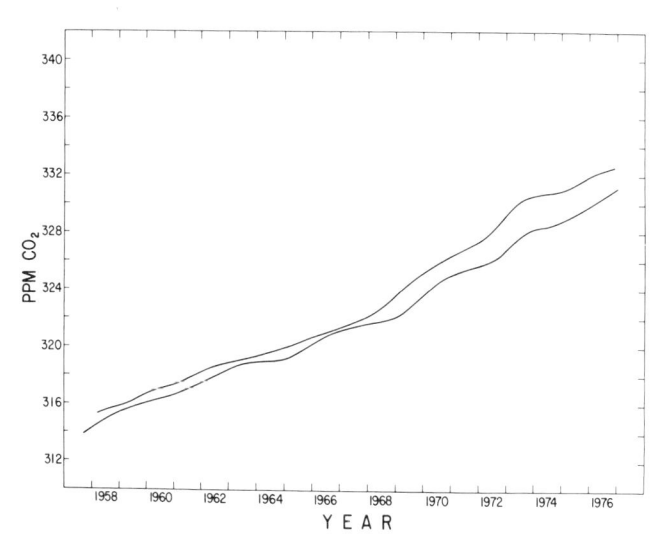

FIGURE 4.5 Comparison of seasonally adjusted trends in atmospheric CO_2 for Mauna Loa Observatory, Hawaii, and the South Pole.

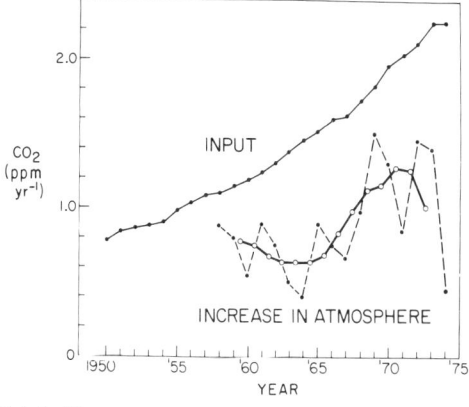

FIGURE 4.6 Upper curve: annual input of fossil fuel CO_2 in the atmosphere versus time, expressed as an annual increase in concentration of atmospheric CO_2 neglecting any possible removal. Lower dashed curve: annual increase of atmospheric CO_2 based on the average of the secular trends for Mauna Loa Observatory, Hawaii, and the South Pole as shown in Figure 4.5. Lower full curve: four-year running mean of the annual increase as plotted by the dashed curve.

flow of carbon dioxide from land plants to the atmosphere may even have occurred in recent years (Whittaker and Likens, 1973).

With respect to uptake by seawater, direct measurements of dissolved carbon, precise enough to reveal secular changes in storage, are possible but have not been undertaken. It is therefore necessary to seek indirect estimates of industrial CO_2 partitioning of which the most reliable is based on comparing radiocarbon CO_2 withdrawal from the air with nonradioactive CO_2 withdrawal. Revelle and Suess (1957), Broecker et al. (1971), and Machta (1973) by this method have concluded that most of the industrial CO_2 leaving the air has been taken up by the oceans, with little net change in the carbon pool of land plants. In contrast, Bacastow and Keeling (1973) predicted substantial biota uptake. The issue is not settled, however, since the radiocarbon data themselves are uncertain.

4.5 PREDICTED FUTURE CO_2 INCREASE IN THE ATMOSPHERE

Success in forecasting atmospheric CO_2 levels over the coming decades and centuries depends on being able to estimate future fossil-fuel consumption and the fraction of fuel-derived CO_2 that will remain airborne. If the world community continues to rely on the energy policies that have prevailed during the past 25 years, fossil-fuel consumption may continue to increase by as much as 4 percent per year for several more decades. On the other hand, if alternative sources of energy are found, the growth in fossil-fuel usage may soon slacken. Except for short periods, the rate will, however, probably not dip below 2 percent until well into the next century (compare Table 1.9 of Chapter 1 with Figure 4.10 below).

As for the airborne fraction, it has probably until now remained nearly constant, except for short-term variations of the kind illustrated in Figure 4.6. This is because the production of industrial CO_2, although steadily rising, still amounts in total to only a minor perturbation to the natural carbon cycle. Until now, no persistent secondary interactions are likely to have complicated the primary mechanisms of atmospheric CO_2 uptake. Indeed, if the adjustments of the natural carbon pools to industrial CO_2 production were all small enough to be precisely linear responses in a mathematical sense, and if the nearly exponential rise in industrial CO_2 input had been precisely exponential for a long time in comparison to the response times of the carbon pools, the airborne fraction would be invariant (Ekdahl and Keeling, 1973). Even as far into the future as A.D. 2000, when the CO_2 input is likely to have reached 50 percent of the preindustrial inventory, predictions based on a constant airborne fraction should still be nearly correct. On this basis, between 375 and 400 ppm of atmospheric CO_2 are expected in A.D. 2000 (Machta, 1973).

Beyond A.D. 2000, the reliability in year-to-year predictions obviously diminishes. The need to predict CO_2 levels correctly is, however, even greater than for this century because of the prospect of far higher CO_2 levels. One approach to making such distant predictions is to fix attention on the ultimate production of CO_2 and to construct simulated histories for the entire era of fossil-fuel exploitation. This requires that some new but not unreasonable assumptions be made.

The most important of these assumptions involves estimating the ultimate amount of recoverable fossil fuel. This amount is uncertain to at least a factor of 3. Hubbert (1969), whose careful estimates are by no means the highest published values (Gillette, 1974), predicts an ultimate fuel production that will yield between five and nine times the amount of CO_2 in the preindustrial atmosphere. But left out of his estimates are shale oil and tar sands, which may become economical sources of fuel when more easily recovered fuels approach exhaustion. With this point in mind, Baes et al. (1976) recently estimated ultimate CO_2 production from fossil fuel at 12 and Zimen and Altenheim (1973) at 13 to 14 times the preindustrial amount of CO_2. Perry and Landsberg (see Chapter 1) advance a "best estimate" for ultimate fuel production slightly below Hubbert's upper estimate but well below the other estimates quoted above.

A second important assumption relates to the rate of fuel exploitation. One cannot determine today whether fossil-fuel reserves will be processed as rapidly as economically feasible or will be stretched out by economic depressions, by wars, or by deliberate measures to conserve energy. A reasonable approach to long-range prediction is to investigate several cases so that we can appraise alternative courses of action.

An additional factor to consider is to anticipate future changes in the airborne fraction. Irrespective of the pattern of fuel use, this fraction will not remain constant during the next century, either with respect to land biota or the world oceans.

First, whether or not the land biota have recently increased their storage of carbon a few percent in response to a 10 percent rise in atmospheric CO_2 level, it seems improbable that the biota will absorb more than a small fraction of the industrial CO_2 near the peak stage of the fossil-fuel era because to do so would require an increase in storage of carbon of the order of the total present land biomass. Because much plant growth is limited by temperature, light, water, or nutrients rather than CO_2, such a large increase seems unlikely even if the land vegetation were otherwise undisturbed by man. But since the major sources of carbon storage on land are extensive forests in Canada, Siberia, and the tropics, and since both human population and per capita consumption of natural resources are expected to increase, most of these forests will be repeatedly exploited for wood products or will be cleared of trees to produce food crops. The overall storage of carbon on land is thus likely to increase only moderately and might even decrease and add still more CO_2 to the atmosphere.

Second, the oceans, although they store much more carbon than the land biota, are not able to accept large additional amounts of CO_2 as easily as they accept industrial CO_2

today. This is because only about 10 percent of dissolved carbon in seawater is presently in the form of ionic carbonate, and it is the presence of this chemical species that principally accounts for the oceanic uptake of CO_2. As dissolved carbonate in the water is used up by reaction with industrial CO_2, the CO_2 pressure builds up in response to shifts in chemical equilibria among carbonate ions, bicarbonate ions, and dissolved CO_2. The water is less and less able to absorb added amounts of industrial CO_2. Furthermore, most seawater below a shallow surface layer exchanges with the sea surface so slowly that only the dissolved carbonate in this surface layer is readily available to react with industrial CO_2. Some additional carbonate may be supplied to seawater by accelerated input from rivers, but historical rates of input (Revelle and Fairbridge, 1957) would have to be greatly increased to hold back the increase in airborne fraction significantly. Finally, accelerated dissolution of carbonate sediments on the sea floor might hold down the airborne fraction as discussed below.

A computer model that takes all these features of the carbon cycle into account has recently been investigated by Bacastow and Keeling (1973 and to be published). The ocean, atmosphere, and land biota each were divided into two reservoirs as depicted by Figure 4.7 and Table 4.2. The principal transfer coefficients for carbon exchange between reservoirs at steady state were derived from radiocarbon data or direct measurements of fluxes and abundances. The model allows for relatively rapid carbon exchange between air and the sea surface but slow exchange (1500-year mass-to-flux ratio) of deep water with surface water. The preindustrial era turnover time for annual plants including detritus is prescribed as 2.5 years; for perennials, 60 years. A biota growth factor relates the average CO_2 uptake rate of plants to the ambient CO_2 concentration. This factor and the effective size of the ocean surface layer were adjusted so

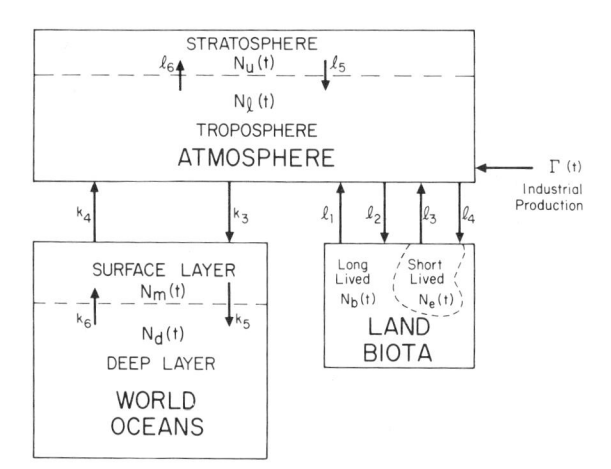

FIGURE 4.7 Six-reservoir model of the carbon cycle of Bacastow and Keeling (1973). The mass of carbon in each reservoir is represented by the $N_j(t)$, and the transfer coefficients between reservoirs are given by the l_j and k_j. The magnitude of the initial values of the $N_j(t)$ and of the time-independent values of the k_j and l_j are listed in Table 4.2.

TABLE 4.2 Six-Reservoir Model Parameters

Initial Values (10^{18} grams of carbon)
N_{uo}, CO_2 in stratosphere	= 0.092
N_{lo}, CO_2 in troposphere	= 0.523
N_{bo}, carbon C in perennial biota	= 1.56
N_{eo}, carbon C in annual biota	= 0.075
N_{mo}, inorganic and organic carbon in ocean surface layer	= 2.46
N_{do}, inorganic and organic carbon in deep ocean water	= 36.2

Perturbation Transfer Coefficients (yr^{-1})
l_1, perennial biota to troposphere = 0
[Set to zero because the perennial plants are assumed to assimilate as well as respire carbon proportional to their mass. See Bacastow and Keeling (1973, p. 94).]
l_2, troposphere to perennial biota = 1/75.7
l_3, annual biota to troposphere = 1/2.50
l_4, troposphere to annual biota = 1/65.6
l_5, stratosphere to troposphere = 1/2.00
l_6, troposphere to stratosphere = 1/11.3
k_3, troposphere to surface ocean = 1/5.92
k_4, surface ocean to troposphere = 1/3.02
[Varies with inorganic carbon in surface ocean layer: this is initial value. See Bacastow and Keeling (1973, pp. 128–133).]
k_5, surface ocean to deep ocean = 1/113
k_6, deep ocean to surface ocean = 1/1500

Additional Model Quantities
$\Gamma(t)$, industrial CO_2 production
  1700 to 1860: assumed exponential increase at 4.35% per year with 1860 production of 9.53 × 10^{13} g of carbon (Keeling, 1973)
  1860 to 1969: based on annual records as cited by Keeling (1973)
  1970 to 1975: based on annual records as cited by Rotty (1976)
  After 1975: as predicted by Eq. (4.2) (see text)
N_∞, ultimate production of fossil fuel = 8.2 ($N_{lo} + N_{uo}$)
β, biota growth factor = 0.266
h_m, depth of surface layer = 265 m

that the model simultaneously predicts the observed airborne fraction based on the average of the CO_2 records for Hawaii and the South Pole (see Table 4.1) and also the rate of depletion of radiocarbon by fossil fuel as deduced from studies of carbon in tree rings. Beginning in A.D. 2000, however, the biota growth factor was linearly reduced in absolute magnitude 4 percent per year, so that after A.D. 2025 the total mass of the biota remains constant. For some of the calculations, specifically noted below, the deep-water reservoir or both the surface and deep-water reservoirs were assumed to be in equilibrium with calcium carbonate since preindustrial times. For these cases, reservoir exchange of dissolved calcium as well as carbon species was taken into account and the biota growth factor readjusted to preserve agreement with the Hawaii and South Pole records. Further justification for adopting such a model is discussed in Appendix 4.A.

The production of CO_2 in the past was computed from historical data (see Table 4.2). The future production of CO_2

was simulated by a modified logistic function that depicts a process that initially grows exponentially in response to an almost unlimited supply of fuel but that, after proceeding long enough to decrease significantly the supply of fuel, is held back as a function of the ultimate fuel reserve remaining.

Specifically, if N is the amount of carbon in fossil fuel that has been combusted to CO_2, and N_∞ is the ultimate amount that will be combusted, then $R(t)$, the relative rate of increase in N (which rate would be constant for exponential growth) is given by

$$R(t) \equiv \frac{1}{N}\frac{dN}{dt} = r\left(1 - \frac{N}{N_\infty}\right), \qquad (4.1)$$

where r is a constant chosen so that $R(t)$ has a prescribed value for the first year of prediction. The CO_2 production rate, dN/dt, simulates the characteristic rise and fall associated with growth in a limited environment. To be able to fit the curve of N versus time, t, to the recent historical period and still permit some variation in the arrival of peak values and in the steepness of rise and subsequent decline, calculations were also performed with a modified equation having an adjustable growth cutoff parameter n such that

$$R(t) \equiv \frac{1}{N}\frac{dN}{dt} = r\left[1 - \left(\frac{N}{N_\infty}\right)^n\right]. \qquad (4.2)$$

Figure 4.8 shows curves that predict possible fuel combustion patterns as a result of choosing different values of the growth cutoff parameter, n. For the ultimate amount of fuel carbon, N_∞, the estimate of Perry and Landsberg (see Chapter 1) was chosen as a standard case. This value is 8.2 times the amount of carbon in preindustrial CO_2. A prediction of the year-to-year sequence of production close to that

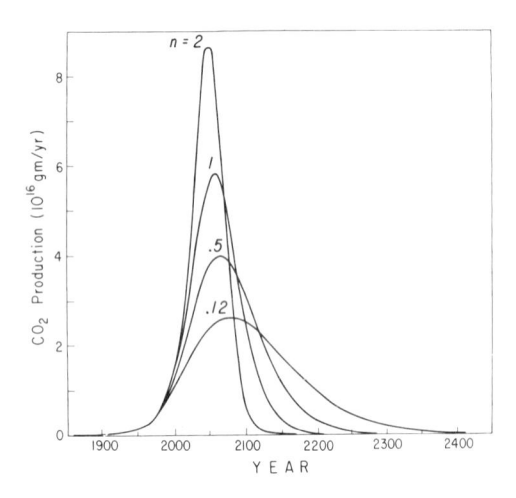

FIGURE 4.8 Industrial CO_2 production in grams of carbon per year for various assumed patterns of fossil-fuel consumption. The ultimate production of industrial CO_2 is fixed at 8.2 times the amount of CO_2 in the preindustrial atmosphere in agreement with the best estimate of Perry and Landsberg (see Chapter 1).

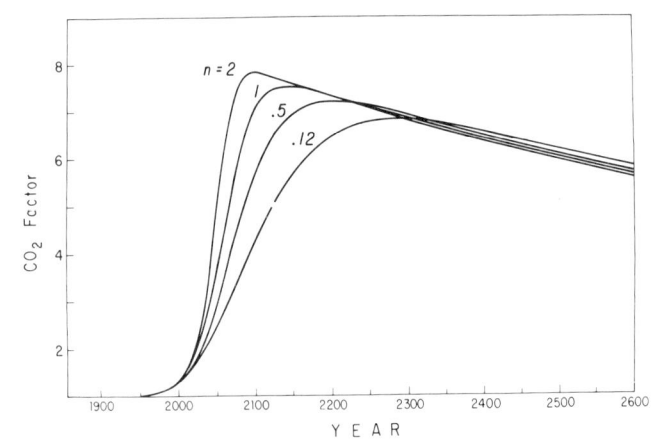

FIGURE 4.9 Predicted increase in atmospheric CO_2 from A.D. 1900 to A.D. 2600. Curves are shown for the four simulated fossil-fuel consumption patterns shown in Figure 4.8. The CO_2 factor expresses the amount of CO_2 in the atmosphere as a multiplier of the preindustrial atmospheric CO_2 abundance.

of Perry and Landsberg was obtained when the growth cutoff parameter was set at $n = 0.5$. The relative rate, $R(t)$, for January 1, 1976 was set at 4.53 percent per year, consistent with the rate of increase in the annual fuel production from 1947 through 1972, a period when very nearly exponential growth prevailed.

This rate, however, is not representative of the rate for the three decades before 1945 when two world wars and a worldwide economic depression markedly slowed growth in fuel usage. The long-term average growth rate given by the product of $1/N$ with dN/dt for 1975 is, indeed, only 3.68 percent per year. Stated another way, setting the rate $R(t)$ equal to 4.53 percent per year and assuming exponential growth implies that 2.22 percent of the fossil-fuel carbon ultimately available (Perry and Landsberg's estimate) has been consumed before 1976, whereas the historic data indicate 2.82 percent. Since it makes little difference to the predicted future trend whether the former or latter fraction is used, we assumed 2.22 percent so as to provide mathematical consistency with exponential growth during the recent past.

The model predicts an initially rapid rise in atmospheric CO_2 level consistent with the recent trend, a peaking sometime between A.D. 2100 and A.D. 2250, and then a period of slow decline (Figure 4.9). The pattern of fuel combustion influences the time of arrival of various levels of atmospheric CO_2, and, as might be expected, the maximum CO_2 level is lower if the period of combustion is stretched out. After the maximum concentration has been attained, the CO_2 level falls at almost the same rate regardless of the combustion pattern chosen.

The model might be criticized for not taking into account recent perturbations in fossil-fuel production. During 1974 and 1975, the acceleration in worldwide production fell abruptly to practically zero in response to a sudden rise in the price of crude petroleum charged by the major oil exporting nations. Preliminary data for 1976 indicate, however, that the production has once again returned to an annual growth rate near 4 percent (Rotty, 1977). This resumption may indicate that the slowdown in growth in 1974 and 1975 was only a minor perturbation to the historical long-term rising trend. For example, according to Eq. (4.2), with parameter n equal to 0.5, the relative growth rate $R(t)$ is predicted to fall from 4.5 percent in 1976 to 4.3 percent in 1990 and 4.0 percent in A.D. 2000. Perhaps the growth in 1976 reflects a strong tendency toward adhering to this prediction. On the other hand, the resumption in 1976 may itself be an aberration with slower growth rates to return in the near future. To examine this latter possibility, we have made projections of fuel use based on the unmodified logistic Eq. (4.1) in which the rate $R(t)$ for 1976 was varied (Figure 4.10). The corresponding atmospheric CO_2 levels are shown in Figure 4.11. Although the curves are somewhat flatter near the maximum than for the earlier cases where the parameter n was varied, the differences in pattern are not very significant. Either set of curves falls well within the limits of predictability of fuel-consumption patterns. Since the assumption of steady growth is consistent with nearly three decades of recent historical record, we prefer to use Eq. (4.2) with parameter n varied from case to case and to ignore the very recent aberrations in trend.

A striking finding of the model is the diminishing effectiveness of the oceans in removing industrial CO_2 as the assumed ultimate amount of fuel rises. For example, the fraction airborne at peak concentration (with growth cutoff parameter $n = 0.5$) is 73 percent for an ultimate consumption, N_∞, fivefold the preindustrial level of atmospheric CO_2. It is 81 percent for tenfold that level, and 84 percent for fifteenfold. The peak CO_2 levels for these three cases are 4.5, 8.9, and 13.4 times the preindustrial level.

The predicted atmospheric CO_2 levels shown in Figure 4.9 are altered considerably if appreciable amounts of sedimentary carbonate on the sea floor are assumed to dissolve and furnish additional carbonate ions to the seawater. In deep ocean water, marine carbonates are likely to dissolve as soon as industrial CO_2 reaches these waters because, even under normal conditions, most solid carbonate particles dissolve on reaching the deeper ocean basins, and their carbonate ions are returned to the water column above by the deep-water circulation. During the industrial CO_2 era, an accelerated dissolution is likely to occur and perturb this otherwise nearly steady-state condition, but this tendency is reduced by the slowness with which deep water exchanges with the ocean surface.

If dissolution of shallow marine carbonates should also occur, the removal of industrial CO_2 from the air would proceed rapidly because of the proximity of the overlying waters to the air–sea boundary. Arguing against this possibility is the generally high degree of supersaturation of surface water with respect to carbonates (Revelle and Fairbridge, 1957; Edmond and Gieskes, 1970; Ingles et al., 1973). Pure calcite will probably not dissolve appreciably in near-surface seawater even at the time of highest atmospheric CO_2 when the pH of the water might drop as low as 7.4. More soluble high-magnesium carbonates might dissolve but probably only near the peak of industrial CO_2 production. The still more soluble carbonate mineral, aragonite, is abundant in shallow tropical seas, and some aragonite is likely to dissolve during the next century. It is doubtful, however, whether there is enough of this mineral available to reduce atmospheric CO_2 levels more than a few percent.

The range of possibilities for dissolution of marine carbonate to remove industrial CO_2 from the atmosphere is illustrated by three predictions, shown in Figure 4.12. If only deep-water dissolution occurs, the CO_2 level for several centuries is only slightly lower than if no dissolution occurs. Gradually, however, the influence of dissolution becomes evident, so that 1500 years from now, the CO_2 level will have fallen to 2.6 times the preindustrial level instead of 3.9 times, as otherwise predicted. The dissolved carbon in seawater will have in-

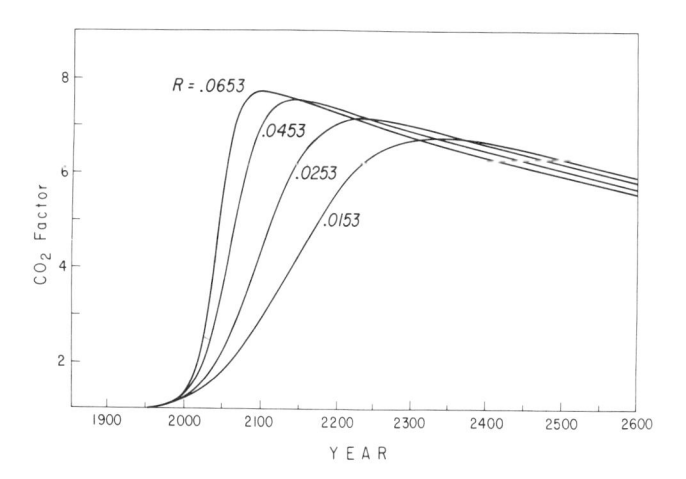

FIGURE 4.11 Predicted increase in atmospheric CO_2 from A.D. 1900 to A.D. 2600 based on simulated fossil-fuel consumption patterns shown in Figure 4.10. The CO_2 factor has the same meaning as for Figure 4.9.

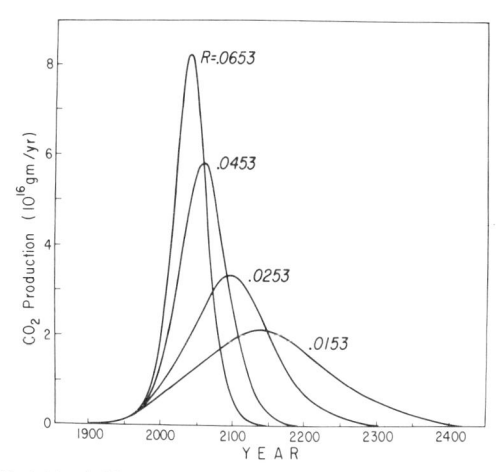

FIGURE 4.10 Additional patterns of industrial CO_2 production, similar to Figure 4.8 except that the growth in rate of fuel combustion, $R(t)$, for 1976, has been varied with logistic cutoff parameter, n, equal to unity, instead of varying n with $R(t)$ fixed. The curve labeled $R = 0.0453$ is the same as that labeled $n = 1$ in Figure 4.8.

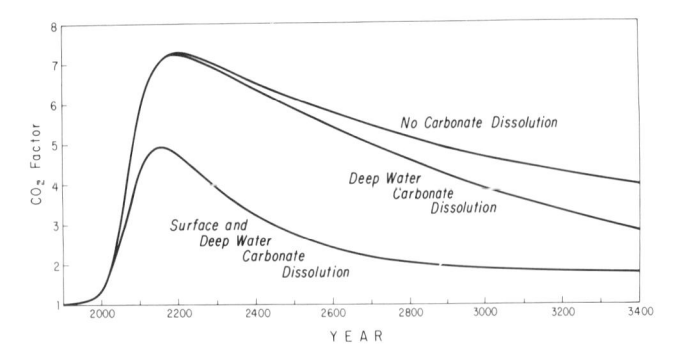

FIGURE 4.12 Predicted increase of atmospheric CO_2 from A.D. 1900 to A.D. 3500 with and without dissolution of carbonate sediments. The pattern of fossil-fuel consumption is as shown by the case $n = 0.5$ of Figure 4.8. The CO_2 factor has the same meaning as for Figures 4.9 and 4.11.

creased by 19.0 percent of the preindustrial concentration (instead of 8.3 percent), and carbonate sediment will have been removed from the ocean floor to a global average depth of about 3 cm.

If dissolution occurs throughout the fossil-fuel era in both shallow and deep water, the atmospheric CO_2 is noticeably lower than otherwise, even as early as the next century. The level in A.D. 3500 will have fallen to only 1.7 times the preindustrial level with most of the falloff before A.D. 2500. The increase in dissolved carbon in seawater in the ocean as a whole will be 22.5 percent of the preindustrial concentration. This amount is not much greater than if dissolution of carbonate occurs only in deeper water (19.0 percent), but in this case, 98 percent of the carbonate that dissolves will have been stripped out of shallow-water sediments. Most of this carbonate, of course, does not remain in surface water: by A.D. 3500 all but 7 percent will have transferred to deep water, principally as dissolved bicarbonate. The average depth of shallow-water carbonate sediment removed from the ocean floor will be the order of 40 cm. Since far less than half of the area of shallow seas contains carbonate-rich sediment, several meters of such sediment on average will have to be stripped of carbonate. This sediment thickness is so great that it seems highly unlikely that extensive dissolution can occur in shallow water within 1500 years even if undersaturation occurs. Thus, the third prediction seems considerably less likely to be fulfilled than the second. Regardless of which prediction is chosen, however, the level of atmospheric CO_2 is predicted to remain near or higher than twice the preindustrial level for a thousand years.

These predictions all assume that the land biota carbon pool is not substantially altered during this long period. If the biomass grows throughout the fossil-fuel era approximately in proportion to the increase in atmospheric CO_2, as assumed in the present model up to A.D. 2000, the predicted atmospheric CO_2 level not only is substantially lower than otherwise near the period of peak concentration but thereafter falls relatively rapidly to near present levels while the biota carbon pool more than doubles. Although this possibility cannot be ruled out on the basis of present limited knowledge of land plant response to CO_2, there is little positive evidence to support it. The subject is further discussed in Appendix 4.A below.

Lesser but substantial uptake of CO_2 by land plants is also possible but still requires a large fractional increase in biotic carbon storage. Since the ultimate amount of recoverable fossil fuel is also uncertain to at least the order of the amount of carbon stored by the land biota, the predicted levels of CO_2 over the next thousand years remain quite uncertain. The present discussion principally emphasizes that the world's oceans cannot be expected to remove a major fraction of the industrial CO_2 from the air for a long time in comparison with the lifetime of human institutions. Although other mechanisms of CO_2 withdrawal may exist, it is probably prudent to expect the concentration of atmospheric CO_2 to persist above twice the preindustrial level for at least several centuries.

If Manabe's prediction of atmospheric temperature rise should be close to correct, even a doubling of the CO_2 level is likely to product climatic change. Manabe's climatic model gives no insight into the importance of the duration of high CO_2 levels, but it seems reasonable that a long period of high CO_2 is more likely to lead to climatic change than would a shorter-term perturbation. As for the ultimate restoration of the ocean and atmosphere to their preindustrial condition, this will require reprecipitation of the carbon added to the ocean water as a result of industrial CO_2 uptake. This is likely to occur only if the world's rivers transport enough noncalcareous alkaline salts to the oceans to neutralize the acidic capacity of the industrial CO_2 already added to the oceans. These salts can only be derived from the decomposition of noncarbonate rocks and soils, and this weathering process probably requires not less than 10,000 years to bring about a major part of the restoration.

REFERENCES

Augustsson, T., and V. Ramanathan (1977). The radiation convective model study of the CO_2 climate problem, *J. Atmos. Sci. 34,* 448.

Bacastow, R. B. (1976). Modulation of atmospheric carbon dioxide by the southern oscillation, *Nature 261,* 116.

Bacastow, R. B., and C. D. Keeling (1973). Atmospheric carbon dioxide and radiocarbon in the natural carbon cycle: Changes from A.D. 1700 to 2070 as deduced from a geochemical model, in *Carbon and the Biosphere,* G. M. Woodwell and E. V. Pecan, eds., U.S. Atomic Energy Commission, pp. 86-135.

Bach, W. (1976). Global air pollution and climatic change, *Rev. Geophys. Space Phys. 14,* 429.

Baes, C. F., Jr., H. E. Goeller, J. S. Olson, and R. M. Rotty (1976). Carbon dioxide and climate: the uncontrolled experiment, submitted to *American Scientist.*

Bischof, W. (1973). Carbon dioxide concentration in the upper troposphere and lower stratosphere. III, *Tellus 25,* 305.

Bolin, B., and W. Bischof (1970). Variations of the carbon dioxide content of the atmosphere in the northern hemisphere, *Tellus 22,* 431.

Broecker, W. S., Y.-H. Li, and T.-H. Peng (1971). Carbon dioxide—man's unseen artifact, Chapter 11 of *Impingement of Man on the*

Oceans, D. W. Hood, ed., Wiley-Interscience, New York, pp. 287-324.

Committee on Impacts of Stratospheric Change (1976). *Halocarbons: Environmental Effects of Chlorofluoromethane Release*, National Academy of Sciences, Washington, D.C.

Edmond, J. M., and J. M. T. M. Gieskes (1970). On the calculation of the degree of saturation of sea water with respect to calcium carbonate under *in situ* conditions, *Geochim. Cosmochim. Acta 34*, 1261.

Ekdahl, C. A., and C. D. Keeling (1973). Atmospheric CO_2 and radiocarbon in the natural carbon cycle: I. Quantitative deduction from the records at Mauna Loa Observatory and at the South Pole, in *Carbon and the Biosphere,* G. M. Woodwell and E. V. Pecan, eds., U.S. Atomic Energy Commission, pp. 51-85.

Flohn, H. (1975). History and intransitivity of climate, appendix 1.2 in *The Physical Basis of Climate and Climate Modelling,* Report of the International Study Conference in Stockholm, July 29-August 10, 1974, GARP Publication Series No. 16, World Meteorological Organization, Geneva, pp. 106-118.

Gillette, R. (1974). Oil and gas resources: Did USGS gush too high? *Science 185*, 127.

Hubbert, M. K. (1969). Energy resources, Chapter 8 of *Resources and Man,* the Division of Earth Sciences, National Academy of Sciences, Washington, D.C., pp. 157-242.

Ingles, S. E., C. H. Culberson, J. E. Hawley, and R. M. Pytkowicz (1973). The solubility of calcite in sea water at atmospheric pressure and 35% salinity, *Marine Chem. 1*, 295.

Keeling, C. D. (1973). Industrial production of carbon dioxide from fossil fuels and limestone, *Tellus 25*, 174.

Keeling, C. D., J. A. Adams, C. A. Ekdahl, and P. R. Guenther (1976a). Atmospheric carbon dioxide variations at the South Pole, *Tellus 28*, 552.

Keeling, C. D., R. B. Bacastow, A. E. Bainbridge, C. A. Ekdahl, P. R. Guenther, L. S. Waterman, and J. S. Chin (1976b). Carbon dioxide variations at Mauna Loa Observatory, Hawaii, *Tellus 28*, 538.

Kelley, J. J., Jr. (1969). An analysis of carbon dioxide in the Arctic atmosphere near Barrow, Alaska, 1961 to 1967, Scientific Report, Office of Naval Research, Contract N00014-67-A-0103-0007, NR 307-252.

Lowe, D. C. (1974). Atmospheric carbon dioxide in the southern hemisphere, *Clean Air,* February, 12.

Machta, L. (1973). Prediction of CO_2 in the atmosphere, *Carbon and the Biosphere,* G. M. Woodwell and E. V. Pecan, eds., U.S. Atomic Commission, pp. 21-31.

Manabe, S., and R. Wetherald (1967). Thermal equilibrium of the atmosphere with a given distribution of relative humidity, *J. Atmos. Sci. 24*, 241-259.

Manabe, S., and R. Wetherald (1975). The effects of doubling the CO_2 concentration on the climate of a general circulation model, *J. Atmos. Sci., 32*, 3-15.

McElroy, M. B., J. W. Elkins, S. C. Wofsy, and Y. L. Yung (1976). Sources and sinks for atmospheric N_2O, *Rev. Geophys. Space Phys. 14*, 143-150.

Newell, R. E., and B. C. Weare (1977). A relationship between atmospheric carbon dioxide and Pacific sea surface temperature, *Geophys. Res. Lett. 4*, 1.

Pearman, G. I., and J. R. Garratt (1973). Space and time variations of tropospheric carbon dioxide in the southern hemisphere, *Tellus 25*, 309.

Ramanathan, V. (1976). Greenhouse effect due to chlorofluorocarbons: climatic implications, *Science 190*, 50.

Reinsch, C. H. (1967). Smoothing by spline functions, *Num. Math. 10*, 177.

Revelle, R., and R. Fairbridge (1957). Carbonates and carbon dioxide, *Treatise Marine Ecol. Paleoecol. 1*, 239.

Revelle, R., and H. E. Suess (1957). Carbon dioxide exchange between atmosphere and ocean, and the question of an increase of atmospheric CO_2 during the past decades, *Tellus 9*, 18.

Rotty, R. (1976). Paper prepared for the Conference "Fate of Fossil Fuel CO_2," Honolulu, January 19-23, N. R. Andersen and A. Malahoff, eds., to be published by the Office of Naval Research, Dept. of the Navy, Arlington, Va.

Rotty, R. (1977). Present and future production of CO_2 from fossil fuels—a global appraisal, paper prepared for Energy Research and Development Agency (ERDA) Workshop on Significant Environmental Concerns, Miami, Florida, March 7-11.

Schneider, S. H. (1975). On the carbon dioxide-climate confusion, *J. Atmos. Sci. 32*, 2060.

Sze, N. D. (1977). Anthropogenic CO emissions: implications for the atmospheric $CO-OH-CH_4$ cycle, *Science 195*, 673.

Wang, W. C., Y. L. Yung, A. A. Lacis, T. Mo, and J. E. Hansen (1976). Greenhouse effects due to man-made perturbations of trace gases, *Science 194*, 685.

Weiss, R. F., and H. Craig (1976). Production of atmospheric nitrous oxide by combustion, *Geophys. Res. Lett. 3*, 751.

Whittaker, R. H., and G. E. Likens (1973). Carbon in the biota, in *Carbon and the Biosphere,* G. M. Woodwell and E. V. Pecan, eds., U.S. Atomic Energy Commission, pp. 281-302.

Yung, Y. L., W. C. Wang, and A. A. Lacis (1976). Greenhouse effect due to atmospheric nitrous oxide, *Geophys. Res. Lett. 3*, 619.

Zimen, K. E., and F. K. Altenheim (1973). The future burden of industrial CO_2 on the atmosphere and the oceans, *Naturwissenschaften 60*, 198.

APPENDIX 4.A: ASSESSMENT OF SIX-RESERVOIR MODEL TO DEPICT FUTURE ATMOSPHERIC CO_2 ABUNDANCES

INTRODUCTION

The global geochemical exchange system recently perturbed by industrial CO_2 production is described above by means of a model with six generalized reservoirs that store and exchange CO_2. It is important to ask whether this model is adequate to the task of predicting future atmospheric CO_2 levels. The following discussion will attempt, at least in part, to answer this question. The six-reservoir model, shown in Figure 4.7 of the main text, will be referred to below as the "6R" model.

This 6R model is representative of a class of models, called "reservoir" or "box" models, which are designed to focus on chemical exchange between various sectors or "pools" of a biogeochemical system. Such models are suited to solving critical problems of chemical transport arising at the boundaries between pools without dealing with mechanisms of transport or storage within these pools. Because of their simplicity, box models are convenient for solving time-dependent problems. The absence of internal complexity promotes freedom to address details of time dependency.

To appreciate the advantages and weaknesses of box models, their most important general properties will now be considered. Afterward, critical aspects of modeling the land biota and ocean water will be discussed in terms of the 6R and selected alternative models.

BOX MODELS

A box model describes a system of reservoirs in which the behavior of a trace chemical substance is described principally or solely in terms of its rate of entering and escaping from each reservoir. A reservoir boundary may be a clearly defined physical surface, such as an air-sea boundary, or it may merely divide regions into convenient subregions, such as shallow and deep-ocean water. Within the volumes described by these boundaries only averages or totals of the properties of the reservoir are considered.

In a box model, the flux of chemical tracer emerging from a reservoir is typically assumed to vary with time in proportion to time changes in the total amount, or mass, of tracer within that reservoir. Specifically, if N denotes the mass of tracer and J the flux out:

$$J - J_0 = k(N - N_0), \quad (4.A.1)$$

where J_0, N_0, and k are constants. The flux and mass constants, J_0 and N_0, often represent steady-state or initial values for the tracer of interest. The ratio N/J is often called a turnover or residence time, and k a box-model exchange coefficient.

For purposes of this discussion, a model in which all the reservoirs are assumed to obey Equation (4.A.1) will be called a "simple box model."

With the great majority of simple box models, it is further assumed that the turnover time is a constant defined by the steady-state ratio N_0/J_0, i.e.,

$$k^{-1} = N_0/J_0, \quad (4.A.2)$$

whence

$$J = kN. \quad (4.A.3)$$

A constant turnover time implies that the distribution of tracer and its circulation within the reservoir (if it circulates) are such that every atom or molecule of tracer in the reservoir has the same chance of escaping at any moment regardless of how or where it entered or where it has resided or circulated within the reservoir. One might say that the tracer is assumed to be randomly distributed or randomly mixed.

This assumption may not be realistic, but it is typically more reasonable than to assume that a tracer is uniformly or well mixed, as is often demanded. Furthermore, if a tracer's actual behavior inside a reservoir is poorly understood, a random escape mechanism may be the most reasonable hypothesis by default. At least, such a model does not assume details that are unknown.

If a random escape mechanism is a poor representation of the real system of interest, Eq. (4.A.3) is likely to be untrue. Nevertheless, Eq. (4.A.1) may still be a good approximation of the tracer transport if the flux and the mass vary so slightly from their steady state values, J_0 and N_0, that these variations are essentially first-order perturbations. To demonstrate this it is only necessary to assume that J is related to N by a function

$$J = f(N), \quad (4.A.4)$$

which possesses derivatives, f', f'', ..., with respect to N. Expanding f in a Taylor's series:

$$J = J_0 + f'(N_0)(N - N_0) + \tfrac{1}{2} f''(N_0)(N - N_0)^2 + \ldots, \quad (4.A.5)$$

where J_0 is equal to $f(N_0)$. Equation (4.A.1) holds whenever the second- and higher-order perturbation terms are sufficiently small to be neglected.

For additions of industrial CO_2 to the earth's natural carbon pools, the fractional increases in carbon in the reservoirs are, up to the present time, only a few percent. Thus Eq. (4.A.1) is likely to be a reasonable approximation irrespective of the true functional relationships of mass and flux within the reservoirs.

Furthermore, if the assumption of random escape is known to be false for a box-model reservoir, the modelist often may be able to subdivide the reservoir to conform better with the known facts. Subdivisions need not even be physically distinct; they may share the same environment. For example, annual and perennial plants exchange comparable amounts of carbon dioxide with the atmosphere, but perennials store many times more carbon. Obviously, the carbon atoms of annuals and perennials have distinctly different chances of returning to the atmosphere. By modeling the land biota as two reservoirs, as was done with the 6R model, this fact can be accounted for even though annual and perennial plants intermingle. Moreover, there is no obligation to divide the biota precisely in terms of annuals and nonannuals, nor need a distinction be made as to whether the short-cycled material is derived solely from short-lived plants or includes material, such as the leaves, of long-lived plants. One can assign any convenient pair of turnover times to define the portions (cf. Keeling, 1973, p. 306).

Reservoir subdivision need not stop at two portions. With respect to the land biota, for example, categories such as annuals, biennials, perennials, deep humus, and peat might all be distinguished. In principle, the modelist could even dispense altogether with a finite number of subdivisions and describe the distribution of the outgoing tracer flux of the entire reservoir as a continuous function with respect to turnover time. Some years ago this approach was proposed by Eriksson and Welander (1956) for the land biota.

Following their lead, let us formulate a generalized box model in which the amount of tracer, N, in a given reservoir (such as the entire land biota) varies with time, t, by an expression of the form

$$N(t) = \int_0^t f(\tau) g(t - \tau)\, d\tau + N_0, \quad (4.A.6)$$

where N_0 denotes an initial unperturbed value of N. The function, $g(t)$, will denote the departure from an initial unperturbed steady state of the tracer flux into that reservoir; and the dimensionless "transfer" function, $f(\tau)$, will describe how molecules or atoms of tracer in that input flux are distributed with respect to the time interval, τ, in which they are expected to remain in that reservoir, i.e., $f(\tau)$ denotes the fraction of molecules or atoms entering the reservoir at time t that are still in the reservoir at time $t + \tau$. The integral of Eq. (4.A.6) thus keeps track of the total increase or decrease in mass of tracer within the reservoir in terms of how much tracer enters and leaves the reservoir at the boundaries after some initial time, $t = 0$.

This integral is a convolution integral. It follows that the Laplace transform of N is equal to the product of the transforms of f and g. The analytic solution of Eq. (4.A.6) is often facilitated by the use of such transforms.

The transfer function for a reservoir of a simple box model has the form (Eriksson, 1961)

$$f = e^{-k\tau}, \qquad (4.A.7)$$

where k is the ratio of change in mass to change in flux defined by Eq. (4.A.1.). For a reservoir such as the land biota, subdivided into two parts with respect to turnover time, the transfer function of the biota as a whole takes the form

$$f = f_1 e^{-k_1 \tau} + f_2 e^{-k_2 \tau}, \qquad (4.A.8)$$

where $f_1 + f_2 = 1$. By a proper choice of k_1, k_2, and f_1/f_2, this function can approximate a considerable range of possible real cases. Analogous expressions apply to subdivision into more than two portions.

More generally, the transfer functions of box models may be of almost any functional form and may even be time-dependent. Models so formulated are not entirely general, however. Equation (4.A.6) requires that there be a linear relationship between the input $g(t)$ and the response $N(t)$, i.e., if $g(t)$ is represented as the sum of very short duration impulses or "spikes" of tracer, then N can be constructed by superposition of responses to these impulses.

Such linearity may not be realistic especially for large perturbations. For example, if the biota should turn over carbon at a different average rate in the next century because of changes in the size of the biota, the biotic response to a single spike would be different than for a series of spikes. In this case, the time dependency of $f(t,\tau)$ cannot be specified in advance since it depends on the character of the source. A convolution integral is then not the most appropriate means of solving the problem of interest. Other approaches, for example involving nonlinear differential equations, may be more suitable.

Another complication arises if several geochemical reservoirs interact, as is the case for the atmosphere, oceans, and land biota. The tracer output flux of one reservoir becomes the input to another reservoir, and more than one reservoir boundary may have to be distinguished.

In modeling the fossil-fuel CO_2 problem, only the atmospheric reservoir unavoidably presents the problem of possessing outputs that are also the inputs to other reservoirs. Fortunately for the modelist, the atmosphere's steady-state internal mixing time is short compared with the characteristic time for industrial CO_2 input. Only small errors arise if industrial CO_2 is assumed to mix randomly or thoroughly in the atmosphere, and Eq. (4.A.1) for the atmosphere holds for large perturbations as well as small ones. The industrial CO_2 input, $g(t)$, to other reservoirs in contact with the atmosphere can therefore be expressed as the product of an exchange coefficient and the globally averaged change in atmospheric CO_2, as in a simple box model. The latter change in turn can be computed in a simple manner knowing the transfer functions $f(t)$ of these other reservoirs. [This is shown in the discussion following Eq. (4.A.13), below.]

If the oceans are divided into a surface and deep layer as in the 6R model, the surface layer presents a separate boundary and a coupled input and output between that layer and the ocean below. If the surface layer is not assumed to be deeper than the physically well-mixed layer of the real oceans (about 75 m), the assumption of a randomly mixed layer obeying Eq. (4.A.3) would be reasonable except that at the air-sea boundary chemical complications arise that require that Eq. (4.A.1) be used. If the surface-layer reservoir is assumed to be deeper than about 75 m, even Eq. (4.A.1) should not be assumed to be a good approximation except for small perturbations. Either a more realistic model should be devised or the approximation demonstrated to be satisfactory for the particular calculation of interest.

With this general view of box models and their capabilities as background, the modeling of the land biota and oceans will now be discussed in relation to the formulation of the 6R model.

LAND BIOTA

The division of the land biota into two carbon pools in the 6R model has already been explained as stemming from a disparity between the fluxes and storage times for long- and short-cycled species. Keeling (1973, pp. 305-309) discusses the limitations of the data base on which this disparity is established. Additional subdivision would have added computational complexity without improving the prediction of future atmospheric CO_2 levels.

Deep-soil humus and incipient geological deposits such as peat were left out of the model since these carbon pools turn over so slowly that they are likely to be negligibly perturbed by any biotic response to changing atmospheric CO_2 levels at least for several more decades. Furthermore, no model based on present knowledge is likely to predict adequately their behavior over longer time periods.

The short-cycled biota carbon pool is assumed initially to contain 4.6 percent of the total biotic carbon but to contribute 54 percent of the carbon exchange with the atmosphere. The remaining flux and mass are attributed to the long-cycled carbon pool. The turnover times of the respective pools are 2.5 and 60 years. No distinction is made between living and dead materials because exchange between these categories is expected to have little influence on the exchange processes of interest.

Because so little carbon is stored in the short-cycled biota, atmospheric CO_2 predictions are almost indistinguishably altered if this carbon pool is neglected altogether as was done by Oeschger et al. (1975). The modeling of the biota as two reservoirs thus has the practical effect of discounting the overall biotic uptake of carbon by about one half while maintaining the assumed biotic storage capacity nearly intact.

The transient increase in biotic uptake of CO_2 in the 6R model is made proportional to the logarithm of the increase in atmospheric CO_2 concentration above the preindustrial concentration. This relationship has been found for individual plants grown in greenhouses (see Keeling, 1973, for original references). The proportionality constant, called a "biota growth factor," was set, as described in the next section, to predict the observed increase in atmospheric CO_2 over the period of direct measurements, 1959 to 1974. The logarithmic function predicts a gradual cutoff of biotic increase at high ambient CO_2 levels but very nearly directly proportional response to the small CO_2 rise that has occurred so far. The model provides that the uptake and release of CO_2 by the long-cycled biota pool depend directly on pool size as well, on the grounds that the leaf area of plants will increase in proportion to total mass, that fixation of CO_2 per unit leaf area will remain constant, and that plants will respire CO_2, and dead materials will decay, in proportion to total mass. None of these assumptions is based on hard evidence. Each merely reflects what would seem reasonable for small perturbations.

For the period 1959 to 1974, and on the assumption either that no carbonate dissolves in the oceans or that carbonate dissolves only in the deep oceans, the 6R model predicts that the biota, short and long cycled combined, increase by 0.63 percent. If dissolution occurs in the surface layer of the oceans, the biota pool size remains almost constant. If the entire past industrial period up to 1974 is considered, the predicted biota increase does not exceed 1.7 percent. These changes are far too small to be directly detectable but, being small, make reasonable the above-mentioned assumptions of the model.

For large atmospheric CO_2 perturbations of the global biota, no trustworthy observations or theories exist to guide the modelist. The calculations described in the main text of this chapter reflect the view that large increases in the biota will not occur because loss of virgin lands to agriculture and widespread increase in density of human settlement with its attendant disturbance of soils will prevent any major future global buildup of organic carbon. In the 6R model, the biotic increase is simply brought to a halt by an external computer program command. For the case of an ultimate industrial CO_2 production of 8.2 times the preindustrial atmospheric CO_2, a logistic parameter n of 0.5, and either no carbonate dissolution or carbonate dissolution only in the deep oceans, the biota carbon pool is predicted to have increased by 5.9 percent in 2025, after which time its size remains constant. If dissolution also occurs in the surface layer, the biotic pool is predicted to decrease very slightly (by -0.1 percent).

Since these small changes in the biota hardly influence the atmospheric CO_2 concentration near or after the time of peak levels, the biota carbon pool could be left out of the 6R model altogether were it not that its presence provides a basis for explaining part of the withdrawal of CO_2 from the atmosphere during the historical record period, 1959 to 1974. With the biota included, the modelist may adjust the model to agree with these observations without being obliged unrealistically to adjust the oceanic parameters to account exactly for the industrial CO_2 uptake.

Only if our understanding of the biota response to industrial CO_2 improves in the coming years will the inclusion of biota reservoirs in predictive models become truly useful. This is especially true for the large biotic carbon pools, such as deep-soil humus, which turn over very slowly under natural conditions.

Finally, it should be pointed out that Machta (1971) and Oeschger et al. (1975) modeled the biota with a transfer function that provides that each molecule of CO_2 entering the biota will escape after exactly the average turnover time of carbon in the reservoir. In the case for which the atmospheric CO_2 concentration rises exponentially, say at a rate proportional to $e^{\mu t}$, their models yield the same prediction for biota growth as the 6R model does, provided that the 6R model growth factor is replaced by their growth factor multiplied by $(1 - e^{-60\mu})$, where 60 years is their assumed value of the biotic carbon turnover time. For the approximate exponential growth in industrial CO_2 since 1945, μ^{-1} is approximately 22 years, and this adjustment factor is about 0.93. Since neither their models nor the 6R model were used to consider biota increases after nearly exponential rise in CO_2 has ceased, it is not worthwhile to discuss their biota model in any greater detail here.

THE OCEANS

The division of the ocean into layers for chemical modeling purposes was first investigated in detail by Craig (1957). The turnover time of the whole ocean, if computed from the steady-state flux of radiocarbon at the air–sea boundary, is about 400 years, whereas the average circulation time based on radioactive tracer studies is known to be greater than 1000 years. As Craig pointed out, such a discrepancy can be accounted for by a simple box model with two tandem oceanic reservoirs. A surface layer of almost any depth up to several hundred meters might be assumed. Craig, however, made a physically realistic choice by fixing the model depth at 75 m to agree approximately with the average depth of the seasonal thermocline above which the water is vertically well mixed. Although the Craig model is a good approximation for this surface layer, it is a gross oversimplification for the water lying below. This is because a transition zone, defined by a steep temperature gradient, divides this surface layer from the relatively homogeneous deep ocean below 1000 m [see, for example, von Arx (1974),

Figure 5.5, p. 131]. An oceanic model could be devised with more than two vertically separated reservoirs to account for the transition zone, but further subdivision, as in the case of the biota, does not necessarily lead to a better model, because available observational data offer a poor basis for establishing the additional transport parameters required. On the other hand, if the ocean is divided vertically into only two reservoirs, the dividing surface is more reasonably located somewhere near the middle of the subsurface transition zone rather than at its top.

As a basis for assigning a depth greater than the physically defined mixed layer, Bacastow and Keeling (1973) forced the 6R model to predict the Suess effect, i.e., the dilution of atmospheric radiocarbon by industrial CO_2 up to 1954 when nuclear-weapons testing began to alter the natural radiocarbon cycle. The significance of this approach to calibration of the 6R model will now be considered in some detail, first in terms of the 6R model and a simplified version of it, the "4R" model, and second in terms of a generalized box model. The discussion will rely in part on a recent investigation of the CO_2 modeling problem by Oeschger et al. (1975).

THE SUESS EFFECT

Because industrial CO_2 contains no radiocarbon, its injection into the air dilutes the atmospheric radiocarbon residing there naturally as a result of cosmic-ray production. Were it not for the ability of the biota and oceans to store radiocarbon and for the chemical reactivity of seawater to inhibit the oceanic uptake of industrial CO_2, the relative decrease in radiocarbon would be almost exactly equal to the relative increase in atmospheric CO_2. Actually, the Suess effect, as observed in tree rings that grew around 1950, is approximately one third of the CO_2 increase at that time.

This is a result of two factors. First, the Suess effect is smaller than otherwise because the land biota carbon pool dilutes radiocarbon. This dilution is nearly independent of small changes in the size of the biotic carbon pool. On the other hand, small changes in pool size may have appreciably increased or decreased the abundance of atmospheric CO_2.

Second, with respect to the chemical reactivity of seawater, a thermodynamic restraint inhibits the oceanic uptake of atmospheric CO_2 while scarcely influencing the radiocarbon distribution. Because the storage of carbon in the oceans is expressed in terms of total dissolved carbon in seawater, whereas gas exchange at the air-sea boundary depends on the partial pressure gradient at that boundary, the steady-state box model exchange coefficient, k_{ma}, for the escape of CO_2 from the surface oceanic layer to the atmosphere must be multiplied by a perturbation factor to account properly for transient exchange. This factor is defined as the ratio of the relative change in CO_2 pressure exerted by the water to the relative change in total carbon in the water. Bacastow and Keeling (1973) calculated its value without approximation from known (if somewhat uncertain) thermodynamic constants for seawater. They called it an "evasion factor" (symbol, ξ). It has also been called the "Revelle factor" after the scientist who first proposed its use and noted that its present value is of the order of 10 (Revelle and Suess, 1957). Taking into account the Revelle factor reduces by this factor the otherwise predicted capacity of the surface ocean layer to store industrial CO_2 but has almost no effect on the radiocarbon distribution. This is because, as with the biota, the oceans serve only to dilute radiocarbon, irrespective of the chemical reactions occurring in the reservoir.

As one might expect when two quasi-independent observations are available, a knowledge of both the observed Suess effect and the observed CO_2 increase permits the CO_2 uptake by the biota and that by the oceans to be separately evaluated. What is less obvious is that this evaluation is almost independent of assumptions as to how carbon circulates in subsurface ocean water and does not require that the Suess effect and the recent CO_2 increase to be known for the same period.

The Suess effect around 1950 is predicted by the 6R model to be approximately -2.2 percent for a surface layer depth, h_m, of 75 m. [Bacastow and Keeling (1973) give exact details.] This prediction is somewhat larger than the average observed Suess effect based on tree-ring studies, a value of about -2.0 percent. If h_m is increased to 120 m, the predicted Suess effect agrees well with the observations. Observed values scatter considerably, however, and, since local industrial CO_2 may have contaminated some of the samples used in the tree-ring studies, the observed value may be systematically too high. Unless all are wrong, however, the Suess effect cannot have been much smaller than the -1.4 percent value predicted if h_m is set at about 700 m. A more likely value for the Suess effect is -1.8 percent. Therefore, the surface-layer depth in the 6R model should probably not be set larger than approximately 700 m, although about 250 m is probably more appropriate.

In the 6R model, in addition to the depth parameter, h_m, there is another adjustable oceanic factor: the average turnover time of the deep layer. Bacastow and Keeling (1973) found (see their Figure 5) that the 6R model yields almost the same relationship for the ratio of the predicted Suess effect to the predicted atmospheric CO_2 increase irrespective of which parameter is varied. Specifically, if h_m is fixed at approximately 120 m and the box-model turnover time of the deep-ocean layer, k_{dm}^{-1}, is varied, the relation between the Suess effect near 1950 and the CO_2 increase for 1959 to 1969 is almost indistinguishable from the relation found when h_m is varied and k_{dm}^{-1} is fixed at 1500 years.

To explain this result, it is important to note, as in the main text of this chapter, that the input of industrial CO_2 has been rising almost exponentially, while past responses of the biota and oceans to this input have probably been essentially linear. Under conditions of linear response to a persistent exponentially rising input, the annual airborne fraction, i.e., the ratio of the annual atmospheric CO_2 increase to the annual industrial CO_2 production, is predicted by the 6R model to approach constancy with time (Ekdahl and Keeling, 1973; Oeschger et al., 1975). The cumulative airborne fraction, i.e., the ratio of the total atmospheric CO_2 increase from the preindustrial value, to the cumulative CO_2 production, also approaches this value.

If the magnitude of the Suess effect is divided by the cumulative industrial CO_2 input, a radiocarbon dilution factor is obtained that is analogous to the cumulative airborne fraction. Like the latter, it is nearly time invariant for a linear response to an exponentially rising CO_2 input. The exchange parameters to predict the dilution factor are, however, modified from those for inactive CO_2, depending on the Revelle factor, ξ, and the biota growth factor (which will be denoted by the symbol β). Since ξ is known from thermodynamic data, one can adjust the unknown magnitude, β, and an additional parameter related to oceanic uptake of CO_2 so that the model predicts both the radiocarbon dilution and the airborne fraction. As already stated, either of two subsurface parameters of the 6R model may be varied to produce the desired predictions.

To explain this point quantitatively, let us define a four-reservoir (4R) model in which the 6R model is simplified by combining its two atmospheric reservoirs and eliminating its short-cycling biota reservoir. Also, the long-cycled biota will be assumed to exchange as in a simple box model, i.e., CO_2 uptake will depend only on changes in atmospheric CO_2 not on biota pool size as well. For this 4R model, as with the 6R model, the assumed depth of the surface layer and the average turnover time of the deep layer are critical for establishing the predicted magnitude of the airborne fraction and radiocarbon dilution.

Specifically for the 4R model, if the Revelle factor, ξ, is assumed constant, and if the industrial CO_2 input to the atmosphere, $g_a(t)$, rises exponentially with time as predicted by the equation

$$g_a(t) = g_0 e^{\mu t} \quad (4.A.9)$$

and if t is long compared to the efold time μ^{-1}, the airborne fraction, is (cf. Ekdahl and Keeling, 1973)

$$r_a = \left(1 + \frac{k_{am}}{\mu}\left\{1 + \frac{k_{ma}\xi}{\mu}\left[1 + \frac{k_{md}}{\mu}\left(1 + \frac{k_{dm}}{\mu}\right)^{-1}\right]^{-1}\right\}^{-1} + \frac{\beta k_{ab}}{\mu}\left(1 + \frac{k_{ba}}{\mu}\right)^{-1}\right)^{-1}, \quad (4.A.10)$$

where the simple box-model exchange coefficients are as defined in Table 4.A.1.

The indices have the following significance: a, atmosphere; b, biota; m, oceanic surface layer; d, deep ocean. The exchange coefficient, k_{ij}, refers to transfer from reservoir i to reservoir j. All k_{ij} are assumed to be constant with time.

This expression is of the form

$$r_a = \left(1 + \frac{A}{1 + B\xi} + C\beta\right)^{-1}, \quad (4.A.11)$$

where A, B, and C have fixed values for a constant value of the efold time, μ^{-1}. The quantities, A, C, and ξ are reasonably well known compared with B and β. Therefore, the latter may be justifiably regarded as the unknowns to be evaluated from the time-dependent observations.

The corresponding radiocarbon dilution factor, for the 4R model, is to a close approximation (cf. Oeschger et al., 1975, p. 181) given by

$$r_a{}^* = \left(1 + \frac{A}{1 + B} + C\right)^{-1}, \quad (4.A.12)$$

where the influence of isotopic fractionation is neglected.

It is now possible to explain the finding of Bacastow and Keeling (1973) that a nearly unique concordance exists between the predicted Suess effect and airborne fraction. If the depth of the surface layer is varied (causing k_{ma} and k_{md} to vary) or if k_{dm} is varied, holding the surface layer depth constant, a change occurs in the value of B. Irrespective of which box model parameter is changed, the change in B has the same effect on the relative values of r_a and $r_a{}^*$.

When the actual year-to-year CO_2 production data were used in the calculation, the nonlinear influence of a variable Revelle factor allowed for, and isotopic fractionation prescribed, Bacastow and Keeling found that the discrepancy between the results varying one parameter or the other was of the order of 0.04 percent in the Suess effect. This discrepancy is too small to be useful in obtaining any additional information about the 6R model over that assuming linearity and an exponential rise in CO_2 production.

What has been gained by taking into account the Suess effect is solely to be able to compute separately the industrial CO_2 input to the oceans and to the biota. This was done by first solving Eq. (4.A.12) for the subsurface oceanic parameter, B, and then solving Eq. (4.A.11) for the biotic growth factor, β.

Although this separate computation of inputs has been shown only with respect to the 4R model, it applies generally. In order to show this as a prelude to considering an alternative model, the significance of the Suess effect will now be reconsidered within the framework of a model with a generalized oceanic transfer function.

THE GENERALIZED OCEANIC BOX MODEL

To keep the analysis of this general model as simple as possible, no changes will be made in the modeling of the atmosphere and land biota. These reservoirs will again be assumed to transfer and store carbon as predicted by a simple box model. As noted earlier, this is a good approximation for the atmosphere. With respect to the biota, a simple box model suffices to explain the significance of the Suess effect even if the numerical values predicted by the model are questionable because of uncertainty in the pool size and the distribution of turnover times of various sectors of the biota.

With respect to the oceans, the essential difference in uptake of industrial CO_2 and radiocarbon should become evident if the subsurface ocean is regarded as being separated from the atmosphere by a thin boundary layer that interacts with the atmosphere in a manner consistent with the known chemistry of seawater and that correctly predicts the transport of carbon to and from the subsurface layer. If the boundary layer is defined to be thin enough to have negligible capacity to store carbon, the transport mechanism within the layer is not important, and a simple box model formulation will almost surely suffice. With this view in mind the generalized oceanic model, as proposed by Welander (1959), will be assumed to possess a simple box-model boundary layer of negligible thickness. This approach is, of course, more rigorous than to assume a boundary layer of 75 m, which in reality may not fully obey the simple box-model equations.

With respect to the ocean below this boundary layer, no restriction will be placed here on the mechanism of carbon transport or storage except to assume that the same mechanism applies to both industrial CO_2 and radiocarbon and that the process is linear.

The box-model exchange constants for interactions between the atmosphere, biota, and oceanic boundary layer will be formulated as before for the 4R model, i.e., the radiocarbon constant, k_{ma}, for transfer from the oceanic surface to the atmosphere will be increased for CO_2 by a factor, ξ, and the radiocarbon constant, k_{ab}, for transfer from the atmosphere to the biota will be modified for CO_2 by a factor, β.

As a first step, general expressions for tracer transport and storage will be derived to show how Eq. (4.A.6) applies to the formulation of the airborne fraction. Afterward, the transfer function for the oceans will be derived on the assumption of a very thin boundary layer. The cumulative airborne fraction and the radiocarbon dilution factor will be compared for this case.

Let n_a, n_s, and n_b denote for the atmosphere, oceans, and biota, respectively, the increase in carbon resulting from input of industrial CO_2. If that input is exponential, as provided by Eq. (4.A.9), the increases, n_i, will sum to the cumulative input, i.e.,

$$n_a + n_s + n_b = \int_0^t g_0 e^{\mu t} dt \qquad (4.A.13)$$
$$= \frac{g_0}{\mu}(e^{\mu t} - 1).$$

For times long compared with the efold time, μ^{-1}, the term $e^{\mu t}$ is much greater than unity. Then, to a close approximation,

$$n_a + n_s + n_b = \frac{g_0}{\mu} e^{\mu t}. \qquad (4.A.14)$$

The atmospheric CO_2 increase is in general [cf. Eq. (4.A.6)]

$$n_a(t) = \int_0^t f_a(\tau) g_a(t - \tau) d\tau, \qquad (4.A.15)$$

where subscript a indicates that the variables refer to the atmosphere. If the atmospheric input, g_a, is given by Eq. (4.A.9),

$$n_a = g_0 e^{\mu t} \int_0^t f_a(\tau) e^{-\mu \tau} d\tau, \qquad (4.A.16)$$

For time, t, sufficiently large, the integral approaches the Laplace transform:

$$\tilde{f}_a(\mu) = \int_0^\infty f_a(\tau) e^{-\mu \tau} d\tau, \qquad (4.A.17)$$

and hence the atmospheric increase is

$$n_a = g_0 e^{\mu t} \tilde{f}_a = g_a \tilde{f}_a. \qquad (4.A.18)$$

The transform, $\tilde{f}_a(\mu)$, has the dimensions of time. For a persistent exponentially rising source it has the form of a transient "turnover time" computed by taking the ratio of the increase in mass of tracer at a given moment in time to the input flux of tracer at that moment.

As already discussed, the recent past behavior of industrial CO_2 in the atmosphere approximates the prediction of a simple box model. Equation (4.A.3) should apply, and the perturbation flux into the oceans may then be expressed as a constant fraction of the increase in atmospheric CO_2 above the preindustrial value, i.e.,

$$g_s = k_{am} n_a. \qquad (4.A.19)$$

From Eq. (4.A.18) it follows that

$$g_s = k_{am} g_0 \tilde{f}_a e^{\mu t}. \qquad (4.A.20)$$

Substituting in an equation equivalent to (4.A.15), except that the subscript s replaces a, and carrying out the same analysis as for atmospheric CO_2, the oceanic carbon increase is

$$n_s = g_s \tilde{f}_s. \qquad (4.A.21)$$

Again this is a close approximation for large values of t. Similarly, if the biota are assumed to take up carbon in proportion to the increase in atmospheric CO_2:

$$g_b = \beta k_{ab} n_a, \qquad (4.A.22)$$

where, however, $\beta = 1$ for radiocarbon. Analogous to Eq. (4.A.21),

$$n_b = g_b \tilde{f}_b \qquad (4.A.23)$$

for large values of t.

The cumulative airborne fraction is defined as

$$r_a = \frac{n_a}{n_a + n_s + n_b}. \qquad (4.A.24)$$

Substituting for n_s and n_b from Eqs. (4.A.21) and (4.A.23) with g_s and g_b replaced according to Eqs. (4.A.19) and (4.A.22),

$$r_a = (1 + k_{am} \tilde{f}_s + \beta k_{ab} \tilde{f}_b)^{-1}. \qquad (4.A.25)$$

Thus, for large t, r_a approaches constancy, as pointed out by Ekdahl and Keeling (1973).

To indicate how the radiocarbon dilution factor, r_a^*,

differs from r_a, it is necessary to specify the dependency of the oceanic transform $\tilde{f}_s$, on the Revelle factor, ξ. Since the Revelle factor applies to the air–sea boundary, it should be sufficient, as already proposed, to specify the transfer conditions as applying to a thin simple box-model boundary layer. The transfer function for carbon below this layer will be denoted by $f_d(\tau)$ but left otherwise unspecified.

For the boundary layer, the preindustrial amount of carbon will be expressed as a concentration, C_{mo}, multiplied by the area of the oceans, A, and by the depth, h_m. To relate this amount of carbon to the amount of CO_2 in the preindustrial atmosphere, N_{ao}, a hypothetical depth, h_a, will be defined to denote a surface layer of uniform depth that would contain N_{ao}, i.e.,

$$N_{ao} = C_{mo} A h_a. \quad (4.A.26)$$

The steady-state rate of exchange of CO_2 at the air–sea boundary J_{amo}, is equal to both $k_{am} C_{mo} A h_a$ and $k_{ma} C_{mo} A h_m$, whence

$$k_{am} h_a = k_{ma} h_m. \quad (4.A.27)$$

Irrespective of the actual thickness assumed for the boundary layer, the product $k_{ma} h_m$ is finite and equal to $k_{am} h_a$. From measurements of the radiocarbon gradients at the air–sea interface, the value of $k_{am} h_a$ is known to be approximately 8 m yr^{-1}.

This velocity is not a good measure of subsurface tracer transport, however, since it implies, as does a simple one-box oceanic model, that the turnover time of the deep ocean is about 400 years. The average transport below the surface must be considerably slower but is not readily evaluated without resorting to a detailed analysis of data and mechanisms not relevant in this immediate discussion. The steady-state input rate of carbon to the subsurface water, J_{mdo}, will therefore be left unspecified by setting it equal to the steady-state flux across the air–sea interface, J_{amo}, multiplied by an arbitrary factor, α. Since the boundary layer is assumed to behave as in a simple box model, J_{mdo} is equal to $k_{md} C_{mo} A h_m$, hence,

$$k_{md} h_m = \alpha k_{am} h_a. \quad (4.A.28)$$

The transform, $\tilde{f}_s$, for the oceans as a whole, by analogy with Eq. (4.A.10), and in view of Eq. (4.A.25), takes the form

$$\tilde{f}_s = \frac{1}{\mu}\left[1 + \frac{k_{ma}\xi}{\mu}(1 + k_{md}\tilde{f}_d)^{-1}\right]^{-1}, \quad (4.A.29)$$

where $\tilde{f}_d$, the transform for the subsurface ocean, replaces $1/(\mu + k_{dm})$ of the simple box model.

Replacing k_{ma} and k_{md} according to Eqs. (4.A.27) and (4.A.28),

$$\tilde{f}_s = \frac{1}{\mu}\left[1 + \frac{k_{am} h_a \xi}{\mu}(h_m + k_{am} h_a \alpha \tilde{f}_d)^{-1}\right]^{-1}. \quad (4.A.30)$$

An analogous expression applies to radiocarbon, except that the Revelle factor is set equal to unity.

Next, letting h_m approach zero:

$$\tilde{f}_s = \{\mu[1 + \xi/(\mu\alpha\tilde{f}_d)]\}^{-1}, \quad (4.A.31)$$

where ξ is equal to unity for radiocarbon.

The radiocarbon dilution factor is, therefore,

$$r_a{}^* = \left\{1 + \frac{k_{am}}{\mu}[1 + 1/(\mu\alpha\tilde{f}_d)]^{-1} + \frac{k_{ab}}{\mu + k_{ba}}\right\}^{-1}. \quad (4.A.32)$$

Given k_{am}, k_{ab}, k_{ba}, and the efold time μ^{-1}, the subsurface ocean transform, $\tilde{f}_d$, multiplied by a factor α may be evaluated in terms of the observed radiocarbon dilution factor. Then with $\alpha\tilde{f}_d$ prescribed, the equation for the airborne fraction,

$$r_a = \left\{1 + \frac{k_{am}}{\mu}[1 + \xi/(\mu\alpha\tilde{f}_d)]^{-1} + \frac{\beta k_{ab}}{\mu + k_{ba}}\right\}^{-1}, \quad (4.A.33)$$

may be solved for the growth factor, β.

In summary, for a persistently exponentially rising input of CO_2 to the atmosphere, the increase in CO_2 and decrease of radiocarbon in the atmosphere are predicted by a linear model to be constant fractions of that input. Given a knowledge of both fractions and the steady-state transfer rates for CO_2 between the atmosphere, oceans, and biota, the uptake of CO_2 by the oceans and biota can be separately evaluated. This evaluation is virtually independent of any assumptions made about the physical processes occurring in the ocean except that they result in a linear response to CO_2 input and that the physical processes below a thin randomly mixed boundary layer effect the carbon and radiocarbon in the same way.

Moreover, the illustration of an exponentially rising input was emphasized only because it very nearly applies to the historic situation of interest. The significance of knowing the atmospheric radiocarbon variation would apply to other patterns of input. For example, if r_a and $r_a{}^*$ were known as functions of time over an extended period and if the input could be resolved into a sum of exponentials of differing efold times, it would be possible to evaluate the modified subsurface transform $\alpha\tilde{f}_d$ for each of these efold times.

Indeed, the mathematical formulation of the perturbation problem may be extended to any arbitrary input pattern. For example, the expressions for n_a, n_s, and n_b given by Eqs. (4.A.18), (4.A.21), and (4.A.23) go over in the case of a general input to

$$\tilde{n}_i = \tilde{f}_i \cdot \tilde{g}_i, \quad (4.A.34)$$

where $\tilde{n}_i(\mu)$ and $\tilde{g}_i(\mu)$ are Laplace transforms of $n_i(t)$ and $g_i(t)$ respectively. Also, the analytic expressions for the airborne fraction found earlier for the 4R and the generalized oceanic model are actually the general Laplace transforms of $f(\tau)$ for these models multiplied by the Laplace transform frequency μ. In the general case r_a is replaced by

$$\frac{\tilde{n}_a}{\tilde{g}_a/\mu} = \mu \tilde{f}_a. \quad (4.A.35)$$

Thus, the analytic expressions derived in this discussion apply with only minor modification to the general problem.

An additional important conclusion arises from the above considerations. Since it is possible to give explicit recognition to the Revelle factor in separating the oceanic and biotic inputs, it is possible with a simple box model solved numerically to make predictions for continued exponential increase in industrial CO_2 input into the future even though the surface oceans will become more and more nonlinear in response, as long as a variable Revelle factor is the only nonlinear oceanic feature involved.

In predicting future atmospheric CO_2 levels, a need for additional information arises because the present exponential rise will not continue indefinitely. As the rate of rise diminishes, subsurface waters will increasingly influence the uptake of industrial CO_2. The value of the modified subsurface transform $\alpha \tilde{f}_d$, estimated from the Suess effect, will become less and less applicable. The modelist must look to other data, especially more direct oceanic data, to establish the nature of subsurface transport and storage of industrial CO_2. We shall now discuss alternative models that consider the problem of subsurface transport and storage.

THE BOX DIFFUSION MODEL

In a recent investigation of industrial CO_2 uptake by the oceans and land biota, Oeschger et al. (1975) approached the problem of subsurface ocean modeling using a "box diffusion" (BD) model in which the vertical distribution of radiocarbon aids in establishing the response of the oceans to atmospheric CO_2 input. The input of radiocarbon to the oceans is known from tree-ring studies to have been more or less constant for thousands of years, and therefore the observed radiocarbon distribution is likely to represent an oceanic steady state, if allowance is made for recent additions of artificial radiocarbon from nuclear bombs to the upper few hundred meters of water. Since an oceanic steady-state distribution reflects a wide range of characteristic response times, this radiocarbon distribution should be a valuable aid in devising a model valid for long-term predictions of industrial CO_2 uptake by the oceans.

The authors, however, have made only limited use of that distribution, because they have assumed a constant eddy-diffusion coefficient below a surface mixed layer of 75 m. As discussed below, salinity and temperature data indicate that it is unlikely that the ocean circulation obeys a constant vertical diffusive principle, and, therefore, the box diffusion model, like previous box models of the carbon cycle, is a highly simplified parameterization of the transfer and storage of oceanic carbon. As with previous models, it must be justified on the grounds that it accords well with time-dependent observational data.

The authors have tested the BD model by noting that it satisfactorily predicts both the uptake of "bomb" radiocarbon produced since 1954 and the Suess effect and airborne fractions resulting from industrial CO_2 production. Bomb radiocarbon as a tracer has the advantage that it was injected into the atmosphere over a short time period and like a single spike allows the free response of the oceans to be revealed. The source, however, is not so well established as that for industrial CO_2, and the observing period since the major weapons testing occurred near 1960 is too short to yield information on characteristic response times longer than a few years. Thus bomb radiocarbon as yet is of limited value in evaluating models to be used for long-range predictions.

With respect to the Suess effect and the airborne fraction, since the box diffusion model is linear, and since the CO_2 injection has been essentially an exponentially rising function of time, the model, for different values of its parameters, predicts a relationship between the Suess effect and the airborne fraction indistinguishable from that for a simple box model such as the 4R model. Specifically, the oceanic transform for the BD model is given by the expression

$$\tilde{f}_s = \frac{1}{\mu}\left\{1 + \frac{k_{am}h_a\xi}{\mu[h_m + \sqrt{K/\mu}\tanh(h_d/\sqrt{K/\mu})]}\right\}^{-1}, \quad (4.A.36)$$

where h_d denotes the average depth of the subsurface ocean, K the vertical eddy-diffusion coefficient, and, as before, ξ is unity for radiocarbon. In comparison, the 4R model has the transform

$$\tilde{f}_s = \frac{1}{\mu}\left\{1 + \frac{k_{am}h_a\xi}{\mu[h_m + k_{dm}h_d/(\mu + k_{dm})]}\right\}^{-1}, \quad (4.A.37)$$

where the depths h_m and h_d may differ from those for the BD model.

If the model parameters K, h_m, and h_d for the BD model are set as preferred by Oeschger et al. (1975) (see Table 4.A.1) and if the 4R model exchange constant for deep water, k_{dm}, is set at 1/(1209 yr) to match the average radiocarbon concentration of subsurface water as predicted by the BD model, the two-model predictions agree for an e fold time of 22 years if h_m of the 4R model is 309 m and h_d is 3491 m. Similarly, for any reasonable value of the BD coefficient, K, values for the 4R parameters, h_m and k_{dm}, can be found which lead to agreement in prediction. As already noted in connection with the 6R model, the actual rate of rise in industrial CO_2 production has been so close to this exponential $e$fold time that no useful information that might distinguish between the predictions of the models can be obtained by considering departures from this input.

The strongest argument of Oeschger et al. (1975) supporting the BD model is that this model consistently describes phenomena with completely different characteristic times, such as bomb radiocarbon uptake and the Suess effect, whereas the simple box model fails to do so. They state

TABLE 4.A.1 Model Parameters

I. *Four-Reservoir (4R) Box Model Initial Values*

N_{ao}, CO_2 in atmosphere	$= 0.6156 \times 10^{18}$ g of carbon
N_{bo}, carbon in biota	$= 1.5600 \times 10^{18}$ g of carbon
F_{bo}, flux of CO_2 between atmosphere and biota	$= 2.6 \times 10^{16}$ g of carbon yr^{-1}
h_a, depth of hypothetical surface layer containing N_{ao} g of carbon	$= 58$ m
h_s, depth of ocean ($= h_m + h_d$)	$= 3800$ m
h_m, depth of surface layer	$=$ variable

Steady-State Transfer Coefficients

k_{ab}, atmosphere to biota	$= F_{bo}/N_{ao}$
k_{ba}, biota to atmosphere	$= F_{bo}/N_{bo}$
k_{am}, atmosphere to surface ocean	$= 1/(7.53$ yr$)$
k_{ma}, surface ocean to atmosphere	$= k_{am}h_a/h_m$
k_{md}, surface ocean to deep ocean	$= k_{dm}[(h_s/h_m) - 1]$
k_{dm}, deep ocean to surface ocean	$= 1/(1209$ yr$)$

Other Factors

ξ, ocean-to-atmosphere CO_2 evasion factor (Revelle factor)	$= 8.8957$
β, biota growth factor	$= 0.198$

II. *Box Diffusion (BD) Model Initial Values*
Same as 4R model except
h_m $= 75$ m

Steady-State Transfer Coefficients

k_{ab}, k_{ba}, k_{am}	same numerical values as 4R model
k_{ma}	same functional dependence on h_m as 4R model
K vertical eddy-diffusion coefficient	$= 3987$ m^2 yr^{-1}

Other Factors

ξ, β same numerical values as 4R model

(p. 168) that "in the box diffusion model the flux from mixed [i.e., surface] layer to deep sea [i.e., subsurface layer] increases with decreasing time constants of the perturbations. This is in [contrast] to [simple] box models where it is essentially independent of the time constants if they are smaller than a few hundred years. Because of this fact our model is valid for predictions of the atmospheric CO_2 response to the various possible future CO_2 input time functions."

Since this view casts serious doubt on the predictive value of the 6R model, it is important to determine, if possible, how different are the predictions of the two models. But to lead up to this point, it is worthwhile first to examine the authors' argument regarding flux dependency on response time.

For an exponentially rising source, the BD model prescribes (Oeschger *et al.*, 1975, p. 178) that the net perturbation in the tracer flux from the oceanic surface layer to the water lying immediately below is

$$\Delta J_{md}(\text{BD}) = \frac{n_m \mu}{h_m} \sqrt{K/\mu} \, \tanh(h_d/\sqrt{K/\mu}), \quad (4.A.38)$$

whereas for the 4R model,

$$\Delta J_{md}(\text{4R}) = \frac{n_m \mu}{h_m} k_{dm} h_d/(\mu + k_{dm}), \quad (4.A.39)$$

where ΔJ_{md} denotes, for the model specified, the difference between the downward and upward fluxes at the boundary between the surface and subsurface layers, n_m denotes the increase in carbon in the surface layer, and where h_m and h_d may differ for the two models.

For large values of the *e*fold time, μ^{-1}, these fluxes approach the same relationship:

$$\Delta J_{md} = n_m \mu (h_d/h_m). \quad (4.A.40)$$

For values of μ^{-1} less than several hundred years, with the magnitudes of K and k_{dm} having values of the order given in Table 4.A.1, the BD model flux is approximately equal to $n_m \sqrt{K\mu}/h_m$, whereas the 4R model flux is equal to $n_m k_{dm} h_d/h_m$.

Thus the essential difference in flux for the two models is that the BD model flux, for a given value of n_m, varies as $\sqrt{\mu}$, whereas the 4R flux is independent of μ. The authors state for the case for which h_m is 75 m for both models (p. 178): "The differences are striking. For a time constant [μ^{-1}] of 35 y... the flux is 3.2 times higher in BD than in [4R]! In the case of ^{14}C even shorter characteristic times are involved and, since the ^{14}C fluxes are essentially the same as those for CO_2, [the] difference between BD and 4R is accordingly greater."

Although this statement is correct, it is not a fair analysis of the differences between the two models. The authors do not consider other possible values of the surface layer depth, h_m, nor the relationship between the flux, ΔJ_{md}, and the cumulative airborne fraction. As can be seen from Eqs. (4.A.25), (4.A.36), and (4.A.37), the dependency of the airborne fraction on μ is quite different from that of ΔJ_{md} and involves the magnitude of μ in relation to the air–sea exchange constant, k_{am}. For μ^{-1} small compared with the exchange time, k_{am}^{-1}, the CO_2 uptake at the air–sea interface, assumed to be the same for both models, becomes the rate-determining process. The airborne fractions predicted by both models therefore approach the same value. Since the models also agree for long *e*fold times, they show a maximum difference in prediction for intermediate times. If h_m is assumed to be 75 m for both models and the other model parameters are set according to Table 4.A.1, the

TABLE 4.A.2 Comparison of Simple Box Model and Box Diffusion Model with $h_m = 75$ m for Both Models

	Models with Biota		Models without Biota		
	Airborne Fraction		Airborne Fraction		
efold Time (μ^{-1}, yr)	4R (%)	BD (%)	3B (%)	BD (%)	Ocean Uptake Ratio (%)
1	92.7	91.2	93.4	91.8	81
2	89.6	86.1	91.0	87.4	71
4	86.1	79.5	88.6	81.6	62
10	79.3	68.3	84.9	72.5	55
22	69.8	56.8	79.9	63.4	55
50	54.9	43.6	71.1	53.3	62
100	40.1	32.6	60.2	44.6	72

TABLE 4.A.3 Comparison of Simple Box Model and Box Diffusion Models with $h_m = 309$ m for 4R Model

	Models with Biota, Airborne Fraction	
efold Time (μ^{-1}, yr)	4R (%)	BD[a] (%)
1	89.5	91.2
10	65.7	68.3
22	56.8	56.8
50	45.8	43.6
100	35.0	32.6
200	24.2	22.7
400	15.4	14.7
800	9.2	9.0

[a] $h_m = 75$ m as in Table 4.A.2.

poorest agreement is found if μ^{-1} is about 20 years (see the final column of Table 4.A.2).

This poor agreement might be a basis for deciding which model is more nearly correct, were there a compelling basis for forcing the 4R model to have a surface layer depth, h_m, of 75 m. As noted earlier, a greater depth is physically more plausible, and any value of h_m between about 100 and 700 m leads to a possibly acceptable value for the Suess effect.

If, instead of 75 m, a surface layer depth of 309 m is assigned for the 4R model so that both models predict the same Suess effect for μ^{-1} of 22 years, the cumulative airborne fraction predicted by both models *for any* μ^{-1} is so similar (Table 4.A.3) that it appears unlikely that any atmospheric measurements during an exponential phase would ever be precise enough to distinguish between the models.

To the extent that the model predictions differ, they are in the opposite sense from what might be concluded from the discussion of Oeschger *et al.* (1975). As can be seen from Table 4.A.3, if the models are adjusted to agree for μ^{-1} of 22 years the BD model predicts *less* uptake of tracer for short disturbances and vice versa. Correspondingly, the BD model predicts greater uptake of industrial CO_2 than the 4R model in the next several centuries after exponential growth in CO_2 production has ceased.

Since prediction of atmospheric CO_2 levels during these centuries is of major interest, we have carried out a series of computations for the BD model similar to those described in the main text for the 6R model. To approximate the industrial CO_2 input, we used historic CO_2 production data and the modified logistic equation with an ultimate production 8.2 times the preindustrial atmospheric CO_2 abundance and a logistic parameter n equal to 0.5. A numerical version of the BD model, as described by Oeschger *et al.*, was used (with 42 subsurface ocean boxes), but instead of fixing the Revelle factor at a value of 10 we varied it in accordance with the thermodynamic constraints of the oceanic CO_2 system as described by Bacastow and Keeling (1973). The other model parameters were set using the BD parametric values listed in Table 4.A.1. The growth factor β was set to the value of 0.198, which, for the BD model, gave the observed airborne fraction for the period from 1959 through 1974 (see Table 4.1 of the main text of this chapter). Computations were also made using the 4R model with values as listed in Table 4.A.1. The surface layer depth, h_m, was set at 284 m, since this value caused the model to predict the observed airborne fraction for 1959 through 1974 using historical fossil-fuel production data.

The results of these computations are plotted in Figure 4.A.1. The BD model predicts 7 percent less atmospheric CO_2 than the 4R model at the time of peak CO_2 concentration (A.D. 2190). Although this difference is associated with a 25 percent greater CO_2 uptake, this considerably greater uptake has only a minor effect on the atmospheric prediction because both models predict that most of the industrial CO_2 will be airborne at that time.

Since a change of 7 percent in prediction is small compared to the uncertainty in ultimate fossil-fuel production, the difference between the models is not striking. The BD

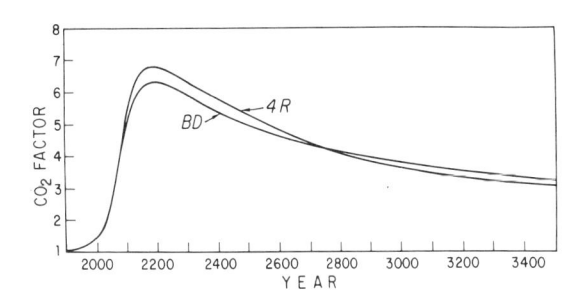

FIGURE 4.A.1 Predicted increase in atmospheric CO_2 from A.D. 1900 to A.D. 3500: comparison of four-reservoir (4R) model and box diffusion (BD) model. The pattern of fossil-fuel consumption is as shown by the case $n = 0.5$ of Figure 4.8. The CO_2 factor has the same meaning as for Figure 4.9.

model has thus been useful in indicating that a model with an analytical formulation distinctly different from that of the 6R model leads to a similar prediction of future CO_2 levels, as in the case of predictions of the Suess effect. If either model is nearly correct, the prediction of atmospheric CO_2 levels in the next 1500 years does not depend strongly on the assumptions regarding transport and storage of carbon in the subsurface water.

ADDITIONAL MODELS

A variety of simple box models with more than two oceanic reservoirs and of continuous models with more parameters than the box diffusion model have been studied in connection with steady-state tracer distributions of subsurface ocean water. Few of these models, however, have been the basis for time-dependent studies, and most involve questionable *ad hoc* assumptions invoked to assign values to some of the model parameters when insufficient data are available. Stuiver (1973) and Veronis (1975) discuss some of the more interesting examples.

Since knowledge of subsurface ocean-water circulation stems largely from studies of the distributions of temperature and salinity, attempts have been made to combine studies of these measured properties with radiocarbon or other tracers having time-dependent properties. Simple box models have not been notably successful in explaining steady-state distributions of salinity and temperature. When small subsurface sources are neglected, both the 6R and the box diffusion models imply zero gradients in both quantities, suggesting difficulties in refining these models.

The simplest steady-state models that effectively use salinity and temperature data are advective-diffusive models. Typically these models assume tracer concentrations to be controlled by unspecified external processes at an upper and lower boundary between which constant vertical diffusion and upward advection prevail. The models have been most often applied to water at depths between 1000 and 4000 m, where the temperature and salinity vary almost exponentially with a vertical scale length of about 1000 m. If K and w denote, respectively, the vertical eddy-diffusion coefficient and upward velocity, the scale length directly reflects this ratio, i.e., K/w is equal to about 1000 m. If the radiocarbon is also considered, separate values for K and w can be computed. From a study of those few ocean stations where prebomb radiocarbon was measured, it is found that K is about 3000 m^2 yr^{-1} [about 75 percent of the value preferred by Oeschger *et al.* (1975)]. Because the observed scale heights vary considerably from place to place, the significance of assuming constant values of K and w seems questionable. Also, the radiocarbon data used in these studies so far admit to considerable latitude in assigning values of K after the scale height is established. Thus the assumption of a constant value of K adopted by Oeschger *et al.* (1975) in their box diffusion model is not strongly supported by salinity and temperature data.

The vertical advective-diffusive models correctly reflect that cold salty water generally underlies warmer less salty water, and they explain in a reasonable way the distributions of salinity and temperature below 1000 m. These models have not, however, explained well the more complicated distributions lying between the surface layer and 1000 m in the region of the so-called intermediate water. As this water is of greater importance in predicting the uptake of industrial CO_2 in the next few centuries than the deeper water, existing advective-diffusive models are not very adaptable to predicting the uptake of industrial CO_2.

More complex two- and three-dimensional models have been proposed to explain subsurface oceanic tracer storage and transport, but these models also have mainly treated water below 1000 m. No model has explained well the more complicated distributions lying between the surface layer and 1000 m in the region of the intermediate water. As this water is of greater importance in predicting the uptake of industrial CO_2 in the next few hundred years than is the deeper water, existing deep-water models are not very adaptable to predicting the uptake of industrial CO_2. Until reliable dynamical models are developed for intermediate water, the simple box and box diffusion models are probably our best tools for predicting future atmospheric CO_2 levels. Actually, the 6R and BD models may not be seriously deficient in this respect. The similarly small oceanic uptake of CO_2 predicted by these models over the next 1500 years suggests that the choice of ocean model may not be crucial in establishing future CO_2 levels.

With respect to predicting the uptake of CO_2 by the land biota over the next few hundred years and longer, the choice of model is much more crucial. In Chapter 10, Revelle and Munk examine this problem in greater detail than previous investigators. In contrast to our exchange model, they postulate a cycling process in which carbon is transferred from the atmosphere to a carbon pool of living photosynthesizing material. This carbon is transferred to a pool of dead and detrital carbon including the wood of forest trees and thence back to the atmosphere without any reverse flow.

Revelle and Munk consider two principal cases. In one case, they assume for the photosynthesizing carbon pool a response to atmospheric CO_2 similar to our model, and for the other carbon transfers, simple box model mechanisms. This first model predicts a damped oscillation in atmospheric CO_2 level similar to a prediction of Eriksson and Welander (1956) (see Figure 10.1 in Chapter 10). For a pattern of production and total ultimate amount of industrial CO_2 similar to that used in our model, their model predicts that atmospheric CO_2 will fall from a peak concentration of 650 parts per million near A.D. 2100 to about 220 parts per million shortly before A.D. 2200. A century later, the concentration returns to near the preindustrial level of 290 ppm. Thus, in this case, with no limit on the biota pool size, the land biota carbon pool has taken up most of the industrial CO_2 within a hundred years after its injection into the atmosphere.

If, on the other hand, Revelle and Munk assume that the

size of the photosynthesizing carbon pool remains fixed and that only the carbon pool of forest wood and detritus increases in size in response to higher atmospheric CO_2 levels, they obtain a prediction that the biota carbon pool will increase by no more than about twice the amount of carbon in the preindustrial atmosphere (see Figure 10.3 in Chapter 10). This second model does not oscillate noticeably, nor does it predict an increase without limit in biota carbon for larger and larger inputs of industrial CO_2.

Revelle and Munk also consider the effect on atmospheric CO_2 of clearing of lands for agriculture and timber harvest. For the model with no limit on biota pool size, clearing has only a small influence on their predictions (see Figure 10.2 in Chapter 10). The model with a limit on the biota, which includes the effect of clearing, would probably be similarly insensitive to its exclusion, although this question was not investigated. Thus the most significant difference between the models almost certainly lies in the differing properties assigned to the photosynthesizing pool.

A possible limitation on the accuracy of these models is their simplified treatment of fluxes and storage times for short-, long-, and very long-cycled carbon species. Revelle and Munk have almost doubled the assumed quantity of organic carbon in the land biota by increasing the amount of soil humus, some of which may have very long cycling times. They have included this additional humus within a single box model reservoir that also includes short-cycled detrital carbon. The latter inclusion scarcely increases the assumed total amount of biotic carbon but approximately doubles the assumed flux to and from the detrital pool. This pool, with its single transfer time (of about 50 years), has a transfer function significantly different from that of a biota model with separate compartments for short-, long-, and very long-cycled carbon. This aggregating of fluxes and masses could result in an overestimate of biotic CO_2 uptake.

On the other hand, the predicted CO_2 uptake probably would be increased substantially by relaxing the requirement that the size of the photosynthesizing pool be constant. In the upper limit, with no restriction on pool size, the total land biota carbon pool is predicted eventually to take up all the industrial CO_2 regardless of the amount. Little evidence exists from studies of plant communities to choose between such different models. Acceptance of a particular model seems therefore to depend mainly on the reasonableness of the prediction of biotic uptake of CO_2. The work of determining the degree of uptake of CO_2 that is reasonable probably must be done by ecologists, botanists, and soil scientists. When that work is done, the models can be adjusted to agree with their findings.

To prepare for this eventuality, we have investigated the result of assuming various amounts of biotic CO_2 uptake using the 6R model with the biota growth factor linearly reduced to zero after A.D. 2000 over different time intervals. These predictions are based on quite arbitrary assumptions, but they reveal a feature of the carbon system that is probably only slightly model-dependent: for biotic uptake less than or of the order of that predicted by Revelle and Munk's model with a limit on the biota, the sum of the uptake by the atmosphere and the biota is nearly the same regardless of the increase in biotic carbon. For example, with a cumulative increase in the biota in A.D. 2250 of 2.6 times the amount of preindustrial atmospheric CO_2, the sum is only 3.7 percent greater than if the biota carbon pool grows negligibly.

Thus the predictions of atmospheric CO_2 increase that we describe above in the main part of this chapter can be reduced by whatever the reader believes ought to be the uptake assigned to the land biota without seriously upsetting the model prediction in other respects. A similar option exists with respect to biota-limited models of the Revelle-Munk type, if assumed mechanisms controlling the photosynthesizing carbon pool are suitably varied.

REFERENCES TO APPENDIX 4.A

Bacastow, R. B., and C. D. Keeling (1973). Atmospheric carbon dioxide and radiocarbon in the natural carbon cycle: Changes from A.D. 1700 to 2070 as deduced from a geochemical model, in *Carbon and the Biosphere*, G. M. Woodwell and E. V. Pecan, eds., U.S. Atomic Energy Commission, pp. 86-135.

Craig, H. (1957). The natural distribution of radiocarbon and the exchange time of carbon dioxide between atmosphere and sea, *Tellus 9*, 1.

Ekdahl, C. A., and C. D. Keeling (1973). Atmospheric CO_2 and radiocarbon in the natural carbon cycle: I. Quantitative deduction from the records at Mauna Loa Observatory and at the South Pole, in *Carbon and the Biosphere*, G. M. Woodwell and E. V. Pecan, eds., U.S. Atomic Energy Commission, pp. 51-85.

Eriksson, E. (1961). Natural reservoirs and their characteristics, *Geofis. Internacional 1*, 27.

Eriksson, E., and P. Welander (1956). On a mathematical model of the carbon cycle in nature, *Tellus 8*, 155.

Keeling, C. D. (1973). The carbon dioxide cycle: reservoir models to depict the exchange of atmospheric carbon dioxide with the oceans and land plants, in *Chemistry of the Lower Atmosphere*, S. I. Rasool, ed., Plenum Press, New York, pp. 251-329.

Machta, L. (1971). The role of the oceans and biosphere in the carbon dioxide cycle, in *The Changing Chemistry of the Oceans*, Proceedings of the Twentieth Nobel Symposium, D. Dryssen, ed., Wiley-Interscience, New York, pp. 121-145.

Oeschger, H., U. Siegenthaler, U. Schotterer, and A. Gugelmann (1975). A box diffusion model to study the carbon dioxide exchange in nature, *Tellus 27*, 168.

Revelle, R., and H. E. Suess (1957). Carbon dioxide exchange between atmosphere and ocean, and the question of an increase of atmospheric CO_2 during the past decades, *Tellus 9*, 18.

Stuiver, M. (1973). The ^{14}C cycle and its implications for mixing rates in the oceans-atmosphere system, in *Carbon and the Biosphere*, G. M. Woodwell and E. V. Pecan, eds., U.S. Atomic Energy Commission, pp. 6-20.

Veronis, G. (1975). The role of models in tracer studies, in *Numerical Models of Ocean Circulation*, National Academy of Sciences, Washington, D.C., pp. 133-146.

Von Arx, W. S. (1974). *An Introduction to Physical Oceanography*, Addison-Wesley, Reading, Mass.

Welander, P. (1959). On the frequency response of some different models describing the transient exchange of matter between the atmosphere and the sea, *Tellus 11*, 348.

The Effect of Localized Man-Made Heat and Moisture Sources in Mesoscale Weather Modification

5

CHARLES L. HOSLER
The Pennsylvania State University

HELMUT E. LANDSBERG
University of Maryland

5.1 INTRODUCTION

Human interference had altered the surface of the earth long before the present era (Thomas, 1956). The first major change started about 7000 years ago when man developed agriculture. This led to systematic changing of forested areas to fields and pastures. Other reasons for deforestation were the needs for structural timber and lumber. In recent times, paper requirements have led to large-scale reductions of forests. Only gradually is a systematic harvesting and replacement policy taking over.

Agriculture and lumbering have undoubtedly led to mesoscale climatic changes, but these are poorly documented, although one can make some approximate guesses at their magnitude. In many instances secondary changes have been more far-reaching. After the clearing, wind and water erosion have washed or blown the top soil away. Bare rock has become exposed, and now far more extreme temperatures and lower humidities prevail where once the even-tempered mesoclimate of the forest dominated. Stretches of Anatolia, the Spanish plateau, and some slopes of the Italian Apennines are silent witnesses to this development.

But by far the most alarming development has been the substitution of rocklike, well-compacted, impermeable surfaces for vegetated soil, a development that is the natural consequence of urbanization. Square kilometer after square kilometer has yielded to the bulldozer and has been converted to buildings, highways, and parking lots. Reservoirs and irrigation also have become important.

5.2 ENERGY BALANCE

Every change in the surface entails as an inescapable consequence a change in the energy balance. It should be recalled here that 80 or more percent of the solar radiation, the principal forcing function of climate, is transformed at the earth's surface into other forms of energy. A look at the basic balance equation tells the story.

$$Q_N - Q_I(1 - A) + Q_{L\downarrow} - Q_{L\uparrow} = \pm Q_H \pm Q_E \pm Q_S + Q_M, \quad (5.1)$$

where

Q_N is the net energy balance;
Q_I is the incoming solar energy;

$Q_{L\downarrow}$ is the long-wave atmospheric radiative energy;
$Q_{L\uparrow}$ is the long-wave terrestrial outgoing radiation ($=\epsilon\sigma T^4$), for which ϵ is emissivity, σ is the Stefan–Boltzmann constant, and T is absolute surface radiative temperature;
Q_H is the sensible heat energy;
Q_E is the latent heat energy;
Q_S is the energy stored in soil or water;
Q_M is heat rejection from man-made combustion and processing; and
A is the albedo.

Only the terms Q_I, $Q_{L\downarrow}$, and Q_H are not directly affected by changes in the surface. All other terms are altered, some of them quite substantially. But even the sensible heat transfer is indirectly affected because all the surface changes alter the surface roughness, which partially governs heat transfer. Scientists engaged in the study of micrometeorology and microclimatology have long been aware of these changes (Geiger, 1961), but these seemingly small-scale phenomena were generally ignored in discussions of climatic change until they reached proportions that began to affect meteorological mesoscale phenomena.

Although the quantitative changes in the physical characteristics may seem to be small, the consequences can be large because of the magnitude of the energies involved. The first notable parameter that changes is the albedo (A), which is the percentage of incident radiation reflected. Table 5.1 gives a small sample of measured albedos of natural and man-made surfaces.

It should be noted, as a generalization, that albedos of cultivated vegetation are higher than those of the forests they replaced. Similarly, much of the urbanized area has a lower albedo than the fields that were there before. The differences in the albedos are immediately reflected in the surface temperatures (Lorenz, 1973a). This, in turn, of course, affects the outgoing long-wave radiation. But it is not only the albedo but also the emissivity that is changed by man's activities. This also is reflected in the long-wave radiation term of the energy equation. A few characteristic values of emissivity are listed in Table 5.2.

Very profoundly affected by man's works are the heat conductivity and heat capacity of surficial materials. Vegetated surfaces usually have considerably lower values of these parameters than building materials such as brick, stone, concrete, and asphalt. Representative values, for example for thermal diffusivity, are

Dry vegetation   0.002 cm^2 sec^{-1},
Concrete   0.02 cm^2 sec^{-1}.

They are an order of magnitude apart. Add to this the fact that man artificially compacts the soil under roads and parking lots to increase the bearing strength and it is easily envisaged that considerably more of the incoming energy from sun and sky is transferred to deeper layers and stored. This will then become available for transfer back to the surface during intervals of net outgoing energy flux. Thus the factor Q_S is the one most profoundly disturbed by urbanization.

Not far behind is the change produced by the evaporation term, Q_E. This is also prominently interfered with in farming operations. Not only is the evapotranspiration of forests different from that of cultivated fields, but the widespread use of irrigation has a profound influence on the local heat balance. Indirectly, shelter belts affect this factor too. Yet, the rural alterations are small compared with those produced in urban areas. There, vegetation, and consequently evapotranspiration, is sharply reduced. But the water, and hence evaporation balance, is even more radically altered by the large amount of impermeable surface, with every effort being made to drain precipitation away as rapidly as possible. The reduced evaporation will, of course, reduce the energy needed for this purpose.

This and the other changes in physical parameters are responsible for a major portion of the urban heat island. The other is the heat rejection from man-made combustion processes, Q_M. However, it is essential to realize that the other elements of the energy balance are quite sufficient to produce a substantial urban heat island. Measurements of surface temperatures by infrared emission techniques from the air (Lorenz, 1973b) show that on sunny days in the summer the urban surface is materially warmer than that of the surroundings. These larger-scale surveys closely resemble infrared thermometer measurements from the surface (Landsberg and Maisel, 1972; Kessler, 1971). These measurements show that (up to a conversion of 50 percent of the surface buildup) the midday temperature difference

TABLE 5.2 Emissivities of Selected Surfaces (in the Long-Wave Spectral Band 8-14 μm)

Surface	Emissivity
Growing crops and grass	0.98 (Fuchs and Tanner, 1966)
Natural freshwater	0.97 (Davies et al., 1971)
Aspalt	0.95 (Lorenz, 1966)
Concrete	0.94 (Lorenz, 1966)
Land (dry)	0.91 (Gorodetsky and Filipov, 1967)
Granite	0.90 (Buettner and Kern, 1965)

TABLE 5.1 Selected Values of Surface Albedos[a]

Surface	Albedo (%)	Surface	Albedo (%)
Fresh snow	85-90	Dry, ploughed field	10-15
Desert	25-30	Wet fields	5-10
Dried grass	20-25	Growing grain	15-20
Deciduous forest	15-25	Stubble field	15
Evergreen forest	7-15	Concrete	12
Granite	12-15	Asphalt	8
Water	7-10	Urban areas	10-15

[a] See Barry and Chambers (1966), Chin (1967), Kung et al. (1964), Nkemdirim (1972), Oguntoyinbo (1974), Stanhill et al. (1966), and Stewart (1971).

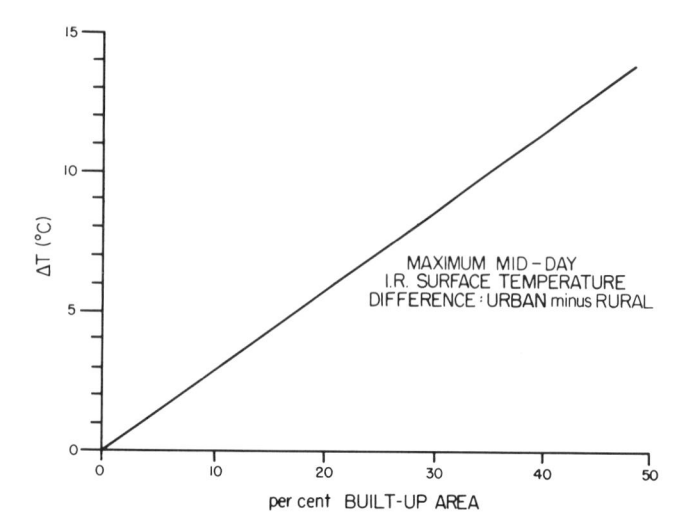

FIGURE 5.1 Midday surface (radiative) temperatures in degrees Celsius as a function of the percentage of built-up area on a sunny day.

ΔT in degrees Celsius on a clear sunny day is directly proportional to the percentage of the area built up, a, approximately by the formula $\Delta T = 0.28a$ (Figure 5.1).

A number of estimates have been made of the total contribution of anthropogenic heat rejection. This varies with season and latitude. In the cold northern climates during winter, anthropongenic heat in metropolitan areas may equal or exceed the amounts received at the surface of the earth from the low and often cloud-hidden sun. But heat production even of large cities, on the whole, at present does not exceed 33 percent of the solar radiation and averages around 20 percent for the year, as documented for Montreal (East, 1971). Yet the amounts of heat produced are quite formidable. In major U.S. cities they amount to 0.03-0.06 cal cm^{-2} min^{-1} (21-42 W m^{-2}) in summer and 0.25-0.30 cal cm^{-2} min^{-1} (174-209 W m^{-2}) in winter. A metropolitan area such as New York has been estimated to produce 2.8×10^{17} cal per year (37,000 MWe) (Bornstein, 1968).

It needs emphasizing, however, that the pronounced heat island effect of calm, clear evenings after sunny days owes its existence to the storage of absorbed solar energy in the urban structures and unvegetated surfaces.

5.3 EFFECTS ON WEATHER

One of us (H.E.L.) followed the development of urban-rural air-temperature differences in the growing town of Columbia, Maryland, in a project studying atmospheric changes.* This gave the opportunity to document the changes over the past seven years, an interval in which the town grew from 2×10^2 to 2×10^4 inhabitants (Figure 5.2). The projected growth over the next decade is expected to see another fivefold increase in population. We can, with reasonable assurance, predict the magnitude of increase in the heat island. The estimated values are well borne out by observations in existing, developed urban areas.

The meteorological literature abounds with accounts of the influence of local heat islands in modifying weather or in generating clouds. The effect of cities on the weather is documented in papers by Landsberg (1972) and more recently in papers by Changnon (1973) and Schickedanz (1974). Among others, Dessens and Dessens (1964) and Stout (1961) have placed on record the influence of localized man-made heat sources in generating clouds. The effect of individual cooling-tower installations in generating cumulus clouds is noted in a photograph by Scorer and Wexler (1968) and similar photographs by Hosler (1971). Recent high-resolution satellite photography in both the visible and infrared regions shows these effects downwind of cities and major industrial centers in a dramatic fashion. In one case, thunderstorms and tornadoes are occurring along a cloud band to a distance of 370 km north of Mobile, Alabama. The cloud band was clearly spawned by the thermal effects of the city. The effects of small islands in midocean where local relief is insignificant but creating temperature anomalies has been noted through the centuries, and it was utilized by early mariners to locate islands. The induced cloud streets and showers often persist for hundreds of kilometers downwind.

The critical matter, however, in answering the question of how important is man's impact on local weather and climate is how these impulses compare with and compete with already existing thermal, stress, and moisture anomalies. Soil and rock type, vegetative cover, soil moisture, antecedent precipitation patterns, surface roughness, and other factors combine with changes in stability and low-level wind structure to give rise to extremely complex patterns of background convergence, divergence, and convection. Many of these natural influences in any given place are of magnitude as great as or greater than the localized heat source induced by man. In most situations these existing disturbances will set the pattern of convection, and man's influence will be lost within it. If, however, the wind tends to be from the same direction during periods when impulses of the magnitude of those induced by man can have an effect, it is reasonable to expect that a precipitation anomaly will be produced downwind. Also, if natural impulses in the area are below the threshold necessary to overcome atmospheric stability, then there may indeed be occasions when man-made heat sources added to the natural impulses can overcome that stability and organize circulations that would not have developed naturally.

Since nature is predisposed to favor existing circulations rather than new ones, subsequent natural changes in the atmospheric stability will provide the energy to feed the pre-existing man-made circulation, thus localizing natural precipitation release downwind of man-made heat sources. Over the open sea in low latitudes or over a flat plain of uniform character one would expect maximal effects of this

---

*Sponsored by the National Science Foundation under Grant GA-2930X.

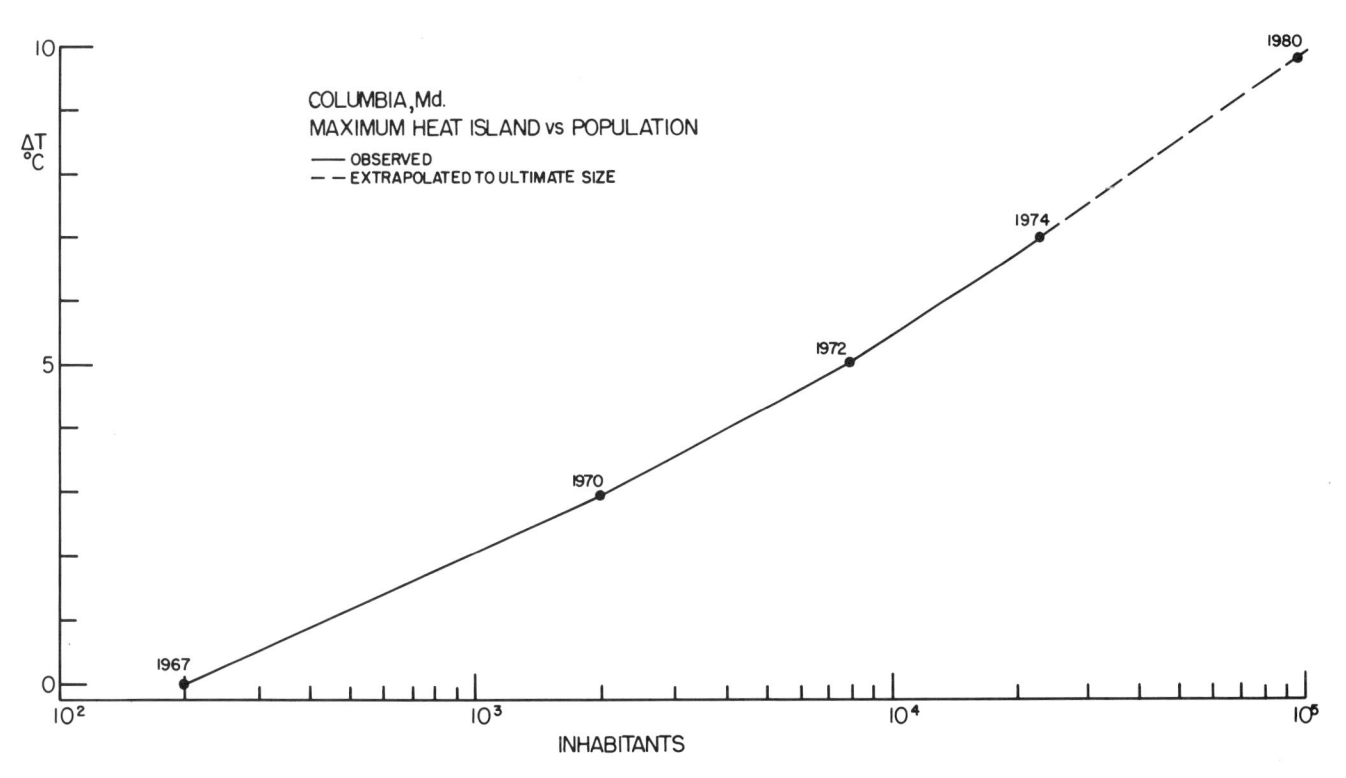

FIGURE 5.2 Maximum value heat island in degrees Celsius (urban–rural air temperatures) on calm, clear evenings at Columbia, Maryland, correlated to population increase.

kind. Where land forms, land use and character are varied, the influence of man-made heat sources may be hard to find and of little consequence. Also, highly variable wind direction and speed in a given locality may spread a small effect over a large area of high natural variability making it impossible to find. The ability of small orographic or thermal sources of lift to produce large effects in producing clouds and precipitation under conditions of marginal stability has been documented by Bergeron (1965) and more recently by Browning et al. (1974). The release of convection may produce rain in cases of deep instability, or it may produce low clouds that feed the precipitation from higher clouds. Table 5.3, produced in part from data given by Huff et al. (1971) and Hanna (1971), presents a comparison of some man-made heat sources and natural phenomena in terms of the megawatt equivalent input into the atmosphere. It is obvious that large cyclonic storms will not be influenced in their overall behavior by man's works. However, we see that the city of St. Louis does produce measurable downwind effects on rainfall, hail frequency, etc. (see Tables 5.4–5.6), and there are indications of similar phenomena occurring downwind of other cities (Changnon, 1973). One might debate the degree to which pollutants, acting to change cloud microphysics, contribute to these effects. Schickedanz (1974) finds evidence of two types of effect, one a result of the heat island and another possibly due to microphysical changes. One can observe the effect of islands such as Aruba both from satellites and with radar, and it is obvious that they produce profound effects on cloudiness and rainfall downwind without a large microphysical component and almost certainly due to heat island effects. It appears from Table 5.3 that it would be reasonable to assume that a 20,000-MW power park (releasing 40,000 MW of heat locally) might produce effects of the same magnitude as St. Louis, Chicago, or the island of Aruba, the latter serving as a heat island in the ocean and the former as heat islands over land caused by a combination

TABLE 5.3 Comparison of Man-Made and Natural Energy Releases

Heat Source	Area (sq. km)	Approximate MW Equivalent Input to the Atmosphere
St. Louis	250	16,100
Chicago	1,800	52,700
Aruba	180	35,000
One-megaton nuclear device (heat dissipated over 1 hour)		1,000,000
20,000-MW power park	26	40,000
Cyclone (1 cm of rain/day)	10×10^5	100,000,000
Tornado (kinetic energy)	10×10^{-4}	100
Thunderstorm (1 cm of rain/30 min)	100	100,000
Great Lakes snow squall (4 cm of snow/hour)	10,000	10,000,000

TABLE 5.4 Summary of Urban Effects on Summer Rainfall at Eight Cities[a]

City	Effect Observed	Maximum Change Millimeters	%	Approximate Location
St. Louis	Increase	41	15	16–19 km downwind
Chicago	Increase	51	17	48–56 km downwind
Cleveland	Increase	64	27	32–40 km downwind
Indianapolis[b]	Indeterminate	—	—	—
Washington	Increase	28	9	48–64 km downwind
Houston[c]	Increase	18	9	Near city center
New Orleans[c]	Increase	46	10	NE side of city
Tulsa	None	—	—	—
Detroit	Increase	20	25	City center

[a] Source: Changnon (1973).
[b] Sampling density not adequate for reliable evaluation.
[c] Urban effect identified only with air mass storms—apparently little or no effect in frontal storms.

of the effects of the city on albedo and the man-controlled heat releases due to intensive energy utilization.

A 1-Mton nuclear device (assuming the heat from it is dissipated over 1 hour) represents a larger impulse, as does a weak cyclonic storm or a Great Lakes snowstorm. A tornado, on the other hand, is a relatively minor phenomenon in terms of its kinetic energy as compared with the energy impact of the large power park to the atmosphere. The tornado is the result of a larger thunderstorm system. The latent heat released in a thunderstorm is at a rate an order of magnitude greater than that of the energy application at a power park; however, it is not unreasonable to assume that the power-park energy applied to the atmosphere is of a magnitude comparable with the initial impulse that started the cumulus cloud that later may have developed into a thunderstorm.

The extent to which heat sources of the size man has created will cause meaningful changes under a wide variety of synoptic situations with varying stability, wind, moisture distribution, and other variables can best be determined by improved physical and mathematical models.

TABLE 5.5 Thunder-Day Summary for Eight Cities[a]

City	Maximum Point Increase (%)	Location
Chicago	38	City and 40 km downwind
Cleveland	42	City and 16 km downwind
Houston	10	City and 16 km downwind
Indianapolis	0	—
New Orleans	26	City
St. Louis	25	City
Tulsa	0	—
Washington	47	City

[a] Source: Changnon (1973).

TABLE 5.6 Hail-Day Summary for Eight Cities[a]

City	Maximum Point or Area Increase (%)	Location
Chicago (Pt)	246	40 km downwind
Cleveland (Area)	90	18–32 km downwind
Houston (Pt)	430	Area of industrial growth 16 km downwind of urban area
Indianapolis	0	—
New Orleans (Pt)	160	City
St. Louis (Area)	150	18–32 km downwind
Tulsa	0	—
Washington (Area)	100	City and 18 km downwind

[a] Source: Changnon (1973).

The state of modeling of circulations created by cities or large energy sources created by a power park does not permit us to answer all of our questions. However, the crude two-dimensional models available tend to indicate that energy sources of the magnitude that we are talking about will create circulations of adequate size and intensity to initiate cumulus development and focus subsequent natural shower development. One would hope that in time model sensitivity will reach a level where the results can be used for prediction. Modeling progress is also limited by the inability of present computers to handle physical processes in sufficient detail and with a sufficient data base. This forces gross simplification of structure and parameterization of important processes. Calculations by Black (1963) indicate that heat sources of comparable magnitude with those of power plants or cities can be produced by asphalting large areas of desert near the seacoast to decrease the albedo. The resulting circulation would be sufficient to overcome the stability on some days, and the result would be induced cloud streets and showers downwind comparable with those found downwind of islands.

A preliminary attempt to determine whether the effects of atmospheric heating by large power plants can be assessed numerically has been made by Dennis Deaven of the National Center for Atmospheric Research. His model uses a form of isobaric coordinates (sigma coordinates) in the boundary layer and isentropic coordinates above. The surface temperature at the central gridpoints was brought up to a value that eventually produced a quasi-steady sensible heat flux of 1340 MW km^{-2}. The quasi-steady structure of the isentropes after 5.6 hours is shown in Figure 5.3, and the streamlines for the flow induced by the surface heating are shown in Figure 5.4.

The model used was two-dimensional and did not include condensation effects. It is to be expected that a three-dimensional process of the magnitude shown would necessarily result in formation of a vortex flow and that the inclusion of moisture effects would result in additional instability. Even so, the disturbance in the flow shown in the diagrams makes it clear that a significant local circulation results when a large power park or city imposes a thermal anomaly of this magnitude upon a marginally stable atmosphere. For more stable soundings the effect is less, and for less stable soundings greater.

During 1960 and 1961, showers in central Pennsylvania were analyzed as to their direction of movement, speed of movement, and persistence (Hosler et al., 1962). For middle latitudes these data probably serve as a clue to how serious downwind effects might be and over what area they might be expected. This study shows that there is a prevalent direction of shower motion from 260° with almost all showers moving between 220° and 280°. Any effects of a heat source on shower development might be concentrated in this case in a direction slightly to the north of east of the heat source. The average shower moved 21 km after the radar echo formed before dissipating. These were summers,

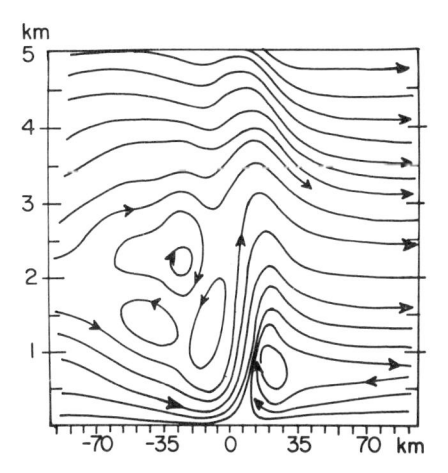

FIGURE 5.4 Streamlines resulting from a 20,000-MWe power park. (Source: Dennis Deaven, NCAR.)

however. In winter, some snow flurry activity may be similarly initiated; however, the natural energy available for convection is so much less in winter that the amplitude of any effect would probably be small.

Because at least 20 minutes is required in the development of precipitation in a cumulus cloud, a shower developing downwind as a result of the influence of a man-made heat island is likely to be at least 8 km downwind before rain begins if it travels at the average echo speed shown in this study in Pennsylvania. This together with the fact that the average echo moved a distance of 21 km during its lifetime would imply that the maximum precipitation increase in Pennsylvania would occur about 29 km to the east northeast but be spread from 8 to 64 km downwind because of the variability of the speed of shower movement. Table 5.4 indicates that this reasoning produces a result similar to what is observed.

These data are for Pennsylvania and are not universally applicable, although Ligda and Mayhew (1954) also found that the showers move with the 700-mbar wind direction, and Clark (1960) found that echos traveled about the same distance as found in Pennsylvania. In some other areas, the direction and speed of movement of showers may be more highly variable. In that event, the effect in generating showers would be spread over a wide sector and thus of small consequence, since the net rainfall is not expected to increase significantly but only to be changed in location on a given occasion.

Because violent weather, such as the occurrence of tornadoes, results from convective phenomena, it is natural to ask whether there would also be an increase in these phenomena. Since severe weather is usually associated with organized lines and bands due to larger-scale convergence in the lower levels, the impact of the man-made heat source may not be so great as upon scattered showers. Any increase would also be 8 to 64 km downwind from any large heat source in an area like Pennsylvania, with the greater distance being more likely. Tornadoes tend to develop from large, intense cells.

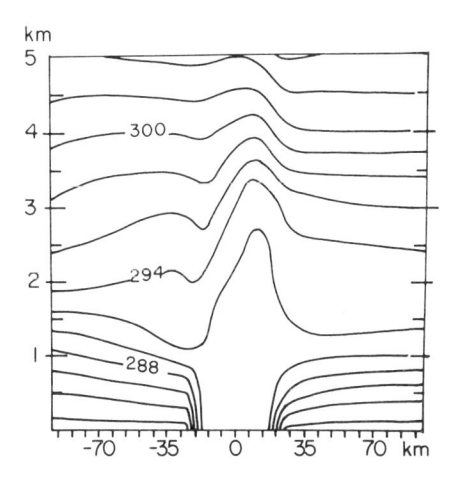

FIGURE 5.3 Disturbance of isentropic surfaces by a 20,000-MWe power park. (Source: Dennis Deaven, NCAR.)

Tables 5.4–5.6, taken from Changnon (1973), summarize the effects noted on rainfall, hail, and thunderstorm frequency. Taken together, there seems to be little doubt that these effects are real.

The question arises when considering the impact of large power parks as to whether it makes a difference in how the heat is dissipated. That is, would there be any difference in the effects of a cooling pond, dry cooling, or evaporative cooling? Certainly when we discuss fogging, icing, or invisible plumes one is concerned about the quantity of water released and the altitude of release. This will be discussed later.

The net effect on downwind convection is apt to be somewhat similar in all cases. The initial vertical velocities would be larger in the case of the dry cooling towers, where sensible heat is added directly. Subsequent tapping of the naturally available latent heat will sustain convection in an unstable atmosphere. In the case of evaporative towers or the cooling pond, the lesser amount of sensible heat, somewhat compensated for by the reduction in air density due to the water vapor addition, may not result in vertical motion that will be as great initially. However, as long as the air reaches the condensation level, the net effect will be the same because the latent heat available in the vapor plume will be released to maintain positive buoyancy and sustain the convective development for some distance downwind. In the absence of any direct empirical evidence to the contrary, it would be a reasonable assumption that the manner of waste-heat dissipation is probably not extremely important in determining the effects of convection during shower periods that occur naturally. Thus we would not expect the power plant itself to have any more effect on weather than the city or industrial center that ultimately consumes the electrical energy generated at the site of the power plant unless the initial area is critical. Eventually the heat from the point source spreads to an equivalent area of the city. If one questions the environmental impact of the power plant, one must similarly question the environmental impact of consuming the energy, which results in sensible heat being added to the atmosphere at the terminal point of transmission line. Improved models and field studies will eventually answer the questions raised above and substitute physics and data for conjecture.

The higher dew point temperature in the vapor plume from evaporative cooling towers is in itself a destabilizing effect as latent heat is released earlier even if natural impulses cause the initial updraft to form a cloud. The air in the plume can thus be rendered unstable by lifting more easily than the ambient air. Also, natural cumulus clouds building into a moist plume will grow faster and bigger because of the reduced influence of entraining cool dry air, the principal enemy of cumulus growth (Austin, 1948).

In addition to the direct effect of the plume on convection, there is also an indirect effect of raising the humidity in a large swath downwind of the plant, which will in turn reduce the evaporation rate for precipitation falling from cloud bases higher than the plume. When a visible cloud exists, the precipitation elements will grow slightly as they fall through. J. R. Hummel of the Pennsylvania State University in an unpublished study has made some calculations that indicate that at low rainfall rates the increase in rainfall may be as much as 10 percent but usually of the order of 1 to 5 percent for reasonable rates of rainfall. These increases are within 2 km of the plant and are thus insignificant.

5.4 FOGGING AND ICING

If evaporative processes are used to dissipate the waste heat in a large power park then on the order of 10^7 g of water per second will be injected into the atmosphere at that point. Under average diffusion conditions this means that 2 km downwind the water vapor content of the air will be increased by on the order of 10 g per m^3 and that increases of over 1 g will extend over many kilometers downwind. Especially in high and middle latitudes in the winter, this will generally tax the ability of the atmosphere to absorb it as vapor and a visible plume will persist downwind. For natural draft cooling towers these will be elevated and of little concern. Downwind plumes may be of concern for mechanical draft towers or cooling ponds. The natural draft plumes have been studied in some detail at the Keystone plant in Pennsylvania, and Table 5.7 shows the frequency with which plumes of various lengths occur at the Keystone plant according to Bierman et al. (1971). This is an 1800-MW fossil-fuel plant and is evaporating only about one tenth of the water vapor to be evaporated from a major power park of 20,000-MW capacity. Figure 5.5 shows an extreme case in which the ambient surface temperature was $-12\,°C$. The plume leveled off at 2000 m and persisted for many kilometers downwind. It is obvious then that long, persistent

TABLE 5.7 Plume Length at Keystone Power Station in Western Pennsylvania when It Did Not Merge with Overcast

Length (m)	<600	>600 to <1500	>1500
Frequency (%)	87.3	10.1	2.6

FIGURE 5.5 Extended high-level plume production at Keystone on a cold February morning.

plumes downwind in cold and humid weather will, if nothing else, have aesthetic effects in a populated area.

Examination of saturation deficit data indicates that plumes 2 km or more long will occur as much as 20 percent of the time if evaporational cooling is used at a large power park in Pennsylvania. If the wind direction is fairly persistent, these plumes may have an impact on solar radiation reaching the ground immediately downwind of the power plant. The question of whether these plumes would affect aircraft has been raised.

Extensive measurements have been made in the plumes from Keystone by flying through them in light aircraft. These measurements reveal that the vertical motions in the plumes above the very large natural draft cooling towers are of sufficiently low magnitude and short duration as to present no significant hazard to aircraft operating in or around them. The liquid-water content and drop sizes are so small and the ability to stay in the cloud is so limited that there is no possibility of significant icing or ice accumulation in the plume. Thus, the only significant effect of the plume on air navigation is to reduce visibility aloft during those few occasions when extended plumes occur. These occasions all correspond to periods when it is almost certain that low clouds and precipitation will be present naturally. The plume from the cooling tower has no special properties that will distinguish it from the natural clouds except its location. On the few occasions a year when the plume levels off immediately below the natural cloud base, there will be the effect of lowering the cloud base by as much as 100 m. Because of the penetration achieved by these plumes, this phenomenon will always occur at an altitude above 500 m and usually above 600 m. Thus in no case would this tend to increase the number of hours when ceilings would be below minimum. It should be emphasized that penetrations made were as low as 50 m above the towers. At these altitudes any condition of concern would be exaggerated as compared with what one would encounter at a height of several hundred meters above the towers.

It does make a great difference depending upon the altitude at which the water vapor is injected as to the severity of the downwind effect from the standpoint of persistent plumes affecting transportation or visibility. Large natural draft cooling towers and the attendant buoyancy of the air and water ejected from them would eliminate the chance of any surface effects. Fogging has not been a problem downwind of these installations and is not likely to be. However, in the case of mechanical draft cooling towers or cooling ponds the water injected at a very low level under high humidity and low temperature conditions will almost certainly produce large amounts of fog downwind. Figure 5.6 shows steel mills in Pennsylvania at which the low-level sources of both water vapor and pollutants cause dense fog beneath a strong inversion. The persistence of this fog into the morning hours represents a considerable weather modification. The existence for some hours of an elevated plume downwind from a power-plant installation if located almost always over the same area would, of course, have an effect on subsequent convection that might occur later in the day due to solar heating. The area under

FIGURE 5.6 Extended low-level plume production by steel mills on an April morning.

the plume having been deprived of solar energy for several hours in the morning will actually tend to dampen convection in that particular area if the convection depends on solar heat absorbed at ground level (Weiss and Purdom, 1974).

Since in all evaporative cooling towers a certain amount of unevaporated water is ejected from a tower in the form of so-called drift (see Hosler et al., 1974) even with the most modern drift eliminators, it appears that for a 20,000-MW power installation, fallout of dissolved salts must be considered. This would normally not be important unless there are particularly sensitive agricultural crops in the area and high concentrations of substances that might damage the leaves of the crops. For example, if seawater were used the chloride might affect tobacco. If seawater were used in mechanical draft cooling towers then concentrations of thousands of kilograms per square kilometer per year may fall within a few kilometers of the plant, representing a serious problem from the standpoint of corrosion or damage to vegetation. For natural draft towers, the drift problem is less serious as a result of the height of injection in the larger area over which the material is dispersed. For mechanical draft towers, however, the local fallout can become of the order of 10^5 kg km^{-2} per year as opposed to 10^4 kg km^{-2} for a saltwater tower of natural draft variety. When freshwater is used, these rates fall to levels of the order of 10^4 kg km^{-2} per year for mechanical draft towers and 10^3 kg km^{-2} per year for natural draft towers. For the natural draft tower, ice deposition is an insignificant problem because of the great height at which water is injected. Most of the drift drops evaporate before reaching the ground. In the case of the lower-level mechanical draft towers, localized ice deposition may occur in the winter.

5.5 CONCLUSION

A look into the future dictates a more comprehensive investigation and understanding of the consequences of man's activities. There will be more power plants with their warm-water discharges into natural waters or cooling la-

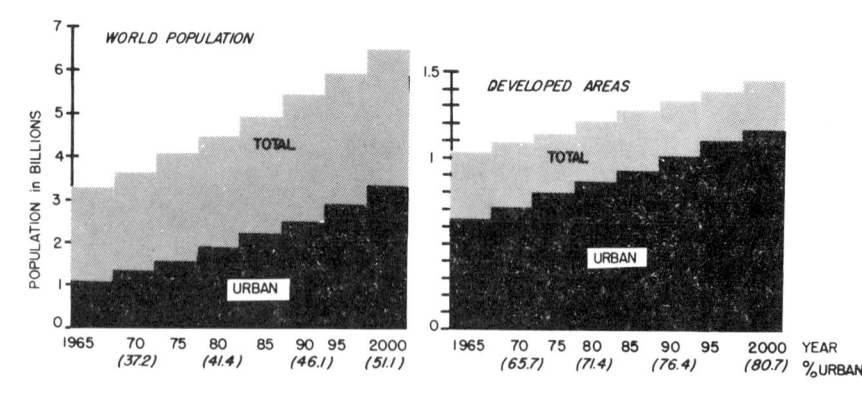

FIGURE 5.7 Projection of world population to the year 2000 in the world as a whole and the developed nations alone and the respective share of urban dwellers, from United Nations sources.

goons or cooling towers. The 1970 anthropogenic heat production was about 8×10^3 GW or 0.01 percent of the heat received at the earth's surface from the sun (Kellogg, 1974). By the end of the century, with an apparently inevitable doubling of the population and increased industrialization and mechanization, a conservative estimate of a tripled heat rejection can be projected. Although this is still only a very small fraction of the solar energy and unlikely to affect the global climate, it has to be considered that most of this heat will be dissipated in the small areas of metropolitan concentration. Here the projections show that by the end of the century 81 percent of the population in the developed nations and 51 percent of the world's population will live in urban areas (Figure 5.7). This will lead to some large megalopolitan regions. There not only local heat rejection but also higher absorption and storage of solar energy will lead to considerable mesoclimatic changes. Not all of these are clearly foreseeable. There will, undoubtedly, be higher-temperature contrasts to the rural environment. There will also be more induction of rainfall. Initiation of severe local storms cannot be excluded, and the downwind effects must of necessity become more pronounced.

Efforts to increase our knowledge of the effects of existing man-made heat and moisture sources should be increased, and studies of the consequences of even larger and greater numbers of sources must be begun.

REFERENCES

Austin, J. M., and A. Fleisher (1948). A thermodynamic analysis of cumulus convection, *J. Meteorol. 5*, 240.

Barry, R. G., and R. A. Chambers (1966). A preliminary map of summer albedo over England and Wales, *Quart. J. R. Meteorol. Soc. 92*, 543.

Bergeron, T. (1965). On the low-level redistribution of atmospheric water caused by orography, *Suppl. Proc. Inst. Conf. Cloud Phys.* Tokyo, pp. 96-100.

Bierman, G. F., G. A. Kunder, J. F. Sebald, and R. F. Visbisky (1971). Characteristics, classification and incidence of plumes from large draft cooling towers, paper presented at American Power Conf., Chicago, Ill., Apr. 22.

Black, J. F. (1963). Weather control: Use of asphalt coatings to tap solar energy, *Science 199*, 226.

Bornstein, R. D. (1968). Observations of the urban heat island effect in New York City, *J. Appl. Meteorol. 7*, 575.

Browning, K. A., F. F. Hill, and C. W. Pardoe (1974). Structure and mechanism of precipitation and the effect of orography in a wintertime warm sector, *Quart. J. R. Meteorol. Soc. 100*, 309.

Buettner, K. J. K., and C. D. Kern (1965). The determination of infrared emissivities of terrestrial surfaces, *J. Geophys. Res. 70*, 1329.

Changnon, S. A. (1973). Urban-industrial effects on clouds and precipitation, in *Proc. Workshop on Inadvertent Weather Modification*, Utah State U., Logan, Utah, Aug. 1973, pp. 111-139.

Chin, L. S. (1967). Albedo of natural surfaces in Barbados, *Quart. J. R. Meteorol. Soc. 93*, 116.

Clark, R. A. (1960). Study of convective precipitation echoes as revealed by radar observation, Texas, 1958-59, *J. Meteorol. 17*, 415.

Davies, J. A., P. J. Robinson, and M. Nunez (1971). Field determinations of surface emissivity and temperature for Lake Ontario, *J. Appl. Meteorol. 10*, 811.

Dessens, J., and H. Dessens (1964). Cumulus artificiel sur la Sahara, *J. Rech. Atmos. 1*, 29.

East, C. (1971). Chaleur urbaine a Montréal, *Atmosphere 9*, 112.

Fuchs, M., and C. B. Tanner (1966). Infrared thermometry of vegetation, *Agron. J. 58*, 597.

Geiger, R. (1961). *Das klima der bodennahen Luftschicht,* Friedr. Vieweg und Sohn, Braunschweig.

Gorodetsky, A. K., and G. F. Filipov (1967). Terrestrial measurements of the atmosphere and of the underlying surface in the spectral region 8-12, *Izv. Atmos. Ocean Phys. 4*, 228.

Hanna, S. R. (1971). Meteorological effects of cooling tower plumes, Contribution No. 48, NOAA Res. Lab., Atmos. Turbulence and Diffusion, Oak Ridge, Tenn.

Hosler, C. L. (1971). *Wet Cooling Tower Plume Behavior,* C.E.P. Tech. Manual No. T40, Cooling Towers, AIChE, p. 1.

Hosler, C. L., L. G. Davis, and D. R. Booker (1962). The role of orographic barriers of less than 3000 feet in the generation and propagation of showers, Rep. to NSF under Grant No. 7363, Dec.

Hosler, C. L., J. Pena, and R. Pena (1974). Determination of salt deposition rates from drift from evaporative cooling towers, *J. Eng. Power,* July, p. 283.

Huff, F. A., R. Beebe, D. M. A. Jones, and R. G. Semonin (1971). Effect of cooling tower effluents on atmospheric conditions in Northeastern Illinois, Circ. 100, Ill. State Water Survey, Urbana.

Kellogg, W. W. (1974). Long range influences of mankind on the climate, paper presented at world conf. on Toward a Plan of Action for Mankind: Needs and Resources—Methods of Forecasting, Inst. de la Vie, Paris, Sept. (typescript).

Kessler, A. (1971). Uber den Tagesgang von Oberflachentemperaturen in der bonner Innenstadt en einem sommerlichen Strahlungstag, *Erdlkunde* 25, 13.

Kung, E. L., R. A. Bryson, and D. H. Lenchow (1964). Study of a continental albedo on the basis of flight measurements and structure of the earth's surface, *Mon. Weather Rev.* 92, 543.

Landsberg, H. E., and T. N. Maisel (1972). Micrometeorological observations in an area of urban growth, *Boundary Layer Meteorol.* 2, 365.

Ligda, M. G. H., and W. A. Mayhew (1954). On the relationship between the velocities of small precipitation areas and geostrophic winds, *J. Meteorol.* 11, 421.

Lorenz, D. (1966). The effect of long-wave emissivity of natural surfaces on surface temperature measurements using radiometers, *J. Appl. Meteorol.* 5, 421.

Lorenz, D. (1973a). Die radiometrische Messung der Boden- und Wasseroberflächen-temperatur und ihre Anwendung insbesondere auf dem Gebiet der Meteorologie, *Z. Geophys.* 39, 627.

Lorenz, D. (1973b). Meteorologische Probleme bei der Stadtplanung, *Baupraxis* 9, 57.

Nkemdirim, L. C. (1972). A note on the albedo of surfaces, *J. Appl. Meteorol.* 11, 867.

Oguntoyinbo, J. S. (1974). Land-use and reflection coefficient (albedo) map for southern parts of Nigeria, *Agric. Meteorol.* 13, 227.

Scorer, R. S., and H. Wexler (1968). *A Colour Guide to Clouds*, Pergamon Press and Macmillan Co., London and New York.

Schickedanz, P. T. (1974). Inadvertent rain modification as indicated by surface rain cells, *J. Appl. Meteorol.* 13, 891.

Stanhill, G. G., G. G. Hofstede, and J. D. Kalma (1966). Radiation balance of natural and agricultural vegetation, *Quart. J. R. Meteorol. Soc.* 92, 128.

Stewart, J. B. (1971). The albedo of a pine forest, *Quart. J. R. Meteorol. Soc.* 97, 561.

Stout, G. (1961). Some observations of cloud initiation in industrial areas, Symp. on Air Over Cities, SEC Tech. Rep. A62-5, Public Health Service, Cincinnati.

Thomas, W. L., Jr. (1956). *Man's Role in Changing the Face of the Earth*, U. of Chicago Press, Chicago, Ill.

Weiss, C. E., and J. F. Purdom (1974). The effect of early morning cloudiness on squall-line activity, *Mon. Weather Rev.* 102, 400.

6
Regional and Global Aspects

RALPH A. LLEWELLYN
Indiana State University

WARREN M. WASHINGTON
National Center for Atmospheric Research

6.1 ESTIMATES OF ENERGY CONSUMPTION

Worldwide energy use has increased at an average annual rate of about 3½ percent over the past 50 years, accelerating to about 5½ percent during the latter half of this period, but with a perceptible slowing beginning in the late 1960's (Darmstadter, 1971; Hubbert, 1971). By 1971, the per capita consumption of energy among the developed countries had reached 164×10^6 Btu/year, having tripled since the beginning of the Industrial Revolution (Cook, 1971).

This dramatic increase was accompanied by an equally significant rise in world population, estimated to have reached 4 billion in March 1976. The combined result of population increases, changing patterns of population densities, and escalating per capita energy consumption has been to concentrate in some regions of the world very large energy fluxes per unit area (energy flux density) when compared with the world average. For example, in 1971, the world average was 1.43×10^9 Btu/km^2 of land area-year, while Japan's energy consumption per unit area was 2.52×10^{10} Btu/km^2-year or nearly 20 times larger. In that year, New York City's Manhattan Island had a total energy flux density of approximately 6.0×10^{12} Btu/km^2-year, very nearly equal to the global average of solar radiation at the surface and roughly 1000 times the average man-made energy flux density for the world's land areas.

The developing countries, on the other hand, consume energy at a much lower rate. The developing nations of Africa, for example, had an average energy flux density of 4.69×10^7 Btu/km^2-year for 1971 (United Nations Statistical papers, 1973), or about 3 percent of the average for the world's land areas and roughly 8×10^{-6} of that of Manhattan. However, data gathered over the past several years indicate the possibility that the situation is slowly but steadily changing. Since 1950, the net population growth rate for the developing countries has risen sharply, largely a result of declining infant death rates. This, coupled with increasing concern and ability on the part of these nations for improving the living conditions of their people, has resulted in rising energy consumption to the extent that the ratio of per capita energy consumption for the developed to the developing countries has remained nearly constant at about 8 since 1960. Said differently, even though the population of the developing world is increasing more rapidly than that in the developed world, the difference in per capita energy consumption has not changed markedly, indicating that the aggregate energy consumption by the developing countries is becoming a larger fraction of the world total as times goes on (Figure 6.1).

This trend, if it persists, will lead to a more rapid increase of energy flux densities in the developing countries than in the developed nations. Indeed, aggregate consumption of all fuels by the developing nations will equal that among the present developed nations in less than 120 years (United Nations Statistical Papers, 1973), if both groups maintain the average annual increase in energy use typical of the past 15 years. No one is suggesting that such trends can easily continue, particularly in the absence of clearly adequate and acceptable alternatives to the fossil fuels; however, pressure on the world's social systems will be to maintain the trends to the degree possible (i.e., continue to improve standards of living via energy expenditures). If anything, the energy conservation strategies being explored in many developed countries may reduce the time to energy equity to significantly less than 120 years.

In suggesting that the upward trend of energy use may persist in the developing countries, we are not assuming a rapid rise in the numbers of automobiles, televisions, air conditioners, and the like throughout the world. The pressure to feed the increasing population will provide the main thrust. Providing India's people with the minimum 3000 kcal/day considered necessary would require that nation to devote more energy to agriculture than it currently uses for all purposes combined. To raise the caloric intake of all the world's people to this level would demand that 80 percent of the world's current total energy consumption be devoted to agriculture (Steinhart and Steinhart, 1974). The problem is monumental, but the effort will probably be made to solve it.

Without concern at this point for possible solutions and the details of new technologies that may contribute to, or the extent to which rapidly depleting fossil fuel reserves may impede, the implementation of potential solutions, it must be recognized that man-made energy flux densities will likely continue to increase, with those in the developing countries increasing more rapidly than elsewhere. These increases seem sure to continue through this century and, with obviously less certainty, well into the twenty-first century.

Our interest here is directed toward what the current trends in energy flux densities may suggest concerning the earth's climate on various scales ranging from convective to synoptic, i.e., from areas smaller than 10^2 km^2 to those 10^6 km^2 or larger (Figure 6.5). As Hosler and Landsberg point out in Chapter 5, the effects of man's activities in modifying the weather on a microclimate scale (areas up to several hundred km^2) are well documented. Microclimate effects due to heat generated by cities and major electric generating stations have been previously noted in the literature. Included among the observed effects have been increments in various atmospheric parameters ranging from cloud cover and temperature to the frequency of thunderstorms and tornadoes. The important point is that changes in these indicators also occur without the benefit of man's intervention; so it is essential to assess carefully whether man-made heat rejected to the atmosphere over small areas, such as cities, really produces significant departures from ordinary fluctuations resulting from natural energy inputs to the atmosphere. The evidence seems substantial that on the convective scale it certainly does (Landsberg, 1956; Hosler, 1971; Changnon, 1973).

Whether the same conclusion is valid for larger areas is another question. In Chapter 2, Mitchell observes that the size of the region over which the climate is affected by a given factor and the geographical extent of that factor are rather closely related. Changnon (1973) noted observable effects on rainfall, thunderstorms, and hail frequency over areas of urban dimensions (St. Louis, Chicago, etc.), which reject heat to the atmosphere with energy flux densities of the order of 10^{11}-10^{12} Btu/km^2-year. For similar effects to occur on a continental or global scale, presumably man-made energy flux densities of similar magnitude would be necessary over the larger region, unless, of course, there are potential positive feedbacks not yet recognized that could become significant at less than macrodimensions. In the absence of such feedbacks, it does not seem likely that man's production of heat energy at current levels will have any influence on climate that can be distinguished from effects arising from natural causes at other than microclimate scales.

The key phrase in the last sentence of the preceding paragraph is "at current levels." As pointed out earlier in this paper and by Perry and Landsberg in Chapter 1, neither current levels of energy consumption nor current global patterns of energy flux density are likely to continue. Although the mix of fuels will change in a way not yet clear, total consumption will undoubtedly continue to rise, and higher energy flux densities will probably become more widely distributed geographically.

As one example of how the energy flux density patterns might change in the future, we have computed flux densities for 1970, 2000, 2025, 2050, and 2075, those after 1970 being projected on the basis of present values and recent trends in per capita energy consumptions and population

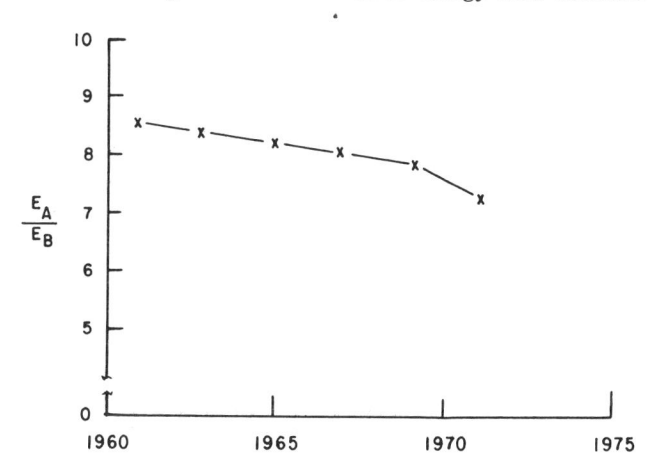

FIGURE 6.1 Ratio of energy use by developed countries (E_A) to energy use by developing countries (E_B), 1960-1971.

densities. Calculations are on a country-by-country basis, except where available data made possible smaller geographical computations. The data are displayed in Figures 6.2, 6.3, and 6.4. The displays, of course, differ from those of Weinberg and Hammond (1970) and Washington (1972), where a uniform per capita consumption (15 kW in the latter) and an ultimate population density based on current values were assumed.

It is not yet clear whether the techniques used herein will result in energy flux density projections of any more value to meteorologists and climatologists than Washington's results. Work is continuing on developing displays with better and more geostrophically precise input data in the expectation that successively more useful outputs will be obtained.

6.2 MODELING CLIMATE IMPACT

SURFACE ENERGY BALANCE

It is possible to suggest mechanisms by which man's activities might influence local weather events (such as clouds and precipitation), but is is difficult to prove that they definitively change regional and global climate (Landsberg, 1970). The problem is in separating natural fluctuations from those possibly caused by man's generation of heat, particularly when the effects of such quantities of heat are relatively small. One promising approach for testing regional and global effects is through the use of theoretical models of the atmosphere and oceans. The basis and development of such models are discussed in Chapter 9. We cover here the preliminary use of such models for estimating the effects, but it may be useful first to review the major factors influencing the surface temperature of the earth.

The earth's surface temperature over continents and oceans is determined by a balance of thermal energy transports. Thus, we can define a surface energy balance equation as

$$F - (1 - A)Q + H + L \cdot E_v + M + l \cdot S_m + E_P = 0, \quad (1)$$

where

$F = F_1 - F_2 =$ net long-wave radiation at the ground,
$F_1 =$ upward long-wave radiation from surface,
$F_2 =$ downward long-wave radiation from atmosphere,
$(1 - A)Q =$ absorbed solar flux at surface,
$A =$ albedo of ground surface,
$Q =$ solar flux arriving at surface,
$H =$ sensible heat flux to atmosphere,
$L \cdot E_v =$ latent heat flux to atmosphere,
$E_v =$ evaporative flux to atmosphere,
$M =$ conduction of heat to subsurface,

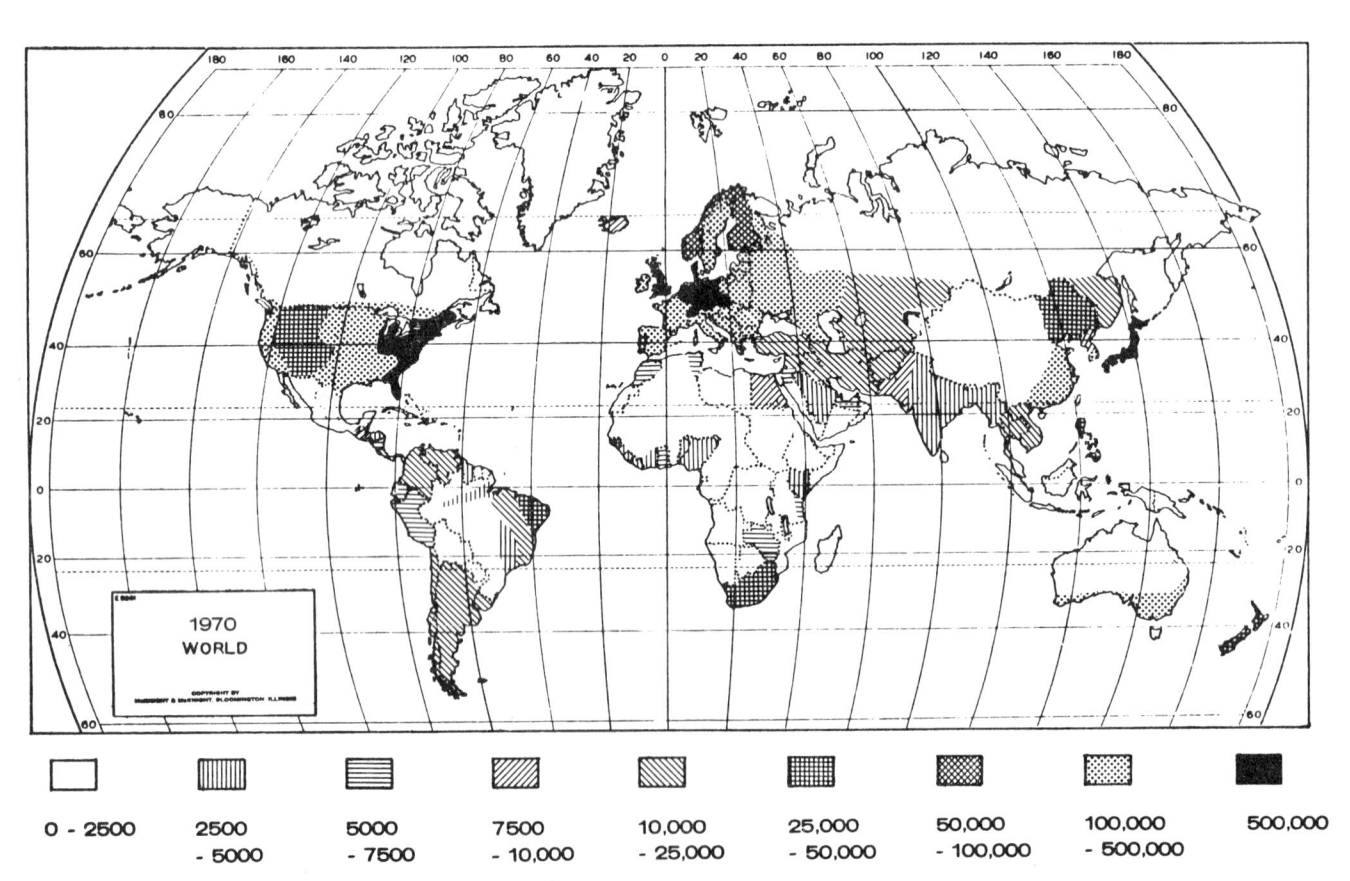

FIGURE 6.2 World energy flux density (kg of coal/km^2/year), A.D. 1970.

Regional and Global Aspects

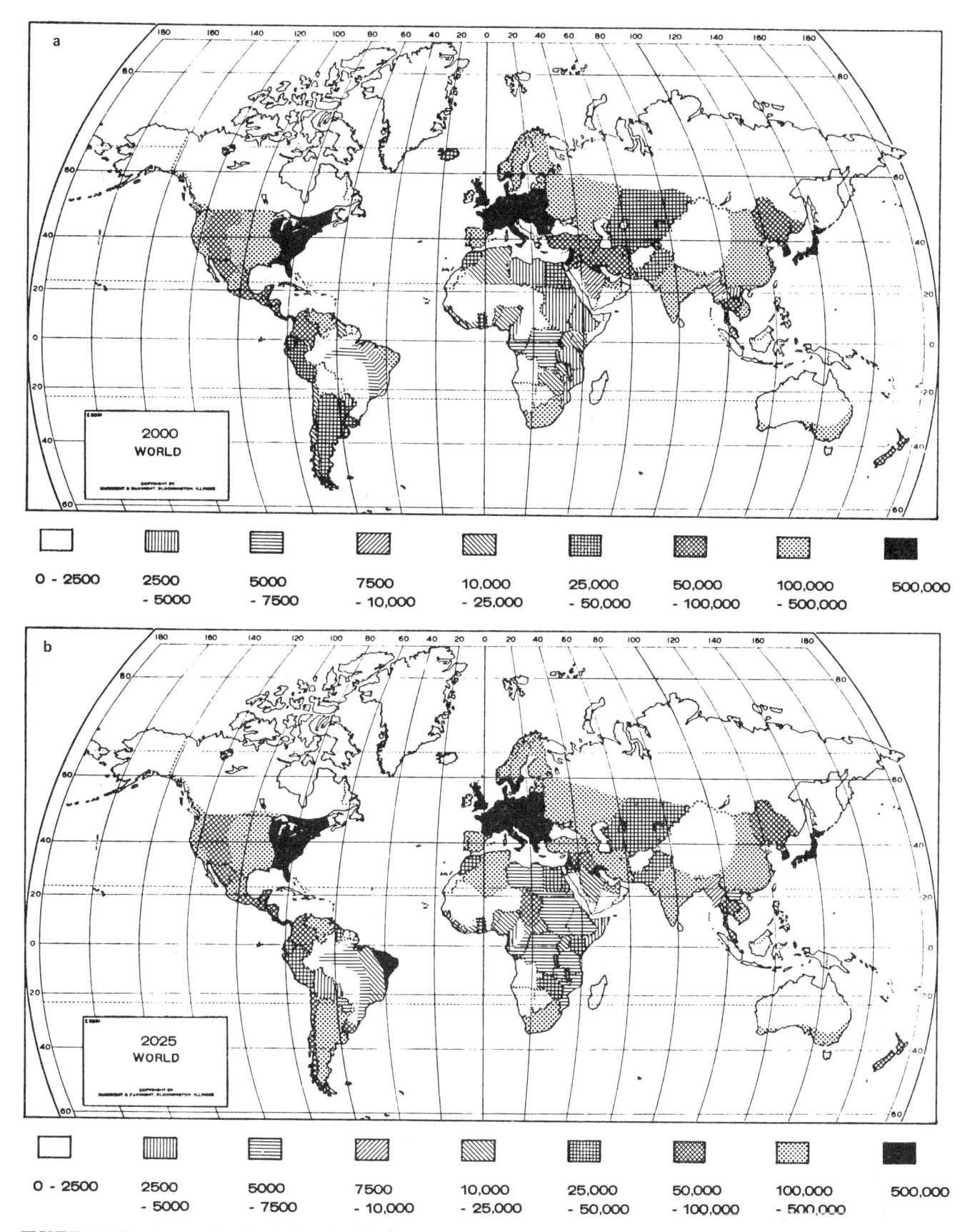

FIGURE 6.3 World energy flux density (kg of coal/km^2/year): (a) A.D. 2000; (b) A.D. 2025.

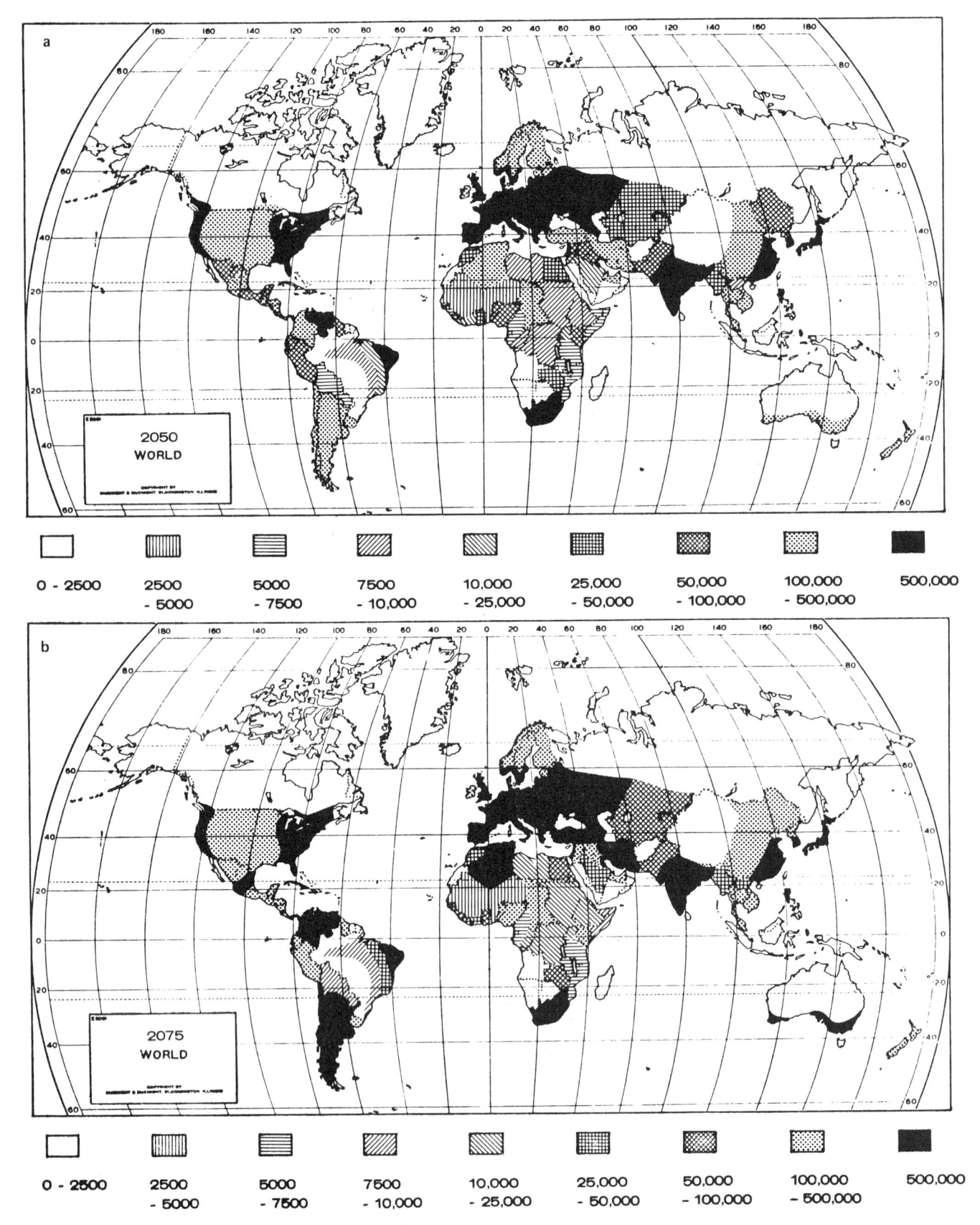

FIGURE 6.4 World energy flux density (kg of coal/km^2/year): (a) A.D. 2050; (b) A.D. 2075.

S_m = melting of snow and ice,
l = latent heat of fusion,
L = latent heat of evaporation L_e or latent heat of sublimation $L_e + l$,
E_P = man-made heat sources.

The effects of horizontal transports of heat have been neglected in Eq. (1). This is a reasonable assumption over continents and ice-snow regions, but, of course, they cannot be neglected over ocean areas—particularly along the western boundary currents such as the Gulf Stream, where enormous amounts of heat are transported poleward. The mechanical generation of heat by wind and tides is also ignored in the definition of Eq. (1). It is assumed in the model studies carried out thus far that thermal heat release E_P is at the surface or in the planetary boundary layer. The proper vertical placement of heat release in the atmosphere may not be a serious problem on regional or global space scales because vertical mixing caused by convection and surface roughness is usually quite efficient in distributing heat in the vertical. There are exceptions where the atmosphere is quite stable as far as thermal or moist convection, and, thus, heat would be trapped in lower layers.

The first terms of Eq. (1) dealing with radiation can be grouped as

$$R = (1 - A) \cdot Q - F. \quad (2)$$

Following Sellers (1965), in Table 6.1 we divide the mean latitudinal values of components of the surface energy balance equation into oceans, continents, and the earth as a whole. In the polar region, the values represent snow and ice cover. The features of Table 6.1 are that R tends to balance latent heat $L \cdot E_v$ and sensible heat H, with the remainder being the transport of sensible heat by the oceans. Vonder Haar and Oort (1973) estimated that ocean transport is 40 percent of the poleward heat flux in the northern hemisphere, suggesting that the role of oceans in heat transport is greater than previously thought. R is positive, except in polar regions, where net long-wave radiation exceeds solar radiation absorbed at the ground. Note that global values of the earth as a whole (including ocean transport) show R as 95 W m^{-2}; later this value will be compared with the present and predicted values of man-made generated heat.

GLOBAL AND REGIONAL CLIMATE IMPACT

It is useful to review model estimates of possible global climatic impacts due to waste heat. Budyko (1969, 1972a, 1972b) and Sellers (1969) used simple steady-state, earth-atmosphere, globally averaged models to estimate the effects. See also the survey of simplified climate models by Schneider and Dickinson (1974). Budyko assumed that the release of waste heat increased by 4 percent each year so that after 200 years waste heat, rather than solar energy, would be the controlling factor in climate. This assumption is not realistic when compared with our estimates. As pointed out, current levels of waste heat probably will *not* continue to increase indefinitely, and, thus, Budyko's assumption appears quite unrealistic. Schneider and Dennett (1975) estimated that current levels of energy consumption are

TABLE 6.1 Mean Latitudinal Values of Components of the Surface Energy Balance Equation (W m^{-2}) from Sellers (1965)

Latitude	Oceans R	$L \cdot E_v$	H	Continents R	$L \cdot E_v$	H	Earth R	$L \cdot E_v$	H
80–90° N							−12	4	−13
70–80							1	12	−1
60–70	30	44	21	26	18	8	28	26	13
50–60	38	51	21	40	25	15	40	37	18
40–50	67	70	18	59	32	28	63	50	22
30–40	110	114	17	79	30	49	96	78	32
20–30	149	139	12	91	26	65	127	96	32
10–20	157	131	8	94	38	55	140	107	21
0–10	152	106	5	95	63	32	139	95	15
0–10° S	152	111	4	95	66	29	139	100	13
10–20	149	137	7	96	54	42	137	119	15
20–30	133	132	9	92	37	55	124	109	21
30–40	108	106	11	82	37	45	106	98	15
40–50	75	73	12	54	28	26	74	70	13
50–60	37	41	13	41	26	15	37	41	15
60–70							17	13	15
70–80							−3	4	−5
80–90							−15	0	−15
Globe	108	98	11	65	33	32	95	78	17

TABLE 6.2 Thermal Energy Generation

Time	Energy Use per Capita (kilowatts/capita)	Population ($\times 10^9$)	Total Energy Use ($\times 10^{12}$ W)
1970[a]	2	4	8
2000[a]	10	7	70
2050[a]	20	10	200
"Ultimate" or[b] "steady state"	20	20	400

[a] From Häfele (1973).
[b] Energy use per capita from Weinberg and Hammond (1971); population from Weinberg and Hammond (1970).

10^{-4} of the solar flux absorbed at the earth's surface and, therefore, are insignificant in the global heat budget. However, they did point out that other industrialization activities may have many times the direct effect of waste heat, e.g., changes in cloudiness, CO_2, and aerosols. These aspects are discussed in Chapter 2 and in Schneider and Mesirow (1976). In Tables 6.2 and 6.3, Kutzbach (1974) compared the Budyko and Sellers global and continental values of energy flux densities with 1/4 the solar constant and net radiation R at the earth's surface. These values can be compared further with those in Table 6.1. If we follow through Budyko's arguments, we would expect the polar ice covers to retreat and eventually disappear. Sellers, with a model similar to Budyko's but applied at 10° latitude bands, spread 20 W m^{-2} of waste heat into these bands—roughly equivalent to the present distribution of large cities. He found changes in surface temperature of the order of 20°C near the north pole and about 11°C in the tropics and concluded that this would eliminate the permanent ice fields. However, models such as Budyko's and Sellers' have highly simplified feedback mechanisms, such as cloud cover, and do not explicitly compute the dynamics of the atmosphere.

Figure 6.5 (Kutzbach, Center for Climatic Research, University of Wisconsin, personal communication) is a graph of energy flux density versus area. Note the solid line showing net radiation balance at the surface, the dashed line representing years circa 1970 and 2050, and the scale of meteorological processes at the bottom. Sources for the estimates of energy consumption appear in the legend. As seen from the slope of the dashed lines, the current energy usage is comparable to the natural energy balance at smaller scales. It will also probably take until 2050 for much effect to be seen on regional and synoptic scales. The general circulation model experiment referred to later is closer to the 2050 estimate.

Without a theoretical model, Sawyer (1974) estimated the effects of sea-surface temperature anomalies on the increase of sensible and latent heat fluxes from the ocean to the atmosphere and compared these with present levels of waste energy fluxes. He concluded that the effects of waste heat fluxes are much less than those caused by natural sea-surface anomalies. From the SMIC report (Matthews

TABLE 6.3 Man-Generated and Natural Energy Flux Densities (W m^{-2})

	Globe	Continents
(a) *Man-generated*		
1970	0.015	0.054
2000	0.14	0.47
2050	0.39	1.4
"Ultimate"	0.78	2.7
"Extrapolation"[a]	(20)	(65)
(b) *Natural*		
1/4 solar constant	350	350[b]
Net radiation, surface	95	65

[a] Extrapolation to circa 2150, based upon continued 4 percent per annum growth rate of energy flux density (Budyko, 1969; Sellers, 1969).
[b] Should be used with caution over continental areas.

et al., 1971), current values of waste heat releases over western Europe and the eastern United States (~10^6 km^2) are 0.74 and 1.11 W m^{-2}, respectively. These small amounts appear insignificant when compared with natural values R, H, or LE in the higher latitudes (Table 6.1). Sawyer, therefore, concluded that a fiftyfold increase is needed before it affects the natural climatic fluctuations. This agrees with the North Pacific anomaly numerical experiments (Chervin et al., 1976), which indicated that for realistic anomaly patterns in the North Pacific the effects did not have much statistical significance except directly over the warm anomaly. They found relatively little downstream effect.

CLIMATE IMPACT EXPERIMENTS WITH GENERAL CIRCULATION MODELS

Some preliminary calculations with a three-dimensional atmospheric general circulation model were carried out with the restriction of a fixed ocean temperature (Washington, 1971, 1972; Murphy et al., 1976). These experiments were conducted to show the atmospheric response to man-made heat uncoupled from ocean changes and should not be regarded as true climate-change experiments. In the first

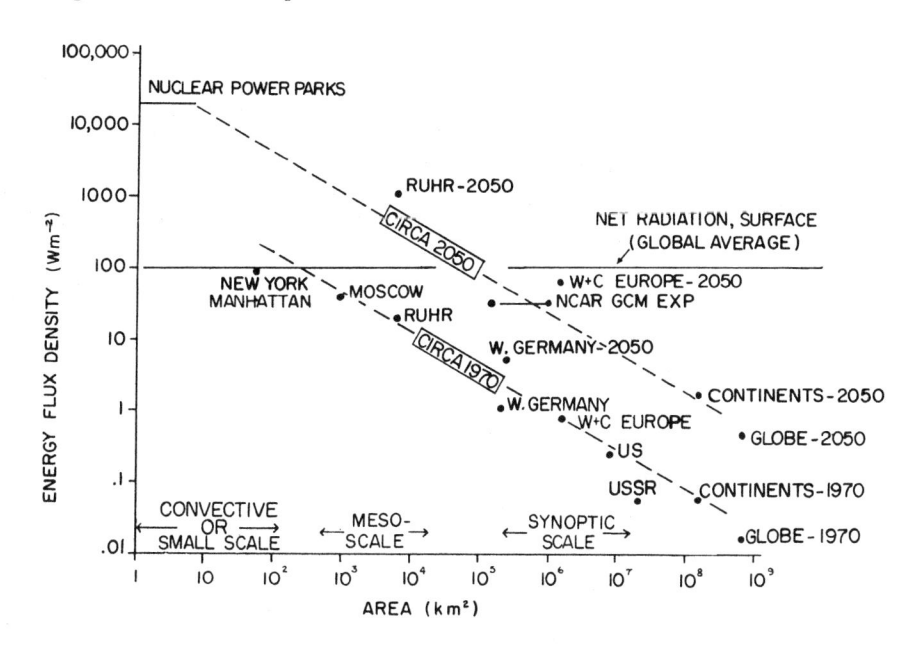

FIGURE 6.5 Energy consumption (W m^{-2}) plotted versus area (km^2) from Kutzbach (1974). Sources: Nuclear power parks, RUHR, W. Germany, continents, globe (Häfele, 1973). Manhattan, Moscow, Europe, United States, Soviet Union (Matthews, 1971). NCAR GCM Experiment (Washington, 1972).

set of experiments by Washington (1971), 24 W m^{-2} of waste heat was added to Eq. (1) over all continental and ice regions—approximately 100 times the heat energy released over the entire United States in 1965, or approximately one third of the surface radiation balance over continental areas. To see how this energy input compares with other terms of the surface energy budget, see Table 6.1 and Figure 6.5. The experiment was compared with a January simulation that did not include waste heat. The surface temperature in the waste-heat experiment rose 1-2° in the tropics and 8° over Siberia and northern Canada within 15 days and thereafter leveled off. The greater response in the northern polar regions than in the tropics was due to a small or zero solar flux in the northern hemisphere winter simulation. In other words, the ratio of thermal heat to net radiation balance (E_P/R) was quite large in the winter hemisphere. The main shortcomings with these experiments, aside from the fixed ocean temperature, were too large an estimate of E_P and assumed uniform distribution. In collaboration with Weinberg and Hammond of Oak Ridge National Laboratory and Hanna of the National Oceanic and Atmospheric Administration, Washington (1972) performed a more realistic set of experiments, still with fixed ocean temperatures. A geographical distribution of thermal pollution was based upon the current population density (this distribution is similar to that shown in Figure 6.2). Following Weinberg and Hammond (1970), Washington assumed a per capita energy usage of 15 kW and an ultimate population of 20 billion. In total energy, this was approximately six times smaller than in the earlier experiments. Figure 6.6 shows the geographical distribution of waste heat. The population assumption ignored the fact that people in the tropics probably require less energy for heating and cooling than those in the polar or midlatitude regions. One million megawatts over a 5° latitude-longitude square in the midlatitudes is equivalent to an energy flux of approximately 6 W m^{-2}. The maximum is approximately 37 W m^{-2} in the most densely populated areas, and the minimum much less than the average 6 W m^{-2}. The maximum is four times smaller than R in the tropics. Four experiments were performed: the first—the control—did not include the effect of waste heat; the second—with waste-heat distribution—was termed thermal pollution. To determine if the differences between the control and thermal pollution experiments were large with respect to the natural fluctuations of the statistics, two other experiments were carried out—one adding negative thermal pollution (i.e., $-E_P$) and one including a small initial random error in the control. Figure 6.7 shows the geographical difference of time means of temperature at the lowest level of the model (1.5 km) between the control and random thermal pollution, control and positive thermal pollution, and control and negative thermal pollution. H denotes warmer than control and C cooler. The temperature changes between the positive thermal pollution and control were up to 10 °C in the northern hemisphere and about 1-2 °C in the tropics. The same pattern emerged for the other experiments. The experiments deviated from the control in the same way, indicating that the thermal pollution effects were not large compared to the noise level for a 29-day, time-averaged January simulation. Put another way, the atmospheric effects of thermal pollution could not be separated from the natural fluctuations of the model for this averaging period. This contrasts markedly with results of the first experiments where the forcing, and therefore response, was much larger. Later Chervin et al. (1974, 1976) investigated the effects of time-averaging on the "noise levels" of general circulation model statistics, so that it is possible to obtain some measure of statistical significance in experiments when a change is made in the model or its boundary conditions.

For additional insight into regional effects and their statistical significance, another January experiment was

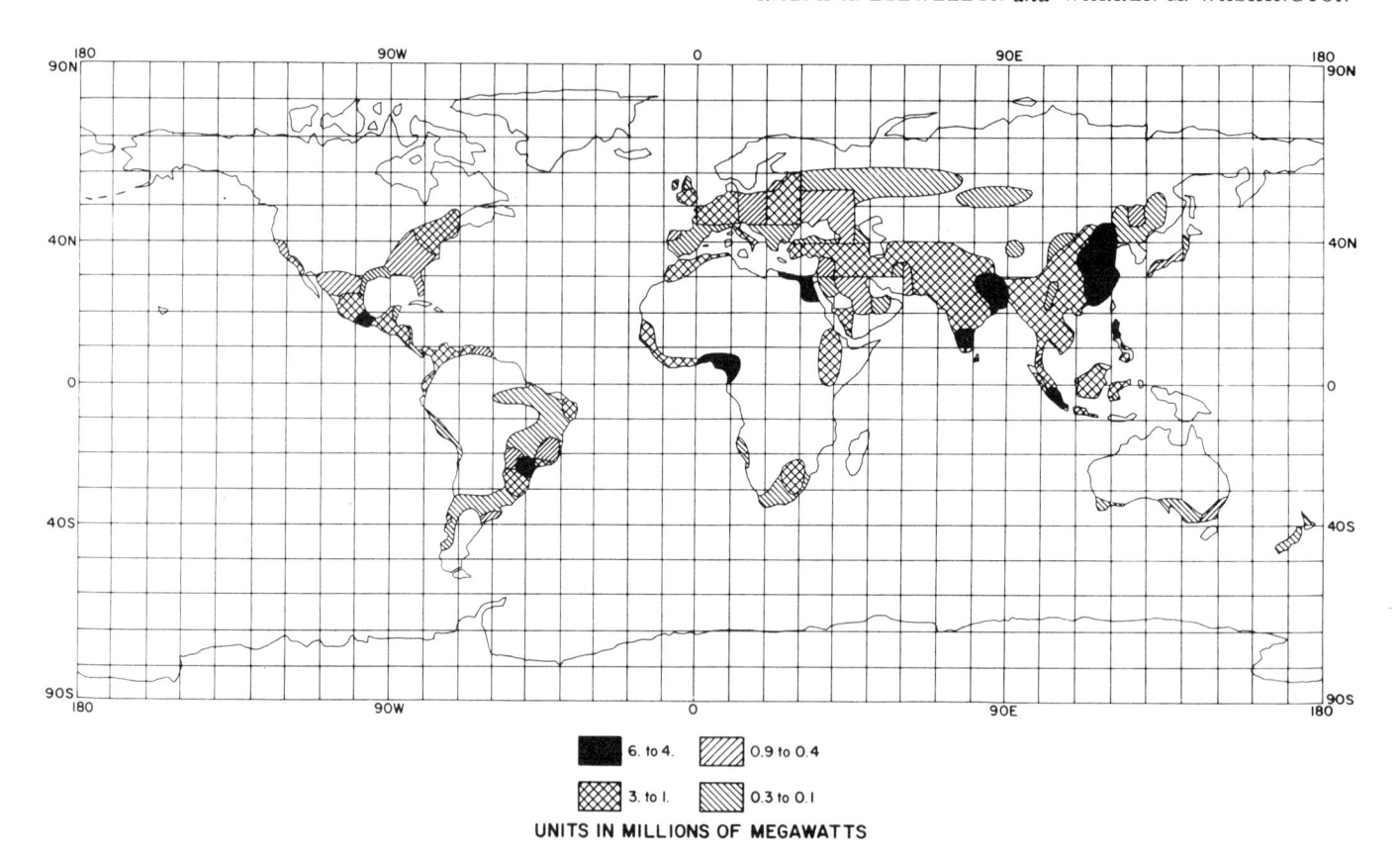

FIGURE 6.6 Geographical distribution of expected levels of man-made heat from Weinberg, Hammond, and Hanna (see Washington, 1972).

performed at the National Center for Atmospheric Research (NCAR). We assumed an urban population distribution for the year 2000 from Pickard (1973; see also Sundquist, 1975), showing a densely populated area from the Atlantic Seaboard to the Great Lakes to Florida, to which we added thermal pollution effects. We further assumed that the energy consumption is equal to that of present Manhattan—approximately 90 W m^{-2} (Figure 6.5). To keep the experiment simple, other regions of the globe were not modified. The experiment was run from days 0-60, averaged over days 31-60, and the results compared with a control experiment without anomalous heating. Figure 6.8(a) shows the time mean temperature difference between the anomaly and control in the boundary layer (i.e., near the ground) $\overline{T}_s$. Note that the temperature differences rise to 12 °C in the vicinity of anomalous heating. Although we show only part of the globe here, no other location has such a high maximum, and it is safe to assume that this is a direct result of the heating. However, it is more difficult to make this point for other levels without statistical analysis.

The temperature difference at 1.5 km, representing a layer thickness of 3 km, is shown in Figure 6.8(b); here the anomalous effect is not obvious. To obtain a quantitative statistical measure, we compute the standard deviation σ of the average January 1.5-km temperature in the model from five separate control runs in which small differences were introduced into the initial data (see Chervin et al., 1976, for details and comparison with observed). Figure 6.8(c) shows the general distribution of 1-3 °C in the midlatitudes and somewhat smaller values over the oceans or in the tropical latitudes. If we define ratio r of the absolute difference of the time means of control and anomaly experiments Δ to the the standard deviation, we obtain

$$r = |\Delta|/\sigma. \tag{3}$$

Following Chervin et al. (1976), we can relate r to the widely used t test in statistics to obtain an objective measure of significance. The significance level is an estimation of the probability that a given value of r could be exceeded by chance (i.e., by merely random fluctuations generated in the model). The confidence level is one minus the significance level. In the experiments here, we have only five samples to estimate standard deviation and, thus, we have four degrees of freedom. The distribution of r in Figure 6.8(d) shows over the eastern United States values of 1 and 2 yielding significance levels of 52 percent and 23 percent, respectively (Table 6.4). Therefore, the differences in this region should be treated with lower confidence than in the boundary layer [Figure 6.8(a)]. In the first case, there is a 52 percent chance of being random, and in the second, 23 percent. Caution should be noted with some of the large values of

Regional and Global Aspects 115

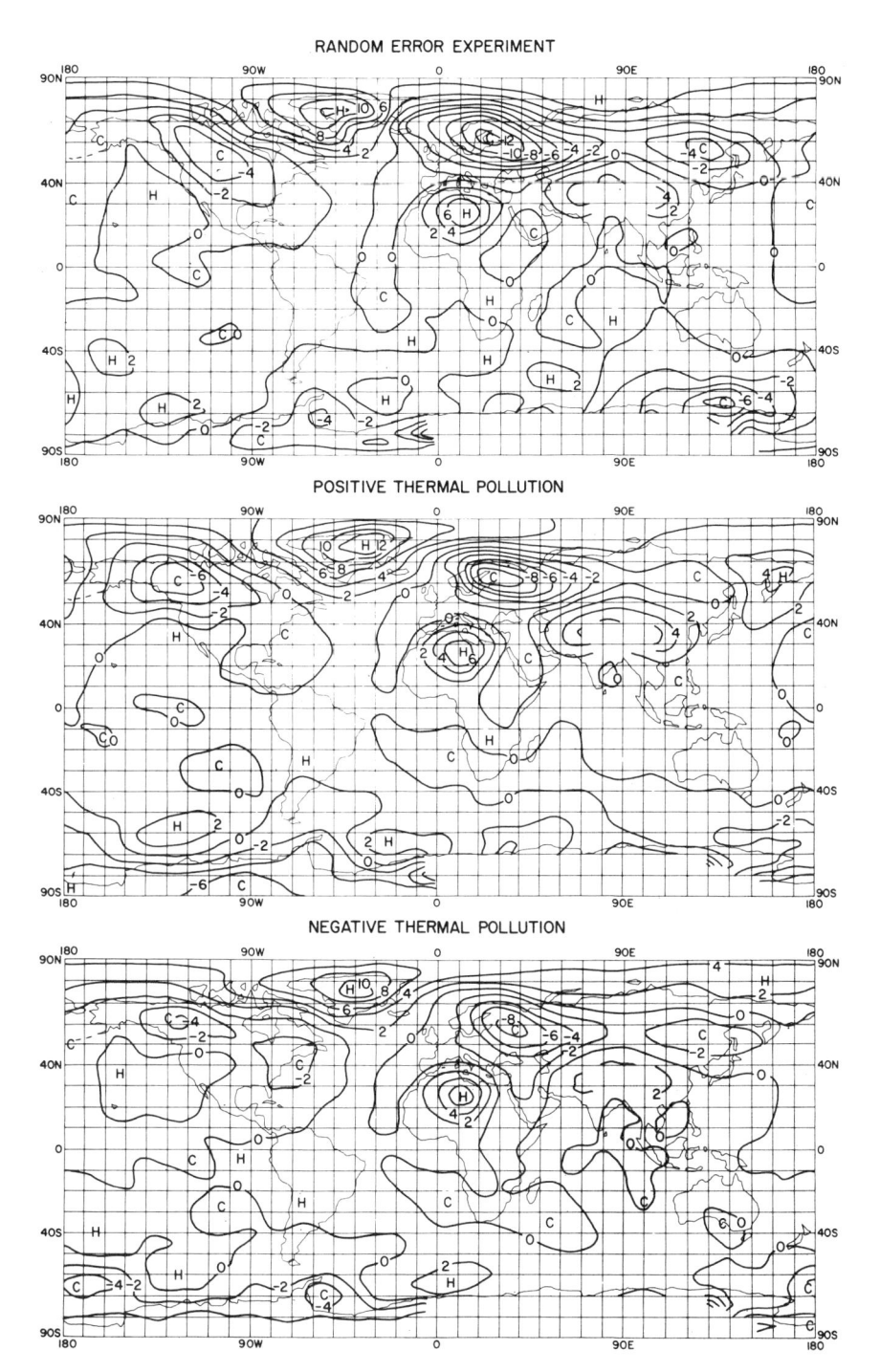

FIGURE 6.7 Geographical difference of time means of global temperature at 1.5 km between control and random initial error experiments (upper), control and positive thermal pollution (middle), and control and negative thermal pollution (lower). *H* denotes warmer than control experiment and *C* denotes cooler; contour interval is 2°C.

r in Figure 6.8(d). These are often caused by unrealistically small values (due to the small sample size) of σ in Eq. (3) and could have been avoided somewhat in Figure 6.8(d) by limiting σ to 0.5°C (a reasonable value from observational data). The major conclusion here is that the changes induced by waste heat are large in the boundary layer but do not seem to have as much influence above it. Remember that because this was a January simulation and the northern hemisphere was mostly stably stratified, the addition of waste heat caused a large temperature change in the boundary layer. In summer over the continents, we expect the effects to be less because solar flux dominates waste heat flux and, because of more convection, the extra heat would not be trapped at the surface as in winter.

In recent January numerical experiments using the United Kingdom Meteorological Office general circulation model,

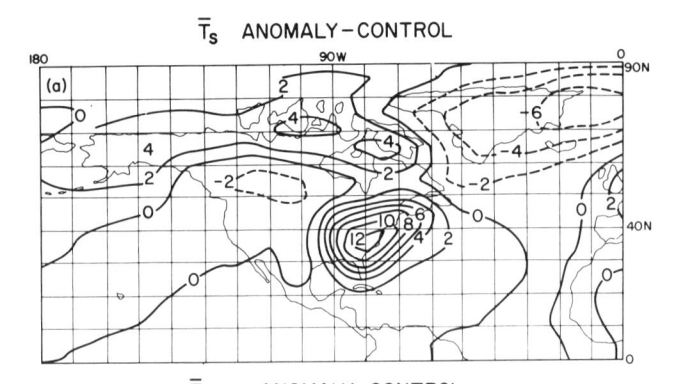

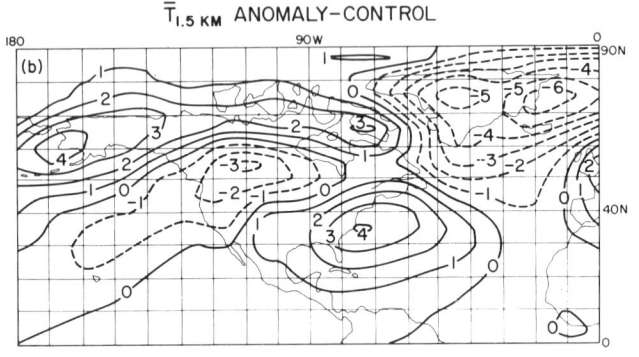

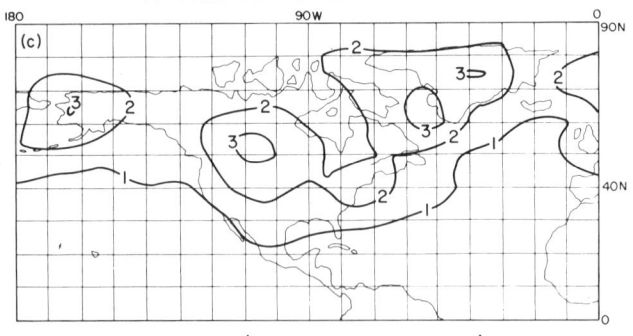

TABLE 6.4 Significance Levels for Temperature Differences

r	Significance Level (%); 4 Degrees of Freedom
1	51.69
2	22.86
3	10.02
4	4.69
5	2.38
6	1.31
7	0.77
8	0.47
9	0.31

Murphy et al. (1976) investigated the effects of waste heat from energy parks. They conducted two experiments, adding 375 W m^{-2} to each over two energy parks with a total area 8 × 10^5 km^2 (approximately four grid boxes in their model). The heat was added to the lowest layer of the model which is 200 mbar thick. This heat is equivalent to the total heat used by Washington (1972) and is approximately four times the energy density of Manhattan (Figure 6.5). In the first experiment, the parks were located in the North Atlantic southwest of England (48° N and 15° W) and in the North Pacific east of Japan (36° N and 148° E). They carried out a second experiment with the Atlantic park located west of Africa (9° N and 22° W), but we discuss here only the first experiment. Two further assumptions were made: (1) all energy was added to four grid boxes in the model (see striped area in Figure 6.9), and (2) the heat is in sensible form only. In future experiments, they will add the park heat in both sensible and latent forms by computing the ocean temperature change in the park region.

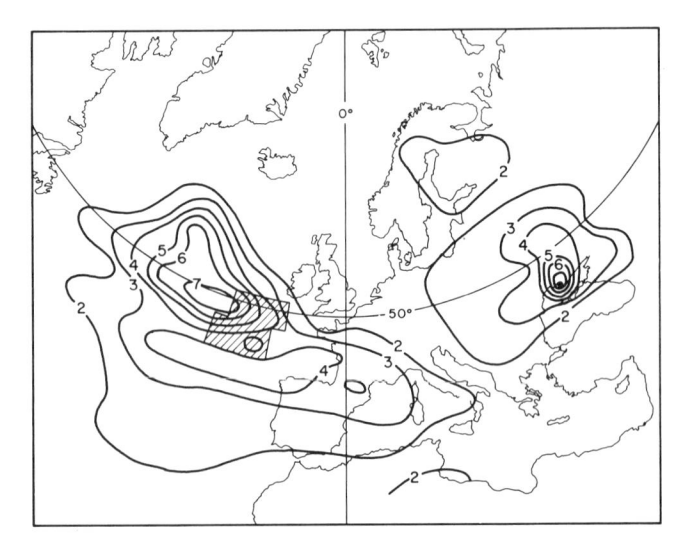

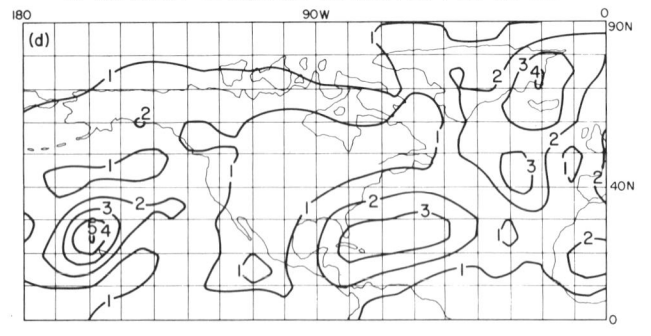

FIGURE 6.8 (a) Difference maps of boundary layer temperature $\bar{T}_s$ between anomaly and control averaged in time over days 31–60. Units are °C. (b) Same as (a) except for $\bar{T}_{1.5 \text{ km}}$. (c) Geographical distribution of standard deviation σ from five Januaries simulated by the NCAR GCM. Units are °C. (d) Ratio r maps of the differences of 1.5-km temperature averaged from days 31–60 to the standard deviation.

FIGURE 6.9 Ratio of pressure differences to one half of the range of sea-level pressures in the four controls over the Atlantic and Europe.

The North Atlantic energy park (4 × 10^5 km^2 in area) experiment produced surface pressure falls east of the park, which extended over the entire western Atlantic with a maximum of −14 mbar, and rises west of the park, extending to eastern Scandinavia with a maximum of +12 mbar. This pattern is consistent with known relationships between atmospheric heating and large-scale pressure patterns. They also found large upstream and downstream effects on precipitation.

Figure 6.9 shows the ratio of sea-level pressure difference between the control and the anomaly divided by one half the range of sea-level pressure in four separate control experiments. This is a type of t test for which values of 3.18 in Figure 6.9 are statistically significant at the 0.05 level (for a two-sided test). With four control experiments, they have three degrees of freedom. We see in Figure 6.9 that increases in sea-level pressure difference over the western edge of the park, as well as south of the park, are highly significant. The differences east of the park are less significant. Over the Crimea, a large ratio occurs, which is not significant because the denominator of the ratio becomes too small. As mentioned earlier with Eq. (3), this can occur because of the small sample size. The statistical significance of the precipitation differences is not shown, although the observed changes are consistent (in a physical sense) with the pressure changes. These effects and those of other energy parks are being investigated in greater detail by a group at the International Institute for Applied Systems Analysis in Laxenburg, Austria (Häfele et al., 1976).

It would be a wrong interpretation of these preliminary experiments to say that thermal pollution can or cannot seriously affect regional and global climate. The experiments described here are extreme. What has been learned is that the effects can be sizable, given levels of waste heat much higher than at present. Model experiments must be performed and averaged over long periods to determine the statistical significance of relatively small effects. These small effects can be important for human affairs. Ultimately, this will require coupled atmosphere–ocean–snow–ice models to account for the total climatic response, and development of such models must continue.

6.3 ACKNOWLEDGMENTS

We thank R. Chervin, S. Schneider, and J. Kutzbach for their comments on the manuscript. We appreciate the editorial assistance of A. Modahl.

REFERENCES

Budyko, M. I. (1969). The effect of solar radiation variations on the climate of the earth, *Tellus 21*, 611.

Budyko, M. I. (1972a). *The Influence of Man on Climate* (in Russian), Hydrometeorological Publishing House, Leningrad.

Budyko, M. I. (1972b). The future climate. *E⊕S, Trans. Am. Geophys. Union 53*, 868.

Changnon, S. A., (1973). Urban-industrial effects on clouds and precipitation, in *Proc. Workshop on Inadvertent Weather Modification*, Utah State U., Logan, Utah, Aug. 1973, pp. 111-139.

Chervin, R. M., W. L. Gates, and S. H. Schneider (1974). The effect of time averaging on the noise level of climatological statistics generated by atmospheric general circulation models, *J. Atmos. Sci. 31*, 2216.

Chervin, R. M., W. M. Washington, and S. H. Schneider (1976). Testing the statistical significance of the response of the NCAR general circulation model to North Pacific Ocean surface temperature anomalies, *J. Atmos. Sci. 33*, 413.

Cook, E. (1971). The flow of energy in an industrial society, in *Energy and Power*, W. H. Freeman and Co., New York, pp. 134-147.

Darmstadter, J. (1971). *Energy in the World Economy*, Johns Hopkins U. Press, Baltimore, Md., pp. 9-35.

Häfele, W. (1973). Energy systems. *Proc. IIASA Planning Conference on Energy Systems*, International Institute for Applied Systems Analysis, Laxenburg, Austria, 17-20 July 1973, IIASA-PC-3, pp. 9-78.

Häfele, W., et al. (1976). Possible impacts of waste heat on global climate patterns, *Research Report RR-76-1, Second Status Report of the IIASA Project on Energy Systems*, 1975, International Institute for Applied Systems Analysis, Laxenburg, Austria, pp. 134-148.

Hosler, C. L. (1971). *Wet Cooling Tower Plume Behavior*, C.E.P. Tech. Manual No. T40. AIChE, p. 1.

Hubbert, M. K. (1971). Energy resources of the earth, in *Energy and Power*, W. H. Freeman and Co., New York, pp. 60-87.

Kutzbach, J. E. (1974). Possible impact of man's energy generation on climate, *Report of GARP Study Conference on the Physical Basis of Climate and Climate Modelling*, Stockholm, Sweden, 29 July-10 August 1974.

Landsberg, H. E. (1956). *The Climate of Towns; Man's Role in Changing the Face of the Earth*, U. of Chicago Press, Chicago, Ill., pp. 584-603.

Landsberg, H. E. (1970). Man-made climatic changes, *Science 170*, 1265.

Matthews, W. H., W. W. Kellogg, and G. D. Robinson, eds. (1971). *Inadvertent Climate Modification, Study of Man's Impact on Climate* (SMIC). MIT Press, Cambridge, Mass.

Murphy, A. H., A. Gilquist, W. Häfele, G. Krömer, and J. Williams, (1976). The impact of waste heat release on simulated global climate, RM-76-79, International Institute for Applied Systems Analysis, Laxenburg, Austria.

Pickard, J. P. (1973). Urbanization and economic change in North American regions, in Sundquist (1975).

Sawyer, J. W. (1974). Can man's waste heat affect the regional climate? Talk presented at the IAMAP Symposia, Melbourne, Australia, January.

Schneider, S. H., and R. E. Dickinson (1974). Climate modeling, *Rev. Geophys. Space Phys. 12*, 447.

Schneider, S. H., and R. D. Dennett (1975). Climatic barriers to long-term energy growth, *Ambio 4*, 65.

Schneider, S. H., with L. E. Mesirow (1976). *The Genesis Strategy, Climate and Global Survival*, Plenum Press, New York.

Sellers, W. D. (1965). *Physical Climatology*, The U. of Chicago Press, Chicago, Ill.

Sellers, W. D. (1969). A global climatic model based on the energy balance of the earth-atmosphere system, *J. Appl. Meteorol. 8*, 392.

Steinhart, J. S., and C. E. Steinhart (1974). Energy and use in the U.S. food system, *Science 184*, 307.

Sundquist, J. L. (1975). Dispersing population: What America can learn from Europe, Brookings Institution, Washington, D.C., pp. 24-32.

United Nations Statistical Papers (1973). World energy supplies, 1968-1971. Ser. J/16, 6-40.

Vonder Haar, T. H., and A. H. Oort (1973). New estimate of annual poleward energy transport by northern hemisphere ocean, *J. Phys. Oceanog. 3*, 169.

Washington, W. M. (1971). On the possible uses of global atmospheric models for the study of air and thermal pollution, in *Inadvertent Climate Modification—Study of Man's Impact on Climate*, W. H. Matthews *et al.*, eds., MIT Press, Cambridge, Mass., pp. 265-276; also in *Air and Water Pollution*, W. E. Brittin *et al.*, eds., Colorado Associated U. Press, Boulder, Colo., pp. 599-613.

Washington, W. M. (1972). Numerical climatic-change experiments The effect of man's production of thermal energy, *J. Appl. Meteorol. 11*, 768.

Weinberg, A. M., and R. P. Hammond (1970). Limits to the use of energy, *Am. Sci. 58*, 412.

Weinberg, A. M., and R. P. Hammond (1971). Global effects of increased use of energy, in *Proc. Fourth International Conference on Peaceful Uses of Atomic Energy*, Geneva, Switzerland, Sept.

III

MONITORING AND MODELING

7

Ocean Dynamics and Energy Transfer: Some Examples of Climatic Effects

D. JAMES BAKER, JR.
University of Washington

7.1 INTRODUCTION

Over the long periods pertinent to climate variability, the ocean's large heat capacity and energy storage temper and modify the atmosphere, which in turn affects the ocean. Cause and effect are not easy to distinguish in the dynamics of this coupled system. Since the ocean responds to external changes more slowly and on smaller space scales than the atmosphere, the interaction and feedback processes are difficult to observe and to model.

These points are noted in Chapters 6 and 9, in which it is pointed out that one of the real shortcomings of the existing climate models is the lack of an adequate parameterization of oceanic processes. In spite of a greatly increased effort in both observation and modeling of the ocean in the past few years, we do not yet have models that could yield the necessary parameterizations.

Thus the oceanographer is faced with an especially difficult task when asked to contribute to the topic of this volume. Oceanographers are now engaged in trying to learn which ocean processes are essential to climate dynamics and how to monitor these in an effective way. We are, therefore, one step removed from the question of how to monitor man's effect on these processes. One particularly good example of our basic lack of knowledge here is given in Section 7.2 below.

The second point to be made is that man's effects on oceanic processes through increased use of energy are apparently negligible, at least in terms of ratios of estimated energy production (see Chapters 1 and 6) to total energy involved in particular natural phenomena. However, we must keep in mind the fact that large-scale fluid flow in nature tends to be unstable: that is, small changes can have large effects. Through man's influence, a relatively small amount of energy could be magnified drastically by the release, via flow instability, of available energy in the system.

In any case, it is clear that the subject of ocean dynamics and the interaction of the ocean with the atmosphere must be pursued vigorously if we are to gain a practical understanding of climate and man's influence on it. Note that the question of the global cycle of CO_2, including the interaction of CO_2 with the ocean, has been covered in Chapter 4.

To illustrate the role of the ocean in global climate dynamics, and to show how ocean processes may be involved in local climate change, I have chosen to describe a few examples below. Man's potential effect is noted where possible. I suspect that the reader will find the resulting story somewhat qualitative and unsatisfactory, especially when compared with the quantitative results quoted in the preceding chapters. If so, I have made my point—that the subject is important, our knowledge is poor, and much work remains to be done.

In view of the existence of other reports (IOS, 1969; Ocean Sciences Committee, 1975), which already suggest a plan of action for study of the role of the ocean in global climate dynamics, I shall not outline here a plan for global ocean monitoring but conclude the chapter with a few general remarks only. The reader is referred to the relevant sections of those reports for details. I also note that the Global Atmospheric Research Program, as it continues into the 1980's, will include a substantial oceanic component aimed at its objective of understanding the physical basis of climate (Joint Organizing Committee for GARP, 1975; U.S. Committee for the Global Atmospheric Research Program, 1975).

7.2 THE OCEAN IN THE GLOBAL ENERGY BALANCE

Radiation data from satellite measurements (Vonder Haar and Oort, 1973 and Figure 7.1) reveal the significant share of heat carried by the ocean in the global energy balance. The ocean transport has been estimated here by subtracting the known atmospheric heat transport from the total required to balance the net loss of heat at the poles. The ocean transport appears to be dominant at low latitudes and midlatitudes even within the measurement uncertainty. Since the transport is a significant component of the global energy balance, its monitoring is an essential element of a climate-prediction system. Unfortunately, we do not yet know enough about the general circulation of the ocean to devise a practical global monitoring scheme. The heat could be carried by mean circulation, wind drift, or eddy processes, but we do not know the relative importance of these. [See Bryan (1962) for a careful discussion of the problems of measuring meridional heat transport in the ocean.]

Our knowledge of the general circulation of the ocean suggests that the heat could be transported northward by the great western boundary currents of the ocean like the Gulf Stream or the Kuroshio current: a stream of water flowing with a speed of 0.5 m/sec, 100 km wide, and 1 km deep represents a potential heat flux of 2×10^8 MW per °C of temperature difference from the surroundings. Only a few such fluxes would make up the necessary total. The variations of these currents could be significant; for example, observations by Niiler and Richardson (1973) show that the seasonal variability of the northward heat transport by the Florida current is 3×10^8 MW (half-amplitude). Even though it is difficult to calculate the magnitude of this variability unambiguously (Montgomery, 1974; Niiler and Richardson, 1974), its importance seems established.

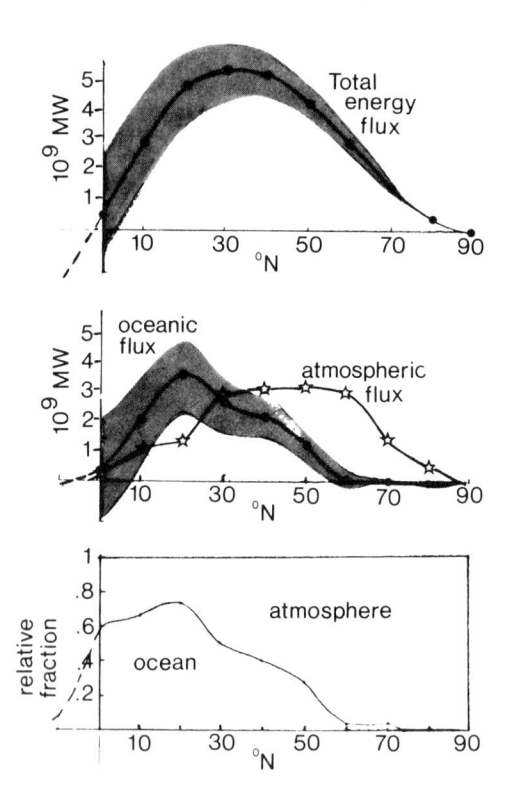

FIGURE 7.1 Variation of net energy transport with latitude over the northern hemisphere [from Vonder Haar and Oort (1973)]. The shading represents uncertainties in the satellite measurements. For reference in magnitudes, recall that the energy input to the atmosphere from a cyclone is about 10^8 MW (see Chapter 5).

If man can influence these western boundary currents, he could affect the overall heat balance. For example, consider a nuclear power park located in a western boundary current for cooling. Since a large power park represents an energy input of about 4×10^4 MW (see Chapter 5), it appears that the change in total heat transport of the western boundary current would be negligible. However, the park could raise the local temperature. For example, a park 25 km² by 100 m deep dissipating 40,000 MW could raise the local temperature by 1 °C in about 3–10 days depending on the rate of flow of cooling water. If this thermal anomaly were advected northward, its eventual communication with the atmosphere in the western boundary current extension regions to the north could lead to local heating and increased cloud formation and precipitation. Local heating could be avoided by using deep cold water as coolant, raising its temperature only to the local surface temperature. Then the current would be changed in mass, and hence total heat transport would be changed also. Some quantitative modeling is required here.

The suggestion that the kinetic energy of these strong ocean currents could be used to generate electric power by the use of turbine arrays has been raised frequently. Von Arx, Stewart, and Apel (Stewart, 1974) estimate the effect on the Gulf Stream near the coast of Florida of extracting energy for electrical power by use of such arrays. They suggest that a reasonable turbine array could yield about 1000 MW, a significant contribution to the local power requirements. Since the total kinetic energy of the current here is about 25,000 MW, the turbine array would extract at least 4 percent of the kinetic energy. Unfortunately, we do not know the effect of such an extraction on the Gulf Stream dynamics. It could be important because of the potential sensitivity of the meandering path of such western boundary currents to local changes (see below) and the possible importance of that path to air-sea exchange in the subpolar regions. Further modeling and observational studies are clearly required.

The fluctuating motion in the ocean (various kinds of "eddies") could also transport heat just as heat is transported by atmospheric eddies. Such ocean fluctuations also include the large-scale sea-surface thermal anomalies and the deeper energetic motions. Newton (1961) has calculated that a single Gulf Stream eddy could transfer as much as 5×10^7 MW from the ocean to the atmosphere. Thus not many eddies are required in order to achieve the necessary heat transfer.

Holland and Lin (1975) have shown in a numerical model that meandering eddies in the Gulf Stream seem to drive large transports in the deep water under the Gulf Stream. These transports result in a heat flux toward the south, but the magnitude of this flux in the real ocean is not known. As mentioned earlier, both observational and modeling studies will be required to establish the importance of these processes.

The large sea-surface temperature anomalies in the ocean described by Namias and others (e.g., see Namias, 1972) are demonstrably linked to climate variability. However, it is unlikely that man-made energy sources will be able to either affect these anomalies or have effects similar in magnitude. For example, a 40,000-MW power park would generate 4×10^{17} J/year; the heat stored in a typical sea-surface temperature anomaly (5000 km × 1000 km × 100 m, 1/2 °C different from surroundings) is about 8×10^{20} J. Locally, such a park has a power dissipation of about 64 W/m^2, which is comparable with the components of the surface energy balance over the ocean [for example, the fluxes of radiation, sensible heat, and latent heat in different regions of the ocean are estimated to lie in the range 0 to 500 W/m^2 (Sellers, 1965)]. However, the small size of a single park compared with the area of the effective heat-transfer regions of the ocean suggests that the global climatic effect will be negligible. Large numbers of such parks could have a more important effect.

We have discussed heat dissipation in the upper layers. A second point is the interaction of the ocean with increased CO_2 in the atmosphere. The positive feedback process, increased CO_2 in atmosphere → increased heating of ocean → less absorption of CO_2 in ocean → increased CO_2 in atmosphere, has been suggested. W. Broecker of Lamont-Doherty Geological Observatory, Columbia University (private communication) has pointed out that the ability of the ocean to absorb excess CO_2 is not dependent on temperature in any significant way. The feedback effect does exist, but its magnitude is apparently small enough to be ignored according to our current knowledge of ocean dynamics. Such a process could have a significant effect if a general warming greatly reduced vertical mixing in the ocean. In that case, the surface layers would be cooled less by the deep ocean and heated more by the atmospheric warming associated with an increase in CO_2.

7.3 THE OCEAN AND THE LOCAL ENERGY BALANCE

The ocean currents can also have a major effect on local climate. Our example here is the Kuroshio current. Figure 7.2 (Robinson and Taft, 1972) shows how this current occasionally takes a path that loops south of the island of Honshu instead of hugging the coast. The periods when the current loops are associated with lower temperatures, decreased rainfall, and poor rice crops in northern Japan (Uda, 1964).

Could or does man have an effect on this variability of the path of the current? The numerical study of the local dynamics of the Kuroshio (Robinson and Taft, 1972) shows that important contributions to the guiding of the path are the bottom velocity and the bottom topography of the local seabed. Relatively minor changes of either could reguide the path.

It is unlikely, however, that man would have a direct effect on either of those two parameters. It is possible that he could affect the path through the influence of local air-sea interaction: increased convection over land could lead to stronger onshore winds. The warm surface water would then tend to be pushed closer to the coast, thus stabilizing the more productive of the two states.

Such arguments must be viewed with some caution, however, if we recall the difficulties of predicting El Niño (warm currents off the coast of Peru), which apparently exhibits links with more global phenomena (Quinn, 1974; Wyrtki, 1973).

7.4 OCEAN HEAT TRANSPORT: THE ARCTIC BASIN

Most climate models (e.g., see Kellogg, 1974) show sensitivity to the size of the Arctic ice pack. This is understandable, because the existence of the ice causes a drastic change in the surface albedo and thus the surface heat balance. But the surface heat balance is also determined by the amount of heat advected in by ocean currents and the amount of ice exported.

As we noted above, the magnitude of the ocean heat transport into the Arctic basin is relatively small, about 7×10^7 MW (Aagaard et al., 1973). However, the input from the ocean is crucial to the size of the Arctic ice pack. The more heat that is brought into the ocean, the thinner the pack.

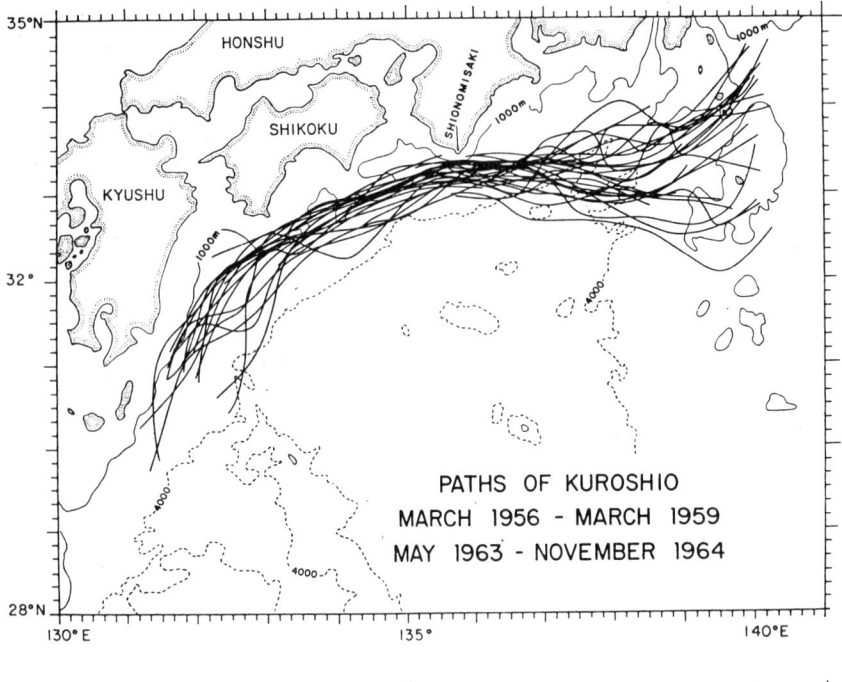

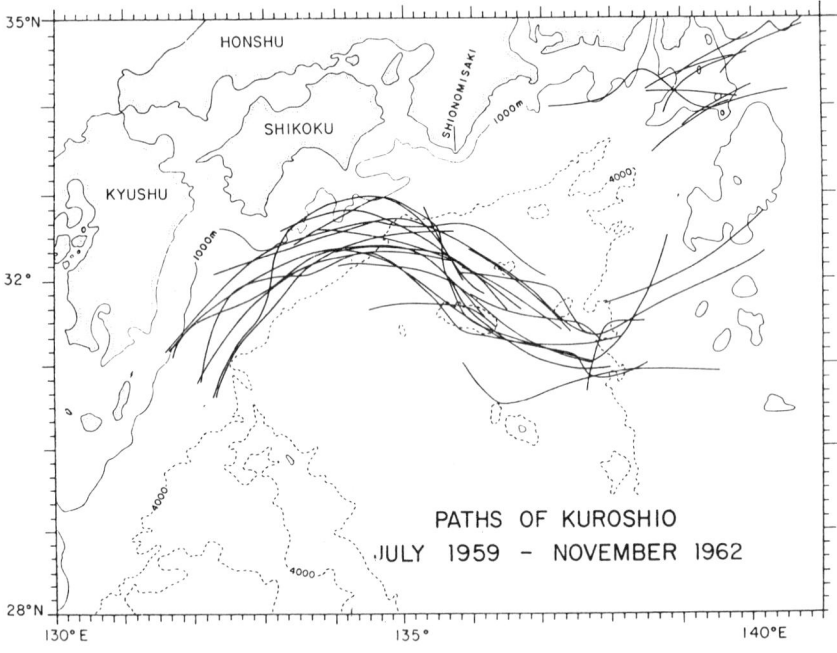

FIGURE 7.2 *Top:* Composite of slope paths for two periods. The 1000-m and 4000-m isobaths are shown. *Bottom:* Composite of meander paths for one period. The 1000-m and 4000-m isobaths are shown.

Most of the heat is transported in through the passage between Greenland and Spitsbergen: a relatively cold (the Greenland) current out, and a relatively warm (the West Spitsbergen) current in. The final contribution to the heat input is the ice export between Greenland and Scandinavia.

A recent study of the water, heat, and ice budgets (Aagaard and Greisman, 1975) based on a year-long deployment of instruments in the Greenland-Spitsbergen passage shows a heat transport larger than that apparently required to melt the Arctic ice pack. The uncertainties from both the measurements and the ice pack models (Maykut and Untersteiner, 1971) make such a conclusion more provocative than definite. However, it does show the sensitivity of this element of climate variability to ocean dynamics and the importance of long-term measurements here.

The influence of man's increased use of energy is not yet clear in this region. The magnitude of heat input from one large power park, about 4×10^4 MW, is still too small to have any apparent significant effect on the total heat input to the Arctic basin. Our knowledge of the basic dynamics and stability of the air-sea-ice system is still too crude, however, to draw any further conclusions.

7.5 RIVER RUNOFF AND THE ARCTIC ICE PACK

Man can produce local climate changes by diverting rivers. In response to chronic flooding of the western Siberian taiga, there have been Soviet proposals to divert southward the Ob and the Yenisei. These rivers discharge onto the shelf bordering the southern Eurasian basin. The magnitude of such an engineering task is enormous, but it is of interest to note potential climatic effects.

Aagaard and Coachman (1975) have pointed out that the relatively small accumulation of freshwater in the southern Eurasian basin is an important feature, because the shallow and weak salinity gradient forms a lid on the warmest and most saline water in the Arctic Ocean. Their heat-budget studies indicate that even under the present conditions of salinity stratification, the upward heat flux from the deep water is an order of magnitude larger in this local region than in the Arctic Ocean as a whole. They note that were the thin veneer of freshwater to be substantially removed in this sensitive area, the high sensible heat content of the Atlantic water could become more readily available for surface exchange. In the absence of other feedback mechanisms, this could lead to prolonged icefree conditions because of the deep-reaching convection in the Arctic Ocean.

The sensitivity of even this simplified system to possible man-made influence shows clearly the need for a correct Arctic Ocean air-sea-ice interaction model.

7.6 OIL SPILLS AND THE ARCTIC ALBEDO

In man's increased use of energy, the potential of oil spills increases. The effect of these spills and of natural seepage has been the subject of much published literature. However, we do not know yet the quantitative effects of this change in surface properties of the ocean on the climate. One example has been suggested by Campbell and Martin (1973). They note that the slow rate of biological degradation of oil at near-zero temperatures has led biologists to suggest that oil spills in the Arctic Ocean might remain there for periods of 50 years or more. Campbell and Martin point out that the dynamics of the ice pack combined with the long life of the oil could allow an oil spill to have a major effect on the albedo in certain regions of the Arctic. They estimate that the transit time of an oil spill on the fringes of the Beaufort Sea around the circumference of the Beaufort gyre would be about 7-10 years. Several mechanisms act both to diffuse the oil and to put the oil on the surface of the ice. Therefore, as the source continued its journey, the area affected by the spill would grow. By the time the original spill site returned to its original approximate geographic coordinates, a considerable area of the Beaufort Sea could have its albedo changed.

The significance of the resultant albedo change for the Arctic heat balance is moot (Ayers et al., Martin and Campbell, 1974), but most scientists agree that the Arctic environment will require special precautions to minimize the risks of accidental oil spills. Man's increased use of resources in the Arctic will be primarily aimed at obtaining oil for energy: the potential effect on the Arctic heat balance due to changing albedo must be studied together with the more basic environmental studies carried on there.

7.7 OCEAN MONITORING FOR EFFECTS OF INCREASED ENERGY USE

As mentioned above, recommendations for global ocean monitoring are listed in the references (IOS, 1969; Joint Organizing Committee for GARP, 1975; Ocean Sciences Committee, 1975; U.S. Committee for the Global Atmospheric Research Program, 1975) and will not be repeated here. It is of interest to note, however, that satellite monitoring (Allison et al., 1975) is beginning to come into its own now and that it will be of use for both local and global measurements. Figure 7.3 is an example of a picture of the Gulf Stream in the visible band, enhanced by special techniques by NOAA scientists. Small-scale structures are clearly visible. Global data on sea-surface temperature from satellites will also be an essential input to climate modeling.

In terms of this report, we note that monitoring of CO_2 and waste heat are paramount. In this regard it is notable that a climatic baseline of the major chemical constituents of the ocean has been established during the past three years by the GEOSECS program. In addition, the Integrated Global Ocean Station System (IGOSS), coordinated by the Intergovernmental Oceanographic Commission (IOC/UNESCO, 1969) is developing a worldwide system of oceanic data collection. The current phase of IGOSS involves an international exchange of oceanographic data; it is planned to extend this to the routine production of oceanic data summaries and predictions.

To summarize the monitoring problem, we can say that a beginning has been made for ocean monitoring, but that we are still too ignorant of basic processes to be able to establish a network that will reveal the significant variability of the ocean climate variables. Part of the reason lies in our ignorance of the dynamics of the system, part lies in the absence to date of the proper instrumentation, and, finally, part of our ignorance arises from the scale of information collection needed; unlike the atmosphere over land, man's activities in and over the open ocean are still infrequent and can never be taken for granted. The cost of comprehensive global oceanic monitoring using present technology is staggering, but oceanographers are hopeful that the use of satellites and the continuing development of other large-scale measurement techniques will help to point the way toward an affordable and efficient monitoring system.

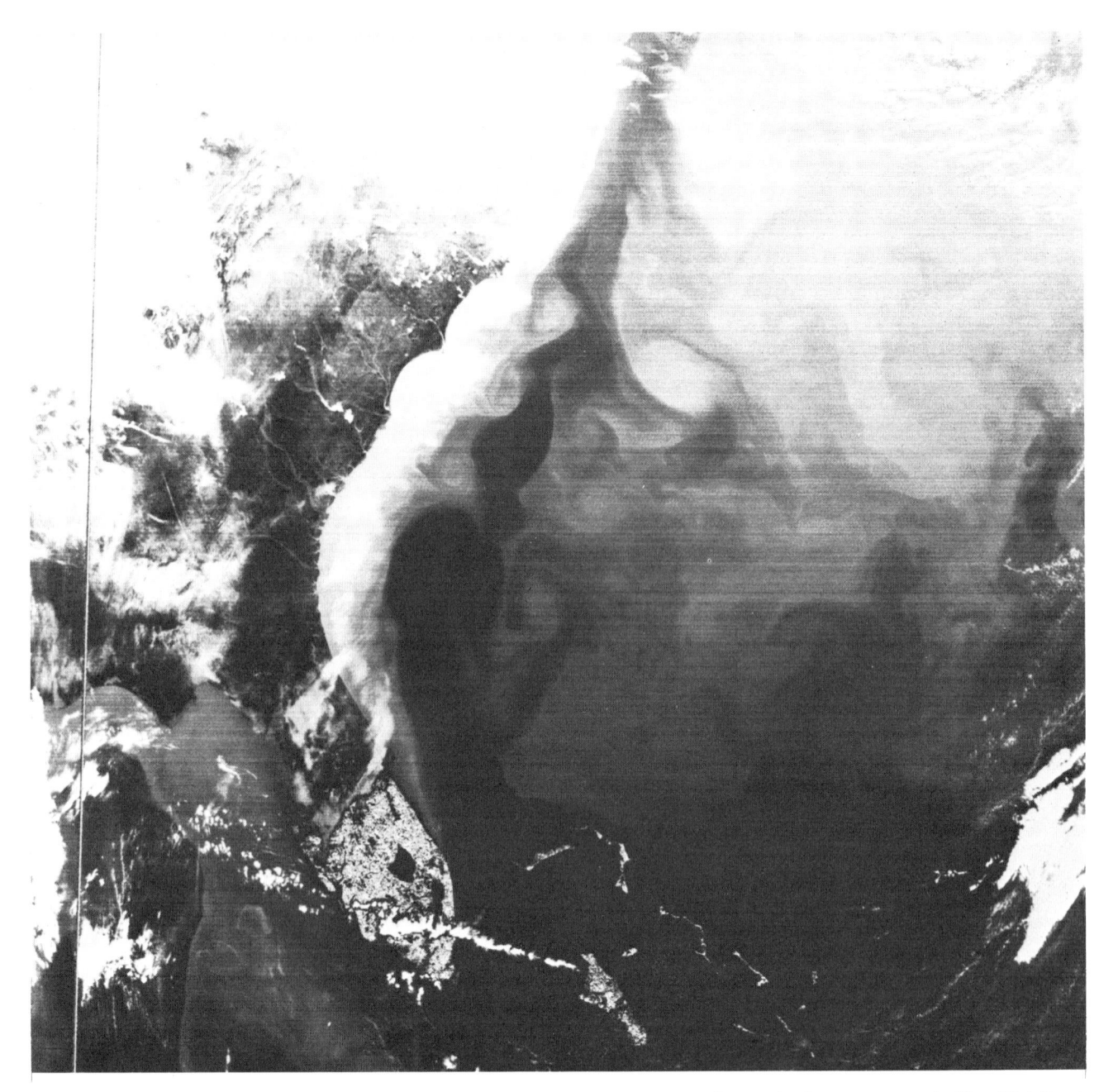

FIGURE 7.3 This infrared image of the sea-surface temperature made by the NOAA-3 satellite on April 1, 1974, uses the Very High Resolution Radiometer, which can resolve temperature changes of 0.5 °C over a spatial distance of 1 km. The warmest water of the Gulf Stream is black, and the coldest inshore water is white. The temperature range for the water is 11 °C. The southern part of Florida is actually hotter than the Gulf Stream, but it was made to appear white for contrast. (Figure supplied by R. Legeckis, National Environmental Satellite Service, NOAA.)

REFERENCES

Aagaard, K., and L. K. Coachman (1975). Diversion of western Siberian rivers towards an ice-free Arctic Ocean, *Trans. Am. Geophys. Union* 56, 484.

Aagaard, K., and P. Greisman (1975). Toward new mass and heat budgets for the Arctic Ocean, *J. Geophys. Res.* 80, 3821.

Aagaard, K., C. Darnall, and P. Greisman (1973). Year-long current measurements in the Greenland–Spitsbergen Passage, *Deep-Sea Res.* 20, 743.

Allison, L. J., A. Arking, W. R. Bandeen, W. E. Schenk, and R. Wexler (1975). Meteorological satellite accomplishments, *Rev. Geophys. Space Phys.* 13, 737.

Ayers, R. C., Jr., H. O. Johns, and J. L. Glasser; S. Martin and W. J. Campbell (1974). Oil Spills in the Arctic Ocean: Extent of spreading and possibility of large-scale thermal effect, *Science 186*, 843.

Bryan, K. (1962). Measurements of meridional heat transport by ocean currents, *J. Geophys. Res. 67*, 3403.

Campbell, W. S., and S. Martin (1973). Oil and ice in the Arctic: Possible large-scale interactions, *Science 181*, 56.

Holland, W. D., and L. B. Lin (1975). On the generation of mesoscale eddies and their contribution to the oceanic general circulation, Parts 1 and 2, *J. Phys. Oceanog. 5*, 642.

IOS (1969). General plan and implementation of IGOSS for Phase I, IOC/UNESCO Document No. SC/IOC/VI/21.

Joint Organizing Committee for GARP (1975). *The Physical Basis of Climate and Climate Modelling*, GARP Publ. No. 16, Geneva.

Kellogg, W. W. (1974). Climatic feedback mechanisms involving the polar regions, in *Proceedings of Climate of the Arctic*, U. of Alaska, College, Alaska, Aug. 15-17, 1973.

Maykut, G. A., and N. Untersteiner (1971). Some results from a time-dependent thermodynamic mode of sea ice, *J. Geophys. Res. 76*, 1150.

Montgomery, R. B. (1974). Comments on "seasonal variability of the Florida Current," *J. Marine Res. 32*, 533.

Namias, J. (1972). Large-scale and long-term fluctuations in some atmospheric and oceanic variables, in *Nobel Symposium 20*, D. Dryssen and D. Jagner, eds., Almquist and Wiksell, Stockholm, pp. 27-48.

Newton, D. W. (1961). Estimates of vertical motions and meridional heat exchange in Gulf Stream eddies and a comparison with atmospheric disturbances, *J. Geophys. Res. 66*, 853.

Niiler, P. P., and W. S. Richardson (1973). Seasonal variability of the Florida Current, *J. Marine Res. 31*, 144.

Niiler, P. P., and W. S. Richardson (1974). Reply, *J. Marine Res. 32*, 534.

Ocean Sciences Committee (1975). Panel on Ocean-Atmosphere Interaction, *The Ocean's Role in Climate Prediction*, National Academy of Sciences, Washington, D.C.

Quinn, W. H. (1974). Monitoring and predicting El Niño invasions, *J. Appl. Meteorol. 13*, 825.

Robinson, A. R., and B. A. Taft (1972). A numerical experiment for the path of the Kuroshio, *J. Marine Res. 30*, 65.

Sellers, W. D. (1965). *Physical Climatology*, U. of Chicago Press, Chicago, Ill., p. 108.

Stewart, H. B., ed. (1974). *Proceedings of the MacArthur Workshop on the Feasibility of Extracting Usable Energy from the Florida Current*, Palm Beach Shores, Fla., Feb. 27-Mar. 1.

Uda, M. (1964). On the nature of the Kuroshio, its origin and meanders, in *Studies in Oceanography*, K. Yoshida, ed., U. of Tokyo Press, Tokyo, pp. 89-107.

U.S. Committee for the Global Atmospheric Research Program (1975). Panel on Climatic Variation, *Understanding Climatic Change: A Program for Action*, National Academy of Sciences, Washington, D.C.

Vonder Haar, T. H., and A. Oort (1973). New estimate of annual poleward energy transport by northern hemisphere oceans, *J. Phys. Oceanog. 3*, 169.

Wyrtki, K. (1973). Teleconnections in the Equatorial Pacific Ocean, *Science 180*, 66.

The Need for Climate Monitoring

8

VERNER E. SUOMI
University of Wisconsin

Many of the papers in this volume express concern over the possibility of inadvertent climate modification due to the activities of man, particularly through the production of large amounts of energy. It is easy enough to see how these activities of man *could* influence climate, but it is quite another thing to prove that man has in fact influenced climate. *On a global scale*, even the most generous estimates of future energy release by man would only represent noise superimposed in the context of the global heat budget. The present uncertainty in the solar constant (1.95 to 2.00 cal/cm^2 min) completely dwarfs many of the concerns raised in other parts of this document.

During the GARP Atlantic Tropical Experiment (GATE) the geostationary satellite SMS-1 over the Atlantic showed an enormous dust cloud coming from the Sahara Desert and extending across the Atlantic. Simple calculations show that the effect on the absorption of solar energy of such sources of pollution completely dominate man-made effects. In a global context, nature's energy sources overwhelm man's sources. However, regional and local scales are quite another matter. It is possible that regional and local changes due to man's activities could trigger a global change in the atmosphere's circulation, but we cannot be certain.

A large part of our difficulty in attempting to assess man's possible effects stems from the fact that we have such a poor data base that we do not yet adequately understand natural climatic phenomena. To answer the questions raised in this volume, we need to understand the dynamics of climate. The purpose of this paper is to explain the need for climate monitoring and to outline a way to carry out such monitoring.

The mechanisms that produce and control the earth's climate are exceedingly complex. One usually considers climate as mainly an atmospheric phenomenon, but it is the atmosphere's interaction with the ocean, the land, and ice masses, together with the sun and space, that controls the atmosphere's behavior. If we are to assess man's possible influence on climate and predict what man's activities might do to our climate, we must first understand the basic mechanisms and physics of climate well enough to model it. This is beyond our grasp at present.

One might argue that if we cannot presently model climate then we should at least try to measure it. Observations of a whole host of parameters describing climate have been collected for centuries. These observations and other indirect but equally valid ones definitely show that climate varies on a short time scale and can change significantly on longer time scales. These facts have been especially well summarized elsewhere in this volume and in two recent reports, *Understanding Climatic Change: A Program for Action* (U.S. Committee for the Global Atmospheric Research Program, 1975) and *The Physical Basis for Climate and Climate Modelling* (WMO/ICSU, 1975).

These documents recommend many new monitoring observations from satellite platforms. Space technology has made truly global observations possible for the first time. Because spacecraft carry the same instruments over different parts of the earth, regional differences and variations should be more easily detected. Despite these important advantages there are also significant limitations. Obviously, spacecraft orbit well above the earth's atmosphere, and certain common climatic parameters such as temperature can only be inferred from the electromagnetic radiation emanating from the atmosphere below. One can hardly expect to achieve the same intrinsic accuracy that an *in situ* thermometer would achieve. On the other hand, other quantities such as the extent of sea ice and possibly even the thickness of sea ice can, at least in principle, be far better determined from space than is economically feasible using observers on the earth.

With these new possibilities for climate monitoring, will we be able to measure man's influence on climate? Perhaps, but most likely not. The difficulty arises from the fact that changes in climate that are significant in their effects on man may be scarcely detectable on a global scale. The situation is not quite so difficult on the regional scale. Regional changes are often larger in magnitude but compensated for by changes in the opposite direction in other areas. Thus small global changes that are difficult to measure may be manifested in regional shifts that we can observe quantitatively.

Even though our ability to monitor the climate of the earth is very much better now than ever before, and even though this new capability will make it possible to obtain a better set of observations of many key climatic parameters, it does not appear possible at present to *measure* how man is changing the *global* climate. On the other hand, it may be possible to measure *regional* climate changes because these changes are larger. We are already certain that we can observe changes in *local* climate because that has already been done without space platforms. We probably can obtain certain local observations even better with them. On the local scale, we may even be able to separate the changes caused by man from those caused by nature.

The most productive approach to devise a climate observing system is through a combination of modeling and monitoring. One of the most successful accomplishments of the Global Atmospheric Research Program (GARP) so far is the clear specification of what is required to describe the initial state of the atmosphere so its short-term transient behavior (i.e., the weather) can be predicted. Numerical atmospheric simulation schemes, often called numerical models, require that the state of the atmosphere be specified at some initial time for several levels over several thousand grid points spaced over the earth. The specification of the atmospheric state can be real (from measurements) or fictitious (from guesses) or both, but the data set cannot be empty or only partly filled. A model that simulates atmospheric behavior is an especially powerful tool not only because it can predict the future weather, but, equally important, it specifies what observations are the *key* ones. Moreover, sensitivity tests with the model can be used to learn how good, how often, and how closely spaced these observations must be. These modeling tools have had a great influence on our meteorological satellite program.

What can be measured from space must not only be novel, it must also be useful. Sensitivity tests indicate just how useful the observations will be. For example, our ability to observe the atmosphere's initial state is now considered promising enough—on the basis of model experiments—to warrant a large international cooperative program—the First GARP Global Experiment (FGGE)—which will be conducted in 1978-1979.

The status of climate modeling in the late 1970's is not so fully developed as weather modeling was in the late 1960's when GARP was first proposed. The problem is that we do not have our theory of *climate* in as good order as our theory of *weather* was at that time. Stated simply, the short-term future behavior of our atmosphere depends on its present physical arrangement and almost unchanging boundary conditions. But the present state of the atmosphere is unimportant for climate because the atmosphere soon "forgets" its present state. Its statistical behavior depends more on the boundary conditions, and the transient behavior of the atmosphere can slowly change the boundary conditions. Such feedback mechanisms can be positive or negative.

In summary, models give us the basis for determining what observations are required for the study of climate and how well they must be obtained. These same questions have been looked into in detail at several study conferences. What follows has been extracted from those conference reports at which the author was a participant. These have been summarized and added to in an attempt to present the latest consensus on the observing requirements. In doing this, the author acknowledges the contributions of the other participants in these meetings (listed in the referenced reports) and assumes full responsibility for any change in emphasis, deliberate or inadvertent, that such a synopsis may incur.

If one takes the time to read all the recent documents on the requirements for a global climate observing system, one comes away with two strong impressions. First, these documents do not read like novels. Secondly, there is a considerable difference between what the modelers want and what has been proposed as feasible by those familiar with how the observations might be obtained. Part of this confusion stems from the lack of a satisfactory theory of climate and part from our ignorance of the best approach to obtain the understanding needed to synthesize such a theory.

The author proposes to use as a framework the latest document on this subject, the report of the October 1975 Tokyo meeting of the Joint Organizing Committee for GARP (ICSU/WMO, 1975). First, it represents in my view the clearest statement of the best strategy to be used in developing an understanding of climate. The strategy, of course, will control the observation program. Secondly, it provides a useful definition of the difference between observations that are strictly climate monitoring and those observations needed to understand and test the models, i.e., their representation of the physical basis of climate. Some detailed proposals for components of a monitoring program will then be made where appropriate.

As a basis for observing climate and its determining factors, we must adopt some definition of the climate system. One conceptual scheme that is widely accepted distinguishes between an *internal* system and an *external* system. The internal system includes those variables whose interrelationships are well enough understood to be modeled quantitatively. Specifically, the following may be termed internal variables:

The atmosphere's dynamic quantities (wind, temperature, pressure, humidity);
Clouds and precipitation;
The motions of the world ocean;
The formation and motion of sea ice;
The hydrological cycle;
The biomass.

In contrast, the external system consists of those quantities that cannot now be modeled and predicted quantitatively. They must therefore be monitored and specified as fixed conditions in numerical models. Some of these are truly external for time scales up to a century or so:

Solar flux;
Surface characteristics such as land or ocean bottom, topography roughness, vegetation albedo, ice-sheet configuration, etc.

Other factors should really be dealt with as part of the internal system but will, for the time being, be considered as external parameters in the climate models either because of inadequate knowledge at present about their proper treatment as internal variables or because such a separation of the problems for the time being appears feasible. These include the following:

Atmospheric aerosols;
Optically active minor constituents in the atmosphere, particularly carbon dioxide and ozone.

It is understood that these working definitions will be modified in the future when our understanding of sources, sinks, and transport mechanisms will allow prediction of the time variations of these factors as part of the internal system.

If we are concerned with the shortest time scale, i.e., annual and interannual variations, it may be sufficient to consider only the uppermost part of the oceans as belonging to the internal system. In such a case even the basic characteristics of the deep-sea circulation would be assumed not to change and to be given as external parameters in the model.

These internal and external elements form a single climatic system. We define the climatic state as the average behavior of this system, as characterized by the statistics of its variables over some specified period of time in some specified domain of the earth-atmosphere system. The time interval is understood to be considerably longer than the life span of individual weather systems and longer than the period over which the behavior of these systems can be predicted.

A climate monitoring system must really observe both the external system *and* the internal system. The reasons for the first have already been given. The reason for the second is that we must monitor the climate *state* to assess how the model is performing and how it might be improved. The documents we have referred to earlier call this category of observations global data sets.

Now if we can refine our specification of both the external parameters and those that account for the climate state into two unique classes, man-originated and nature-originated, we have a basis for separating the inputs and thus for determining the magnitude of change in the outputs of our model. These outputs might be very sensitive to man-made inputs if there is positive feedback in the model (and in nature), or man-made inputs might result in a trivial change in output if man's contribution compared with that of nature is small or if there is negative feedback in this part of the model mechanism.

There are several ways in which man's input can be monitored. One can add a *source inventory* to the data base. This approach has been successful in pollution control. In other instances, trace gases or particulates in the atmosphere are uniquely man-made and can be directly measured. The global inventory is still difficult in this latter case because of the enormous dilution of the atmosphere. In still other instances one can institute patrols from space platforms. Forest fires, industrial particulate plumes, thermal pollution in waterways, and other small-scale sources might be detected from space because concentrations are so large and meaningful signals can be obtained even from the distances needed for space surveillance. The idea here is to measure the parameters *before* they are diluted by the atmosphere. The advantage of the patrol approach is that the signals are large on the regional and local scale, but they are exceedingly small on the global scale.

An extremely important aspect of the entire climate monitoring activity is the data-processing effort required. It is possible for the secrets of nature to be hidden in a flood of data as well as in nature. Clearly, we need *information* more than we need data.

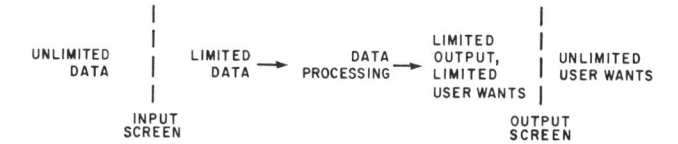

FIGURE 8.1 Design of a typical environmental data system.

Even a superficial assessment of the data presently needed for adequate climate monitoring is staggering. A single day's worth of images from only *one* geostationary satellite will yield 10^{11} bits of information. This number is large enough to record the name and address of every person on earth with enough space left over to give each one a telephone number besides. Figure 8.1 shows the design of a typical environmental data system that is used to screen data gathered mainly for other purposes, i.e., as weather forecasting.

In the design of a climate monitoring system there are two things wrong with the diagram. First, in typical use the flow is from left to right. However, in the *design* of a system, *the task we are considering now*, the order should be just the reverse. What is *desired*, i.e., the *output* requirements of the system, should be considered first. Then and only then will it be possible to say what data are required to get needed information. Designing a system in the form of Figure 8.1 does not provide any assurance of getting the needed information.

Secondly, the data-processing system cannot be independent of the data-collection system. In the case just discussed, an imaging geostationary satellite, there are extremely large quantities of data with low *information* density. However, for most purposes, small amounts of data having high information density are required. To assemble such subsets of data, it is necessary to sort through the entire data set, retaining information along a certain "path" through it. Obviously the various paths that can be taken to increase the information density depend on the information desired, i.e., time, space, parameter. In fact, as will be shown later, knowing what data are desired can greatly simplify the climate monitoring systems. In the system depicted in Figure 8.1, one typically uses large computers and batch processing. In a well-designed data-processing system, a small minicomputer or microprocessor can be an integral part of the data system. The cost and usefulness differences between these two routes can be enormous. The cost ultimately controls what is actually possible. Requirements for a viable climate monitoring system must take account of this fact. Debate and negotiation between the modelers who want the information and the observers who will design and operate the system that will collect it is absolutely essential. Some requirements may since have been relaxed, while others may have been strengthened.

We have made these remarks to indicate the formidable task that faces us. We do so to warn the technologists who are anxious to get on with the task and to gain some sympathy and understanding from the modelers who are so anxious for the information. Clearly, the dialogue that has started in the climate program planning sessions must be continued; the interface between what modelers desire and what technology can provide is not sharp and clear. In some instances, it is foggy and some might even say murky. Members of each group must make the effort to reach some distance into the other's area. Only if effective negotiation goes on between the two basic groups will there evolve a system that meets the needs of the program and that can be held within the bounds of available resources. We have already demonstrated a capability to do this in preparation for the FGGE. The requirements for a monitoring system are exceedingly complex. This complexity is demonstrated in Table 8.1, which was adapted from the summary of the Joint Organizing Committee (JOC) on the formation of data sets needed for studies in climate dynamics. We have simplified the table to show which data need be included in the global data sets and where in GARP Publications Series No. 16 (WMO/ICSU, 1975) these requirements and possible solutions can be found.

TABLE 8.1 Adapted from Summary of the JOC Recommendations on the Formation of Data Sets Needed for the Climate Dynamics Subprogram

Variable	Reference to Tables in GPS No. 16[a]
Total solar flux	6.2
Solar uv flux	6.2
Net radiation budget	6.2
Cloudiness	6.2
Sea-surface temperature	6.3
Surface albedo	6.3
Precipitation over oceans	6.4
Soil moisture	6.4
Water runoff	6.4
Heat content of the upper layer of ocean	6.5
Wind stress	6.5
Sea level	6.5
Near-surface currents	6.5
Deep-ocean circulation	6.5
Extent of snow	6.6.1
Extent of sea ice	6.6.1
Sea-ice melting	6.6.1
Drift of sea ice	6.6.1
Thickness of polar ice sheets	6.6.2
Deformation of polar ice sheets	6.6.2
Change of boundary of polar ice sheets	6.6.2
Water vapor	6.7
CO_2	6.7
Ozone distribution	6.7
Tropospheric aerosols	6.7
Atmospheric turbidity	6.7
Stratospheric aerosols	6.7

[a] WMO/ICSU (1975).

Some of these parameters that may have a man-made component and a few ideas on how they could be measured are discussed below.

Net Radiation Budget All components of this can be measured accurately from space. To detect variations significant for climate, emphasis should be placed on stable instruments and vehicles with long useful lifetimes. Man contributes to the radiation budget by release of heat and by changing surface characteristics, topics treated elsewhere in this volume. These can be monitored to some extent from space. Although trivial in terms of energy, artificial light may be statistically related to total energy release with sufficient reliability to be a useful monitoring tool, since artificial light can be easily detected from space at night. This approach might provide a simple means for monitoring changes in human energy consumption patterns.

Surface Albedo Man changes surface albedo through deforestation, urbanization, grazing, agriculture, etc. These changes also affect evapotranspiration and surface roughness. All of these factors can be monitored by multispectral remote sensing.

Soil Moisture This can be roughly determined from space by passive microwave radiometry. Irrigation, agricultural practices, and large-scale hydrological works of man have significant impact on soil moisture, which in turn affects the albedo, surface temperature, and moisture flux relevant to climate. Both human activities and their consequences in terms of soil moisture should be monitored.

Water Runoff River flow statistics are sensitive indicators of climate variations over the continents and also are influenced by man-made changes in the land. It is difficult to acquire such data directly from space, but satellite communication may make feasible the collection of data from isolated locations.

Carbon Dioxide It is not possible to identify uniquely the contribution of man except by source monitoring and spectral analysis. Of particular value in carbon dioxide monitoring will be measurement of the vertical distribution on a global basis with a view to identification of sources, sinks, and transport mechanisms. This knowledge might clarify the roles of the land biota and the oceans in the atmospheric carbon cycle.

Ozone There is no significant contribution of ozone by man on a global scale. However, ozone is critical in the human environment because of its role in screening out damaging ultraviolet components of the solar system. Other trace gases can act as catalysts to reduce ozone concentration. It is therefore important to monitor this gas.

Tropospheric Aerosols These represent a mixture of natural and man-made particles and can be monitored from space. Some estimates of the proportion due to man may be possible by relating contrast changes due to aerosol loading with known patterns of human activities.

Atmospheric Turbidity This can be monitored through the depolarization of sunlight by aerosols [see, for example, page 208 of *Remote Measurement of Pollution* (NASA, 1971)]. It is also possible to monitor turbidity from the changes in apparent contrast of surface targets of known intrinsic contrast [McLlellan in NASA (1971)].

Stratospheric Aerosols Concern has been expressed on possible increases in stratospheric aerosols due to high-altitude aircraft operations. Limb-scanning techniques in the infrared can be used to search for features due to aerosols. Measurements of the solar disk and the aureole in two wavelengths can give both the real and the imaginary parts of the refractive index, together with particle size.

Trace Gases Limb scanning with an interferometer-spectrometer as done by Rudolf Hanel of the NASA Goddard Space Flight Center (personal communication) in planetary investigations can provide information.

REFERENCES

ICSU/WMO (1975). Report of the Eleventh Session of the Joint Organizing Committee, Tokyo, Oct. 1-8, 1975, Global Atmospheric Research Programme.

National Aeronautics and Space Administration (1971). *Remote Measurement of Pollution*, Scientific and Technical Information Office, NASA, Washington, D.C.

U.S. Committee for GARP (1975). Panel on Climatic Variation, *Understanding Climatic Change: A Program for Action*, National Academy of Sciences, Washington, D.C.

WMO/ICSU (1975). World Meteorological Organization and International Council of Scientific Unions, *The Physical Basis of Climate and Climate Modelling*, GARP Publ. Series No. 16, Geneva.

Modeling and Predictability

9

JOSEPH SMAGORINSKY
Geophysical Fluid Dynamics Laboratory, NOAA; Princeton University

9.1 INTRODUCTORY REMARKS

It is generally the aim of physical science to construct models that are capable of reproducing observational facts. One then has some confidence that the body of physical laws that constitute the model can be used to predict states for differing circumstances, that is, for different initial or boundary conditions, or for different values of the parameters of the model. In principle, the more fundamentally constructed the model, the broader its range of validity in parameter space. These considerations are directly relevant to the problem of climate and climatic variation, within the inherent deterministic limits of the problem. But even the very question of determinism (or intransitivity) can, in principle, be investigated with such models.

One would like to have the capability of answering in some usefully reliable way the "what if?" questions raised by various energy utilization alternatives.

- How sensitive is climate to the release of particulates, gases, and heat and to changes in the characteristics of the earth's surface that may result from a particular mode, or some combination of modes, of energy utilization?
- Is a megalopolis source better or worse than a more uniform distribution?
- Would the climate today be materially different if the industrial revolution had never happened?

Which brings one to the question: Do we really understand "natural" climate and its many time scales of variability spanning from month to month to year to year and beyond? This last question is fundamental to a large class of climate-related problems of which energy use is one important example. Others include the optimization of food and water resources and the impacts of supersonic transports, chlorofluoromethanes, or even nitrate fertilizers.

Underlying these questions is a scientific problem: to develop a fundamental understanding of the mechanisms and dynamics of climate. As such, it has, in the past six years, received extraordinary attention by national and international bodies. These bodies have assessed the current scientific base to determine what can be done to accelerate its development. There are at least three such committees in the National Research Council. Recently, the U.S. Domestic Council (1974) has addressed the question. Among international conferences were the Stockholm Conference on the

Study of Man's Impact on Climate (Matthews et al., 1971) and a comprehensive scientific study conference on climate (GPS No. 16, 1975) that was organized by the Joint Organizing Committee for GARP and sponsored by the World Meteorological Organization, the International Council of Scientific Unions, and the United Nations Environment Program. The conclusions of these bodies generally coalesce about the following points:

Climate modeling seems to be the most clearly promising path.

We are fortunately already well along the way as a result of scientific initiatives of two decades ago, but models are still too simple to answer subtle questions.

One can identify many if not most of the missing pieces of the jigsaw puzzle.

New observational data are required to inspire, advance, and verify improved models.

There is a need for technological innovation, faster computers, and possibly new institutions.

It will be useful to consider the physical basis for climate modeling, to expose some critical problems impeding model development, particularly those relevant to energy utilization, and to show some results of a recent simulation experiment as a vehicle for discussing common misconceptions.

9.2 ON THE INGREDIENTS OF A CLIMATE MODEL

An exposition of the elements that enter into a comprehensive climate model is treated at length elsewhere (Smagorinsky, 1974). Our present purpose is to outline briefly the degree to which we are presently capable of modeling each of these physical processes (see Figure 9.1).

The most extensive experience lies with modeling the large-scale three-dimensional hydrothermodynamics of the global atmosphere. Although constant improvement is still being achieved, it is no longer a critical factor in general circulation modeling. Smaller-scale convective transfer, although a current subject of concentrated research activity, is well enough at hand in general circulation models that simulations with such models of the long-term dispersive characteristics of inert tracers in the atmosphere show reasonable correspondence with observation (Mahlman, 1973). Radiative-transfer theory for a given distribution of the radiatively active constituents carbon dioxide, ozone, and water vapor seems to be adequate. However, arbitrary distributions of clouds and other aerosols still cannot be dealt with in full generality.

The elements of the hydrologic cycle have been modeled with moderate success. The ability to predict the atmospheric water-vapor distribution seems to be sufficient for calculating infrared radiative absorption (the greenhouse effect) but not for determining the formation of clouds. Simple engineering parameterizations seem almost adequate for determining continental water storage, that is, soil moisture. Also, an ability to model variations in continental snow cover gives a reasonable first approximation. This is particularly important in determining changes in surface reflectivity (albedo).

The oceans play a key role in virtually all questions of climatic interest. Coupled ocean–atmosphere models are still in a crude state of development since their first construction in the mid-1960's. An understanding of the mechanisms governing sea-ice variations and how this alters the transmission of heat between ocean and atmosphere is still to be adequately modeled.

As indicated above, an ability to predict the cloud stage with adequate precision to determine the radiative consequences remains one of the most difficult problems in climatic modeling.

The above elements are all essential to a model presumably capable of assessing the sensitivity of climate to thermal pollution. Furthermore, if one wishes to assess the consequences of other industrial effluents, that is, particulates and carbon dioxide, additional elements are required to be determined by models. The CO_2 buffering mechanisms in the ocean and biosphere are not yet fully understood. More-

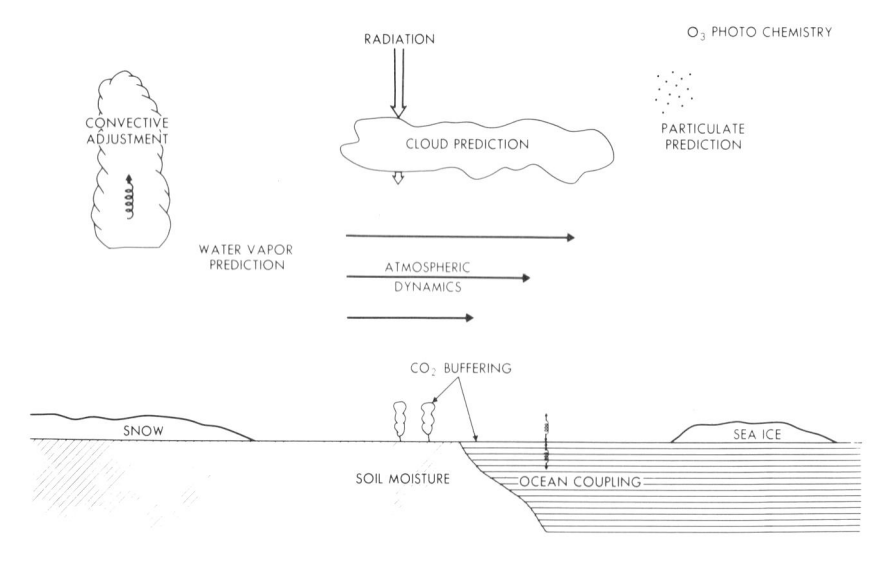

FIGURE 9.1 A schematic representation of the elements that enter into a model of the "climate system."

over, the kinds of particulates, their optical properties, and their source–sink mechanisms are yet to be determined. However, if the modeling requirements to assess effects of thermal pollution are met, a first estimate of climate sensitivity can be made by assuming a discontinuous change in the CO_2 or in the particulate distribution, without trying to understand how it can be maintained by the "climate system."

Finally, the indirect influence of supersonic transports and chlorofluoromethanes on climate through ozone sensitivity depends on a sufficiently correct understanding of ozone photochemistry. This is a rapidly maturing field.

9.3 A CLIMATE EXPERIMENT—CIRCA 1971

One is tempted to presume that climatic change is primarily the result of extraterrestrial influences. For example, the most spectacular of all sensible climatic changes is the seasonal cycle in response to solar radiation changes as an orbital consequence. Nevertheless, the complex earth–atmosphere–ocean–cryospheric system (the "climatic system"), because of its highly interactive nonlinearity, could conceivably be responsible for all past evidences of climatic change, including ice ages. This could be the result of a subtle interplay of positive and negative feedbacks with a variety of relaxation times. This would mean that for fixed external boundary conditions, a unique stable statistical equilibrium does not exist. This possible intransitivity is only suspected now; hard experimental substantiation is yet to be achieved. We shall return to this question later.

Nevertheless, one might ask what would be the result, in the so-called statistical equilibrium, if one of the external conditions were to be changed discontinuously? Recognizing the risk in prematurely trying to ask such a question, let us consider an example—an experiment conducted several years ago by Manabe and Wetherald (1975).

What is the climatic consequence of a CO_2 increase of a factor of 2, in a model in which CO_2 is specified, that is, not self-determined? First, let us note that the observed increase of CO_2 in this century has been 10 to 15 percent. It is estimated that approximately this much has also been buffered by the oceans. In the calculations we shall discuss, it has been assumed that the solar constant and the distributions and amounts of cloud and ozone do not change. It is assumed that the CO_2 concentration in two different simulation experiments (a control and a perturbation experiment) is doubled from 300 to 600 ppm.

The experiment was performed with what is already a physically very sophisticated three-dimensional model, which has succeeded in simulating, with actual geography, many of the details of the contemporary general circulation and climate (Manabe et al., 1974). This model accounts for the complete water-vapor thermodynamic interaction (including radiation as well as released heat of condensation). It also assumes an idealized distribution of oceans and continents but with no heat storage by continents or oceans (the swamplike ocean has water available for evaporation) and with no transport by the ocean. Snow and soil moisture are self-determined over continents, and sea ice is self-determined over the ocean. Reflectivity (albedo) depends on the nature of the earth's surface according to empirical criteria.

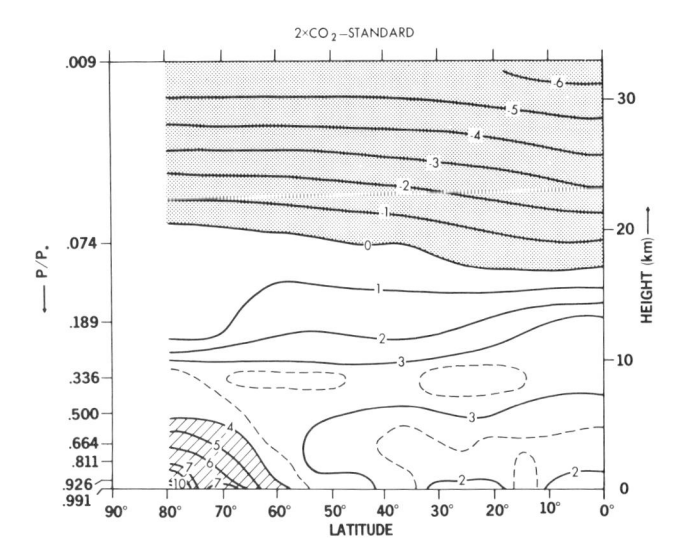

FIGURE 9.2 The latitude-height distribution of the zonally averaged temperature difference (°C) between the statistical equilibrium of a general circulation simulation with the CO_2 concentration set at 600 ppm and one set at 300 ppm. This is a fully three-dimensional model with idealized geography (after Manabe and Wetherald, 1975).

When a new quasi-equilibrium is established after doubling the CO_2 (Figure 9.2), there is general warming in the model troposphere because of an increase of the CO_2 greenhouse effect. Particularly, there is a 2.9 °C global average surface temperature rise with the main increase poleward of 60° latitude. On the other hand, the stratospheric temperature decreases. Incidentally, the reduced meridional temperature gradient resulted in a reduced intensity of baroclinic instability or storminess, the maximum of which moved poleward with the receding perimeter of snow cover.

We must remember that seasonal variability and the effects of the oceans and cloud reaction have not been accounted for and could materially alter, if not reverse, the conclusions.

This model already possesses the mechanisms for both a water-vapor greenhouse positive feedback and a snow-cover positive feedback, which have enhanced the CO_2 greenhouse effect. It is useful to discuss qualitatively the physical nature of these destabilizing mechanisms.

The water-vapor greenhouse feedback chain can be reasoned purely in terms of the interaction between the infrared radiation field, the water-vapor content, and the convection in the presence of buoyant instability. Everything else being equal, one can reason that

• An increase in temperature throughout the troposphere results in an increase in water vapor content,
• Which increases the absorption of infrared (ir) radiation from below (the "greenhouse"),

- Which in turn increases the *in situ* temperature and therefore the back-ir radiation,
- Which yields an increase of temperature in the lower troposphere and the boundary below—a positive feedback,
- Which gives rise to convective instability,
- Which distributes the heat throughout the troposphere but which reduces the magnitude of the surface temperature increase (a partially compensating negative feedback).

The snow-cover feedback involves only the interaction between snow cover, its effect on reflecting solar radiation, and the net effect on the resulting temperature regime. Reasoning as before, we have that

- An increase in atmospheric temperature
- Decreases the area of snow cover,
- Which decreases the albedo of the earth's surface and therefore increases the absorbed solar radiation,
- Which further increases the atmosphere-earth temperature at the latitude of reduced snow perimeter.

This argument is generally reversible for a temperature decrease.

9.4 A CRITIQUE ON WHAT WAS MISSING

Many of the interactive degrees of freedom of the actual terrestrial climatic system were not included in the above model. To name some of the more obvious mechanisms:

CO_2-OCEAN BUFFERING

An increase of atmospheric CO_2 increases the surface temperature of the sea and at the same time will reduce the ocean's capacity to buffer CO_2. Hence, a given CO_2 source rate of increase would yield a greater rate increase of atmospheric CO_2 concentration and hence surface temperature. We do not yet have a parameterization to represent this interaction adequately (see Chapter 4).

CLOUD REACTION

The cloud stage is generally negligible in the water-vapor transition to precipitation, as far as the water and released latent heat budgets are concerned. The importance of clouds comes from their role in reflecting and attenuating solar radiation. Although clouds can be accounted for by simple parameterizations drawn from contemporary terrestrial observations, they must be inadequate to describe systematic large changes of cloud type and distribution that may be associated with large excursions of climatic regime. Yet, the magnitude of seasonal change (or alternatively the interhemispheric difference in January or July) of surface temperature is of the same order of magnitude (about 10 °C) and, therefore, within the scope of actual observation. There is, therefore, some independent empirical check of a cloud parameterization over a significant span of parameter range but probably not enough to cover the extremes of *complete* glaciation to *complete* deglaciation.

A general radiation algorithm for an arbitrary distribution of clouds has yet to be developed. Equally important is the development of a means to predict the large-scale distribution and variability of clouds as the transitional stage between unsaturated water vapor and falling precipitation.

Empirically, stratiform cloud amount varies proportionately with relative humidity (Smagorinsky, 1960), the rate being greater for high clouds. In this sensitivity experiment the intensity of the hydrologic cycle (precipitation and evaporation) increased with increasing CO_2. The relative humidity at the low level increased by about 2 percent, at the middle level decreased by about 1 percent, and at the high level decreased by about 2 percent (Figure 9.3).

However, let us note that an increase in middle and low clouds

- Increases the atmospheric albedo,
- Decreases net downward solar radiation,
- Cools the atmosphere-earth-ocean system,
- Decreases surface temperature.

On the other hand, an increase in high clouds because of its low albedo and low emission temperature

- Increases the absorption of solar radiation,
- Decreases the net outgoing radiation,
- Heats the atmosphere-earth-ocean system,
- Increases surface temperature.

Therefore, the increased low-level relative humidity and the decrease at high levels each contributes to *decreasing the surface temperature*—qualitatively a negative or stabilizing feedback.

A possible parameterization of stratiform cloud amount based on the relative humidity-cloud correlation is very

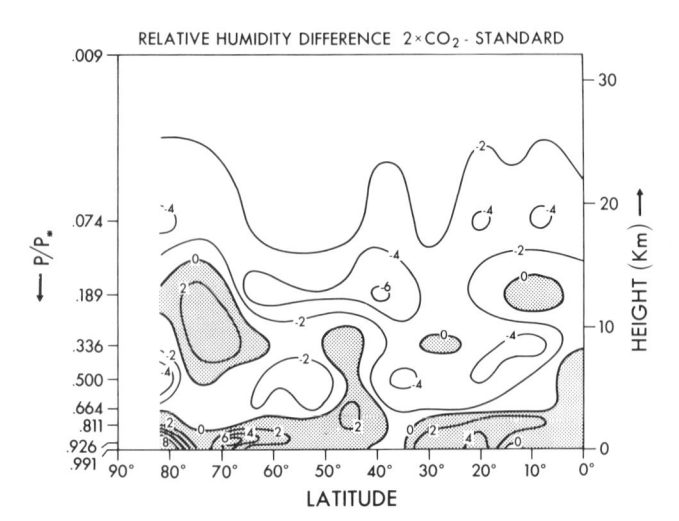

FIGURE 9.3 The latitude-height distribution of the zonally averaged relative humidity difference (%) between the statistical equilibrium of a general circulation simulation with the CO_2 concentration set at 600 ppm and one set at 300 ppm. This is a fully three-dimensional model with idealized geography (after Manabe and Wetherald, 1975).

sensitive. The parameterization (based on Smagorinsky, 1960; Manabe, 1970) yields the following radiative-convective equilibrium surface-temperature dependence on stratiform cloud amount: low clouds −0.82 degree/percent, middle clouds −0.39 degree/percent, and half black high clouds +0.04 degree/percent. Therefore, a 1 percent error in the low-level humidity would give an error in the equilibrium surface temperature (for a surface with no heat capacity) of over 2½ degrees, which is comparable with the magnitude of significant climatic change. Nevertheless, this parameterization may be used to provide an order-of-magnitude estimate of the second-order correction in the Manabe-Wetherald experiment that is due to stratiform clouds: −10 °C due to low-level clouds, +1 °C due to middle clouds, and −0.1 °C due to high clouds. This suggests an overriding compensation by low clouds. Hence the cloud interaction is highly nonlinear and may completely damp the snow-cover and water-vapor greenhouse instabilities resulting from a twofold increase of CO_2. On the other hand, if the sea-surface temperature were specified, implying a boundary of infinite heat capacity of half of the area in this model, the effect of variations in cloudiness as well as CO_2 would be almost insignificant.

Finally, one must reiterate the absence in the above line of argument of how cumuloform clouds might react. One would expect the height of such clouds to increase with the increased intensity of the hydrologic cycle, thus reducing the upward terrestrial radiation at the top of the atmosphere and raising the surface temperature. It all points to the fragility of quantitative and even qualitative conclusions.

AEROSOL INTERACTION

The role of aerosols, such as dust, sea salts, or sulfates, is largely unknown. We have yet to monitor adequately the distribution and variability of the complex of aerosols and to understand the processes that determine the distribution and variability of each of the constituents. In addition to the transport properties of the atmosphere, one needs to know the processes responsible for the nonconservatism of each aerosol's life cycle: the sources and sinks and the phase or chemical changes. Furthermore, upon identifying the aerosols, their optical properties must be determined and appropriate radiative algorithms devised.

Although climatic impact can sometimes be determined empirically, such as in the case of volcanic dust during massive volcanic eruptions (Newell, 1970), one probably will need to resort to simulation techniques in general.

It might very well be that their indirect effects on cloud formation may be aerosols' most important climatic consequence.

OCEANS VERSUS CONTINENTS

Assuming the earth were all ocean-covered with known sea-surface temperature implies infinite heat capacity of the lower boundary, which greatly damps the response of the ocean-atmosphere system to changes in solar radiation, atmospheric composition, or albedo at the lower boundary (see, for example, Washington, 1972). On the other hand, assuming a continent-covered earth with infinitesimal (if not zero) heat capacity provides an earth-atmosphere system that is hypersensitive to changes in solar radiation, atmospheric composition, or albedo at the lower boundary (as is the case in the Manabe-Wetherald experiment).

The real terrestrial situation is somewhere in between. There are oceans and some continents, each with finite but greatly differing heat capacities. The relatively long thermal relaxation time of the oceans is of particular importance with respect to the forced seasonal heating-cooling cycle.

Much of the heat that impinges on the oceans is used to evaporate water rather than to raise the sea-surface temperature. This was taken into account in the Manabe-Wetherald experiment. On the other hand, one must also point out a compensating asymmetric response to heating and cooling. Buoyant instability is activated discontinuously in the atmosphere or ocean when either is sufficiently heated from below or cooled from above. For this reason, everything else being equal, it is easier to decrease the surface air temperature over a continent; and because the heat capacity of the ocean is much larger than that of the atmosphere, it is easier to increase the interfacial temperature over the sea, an effect not accounted for in the Manabe-Wetherald model.

An examination of the available observed seasonal interhemispheric temperature variation (Van Loon et al., 1972) shows that both the annual mean and especially the July-to-January range of midlatitude surface temperatures are smaller in the southern hemisphere. Furthermore, the summertime difference between the hemispheres is about twice as great as the wintertime difference. Since there is more ocean in the southern hemisphere, one would, therefore, conclude that in the absence of other considerations the evaporative effect dominates the compensating effect of buoyant instability. These considerations are further complicated if the earth's surface can become snow or ice covered, especially the sea, for then the communication for heat exchange between the hydrosphere and atmosphere can be all but cut off.

Finally, the oceans transport heat horizontally. Estimates (Vonder Haar and Oort, 1973) indicate that the poleward transfer by the oceans is comparable with that by the atmosphere in meeting the total radiative requirement.

One, therefore, needs coupled ocean-atmosphere models to consider reaction times beyond a month. The reacting depth of the ocean increases with increasing time scale. Such models are under various states of development at a number of research institutions (see, for example, Manabe et al., 1975; Bryan et al., 1974). The fundamental difficulty in constructing ocean models is that the main baroclinic eddies are small in their horizontal dimensions compared with those in the atmosphere, and the computational detail to deal with them explicitly is prohibitive. One has yet to construct adequate means for dealing with their dynamical properties statistically, that is, to devise closure schemes for their parameterization. Conversely, the oceanic time scales are much larger than that of the atmosphere, so that, if one wishes to resolve characteristic atmospheric energetic cycles, long climatic simulations for even hundreds of years require prohibitive amounts of computation time. However, in this, some progress has been already made.

In general, the oceanic heat storage and transport properties can be expected to stabilize climatic sensitivity, although there may be important counterexamples, especially for shorter periods [e.g., Namias's work (1970) suggests some positive feedbacks].

SEASONAL VARIATION

Missing in the CO_2 calculations was the seasonal cycle. The annual average solar radiation was assumed. This is of fundamental importance since the position of the snow perimeter is the net result of winter snow fall and summer melting. These seasonal fluctuations, in turn, are much greater when there are continents at the higher latitudes. Thus there is a fundamental difference between our northern and southern hemispheres. However, it is not yet clear whether the seasonal variation taken together with a correct account of the earth's ocean–continent distribution would intensify or diminish the climatic response to a carbon dioxide increase.

9.5 PREDICTABILITY OF CLIMATE

The Manabe-Wetherald CO_2 "sensitivity experiment" is typical of a class of experiments that asks what the result is of a discontinuous controlled or "external" change in boundary conditions (such as solar radiation, heat from sources below, or albedo) or in composition (such as CO_2 and particulates). It assumes that a new unique quasi-equilibrium "climate" is established. However, since the real atmosphere, as well as a model, is highly nonlinear, the validity of this assertion of transitivity is suspect and is, in itself, an important problem in establishing the limitations on the predictability of climate. One must ask, does a statistical equilibrium ever really exist, or if one waits long enough will a new state evolve; and what is long enough? It should be kept in mind that the magnitude of critical sensitivity to climate change (e.g., several degrees) is within the noise level of the natural variability of the atmosphere and of the present simulative precision of models.

9.6 SOME COMMENTS ON SIMPLE CLIMATE MODELS

In discussing the state of climate modeling, it was implicitly assumed that so-called simple climate models will, in principle, be derived from the knowledge obtained from more complex models. This is largely dictated by our inadequate observed data base on climate within the context of the general circulations of the atmosphere and oceans. In a sense, we use the results of comprehensive model simulations, where necessary, as a substitute for observation in gaining insight into the dominant operative mechanisms.

By this procedure, we can decide when and how we can simplify the models and still have meaningful and useful predictions.

In any case, simple models are didactically useful in isolating the nature of the interaction of particular sets of processes. However, unless it can be satisfactorily shown that other candidate processes are negligible in the parameter range of interest, one may not extrapolate to quantitative or even qualitative predictions of the real complex geophysical medium.

One looks for maximum conceptual simplicity to reduce the computation time for very long climatic simulations. In this respect, the major impact comes from reducing computational resolution; physical simplifications are relatively less effective, e.g., a factor of 2 of resolution in each of the three space dimensions is worth a factor of 16 in computation time.

It is, therefore, natural to look for a means to eliminate one of the spatial dimensions of the physical system by a suitable parameterization and thereby to increase greatly the time step. A favorite candidate is to parameterize the atmospheric baroclinic energy cycle, which has a characteristic time of tens of days, thus eliminating the longitudinal dimension.

It is still not certain whether physically consistent closure schemes are possible in principle. But even if they are, consistency of modeling precision may limit the admissible sophistication of the operative physical processes (e.g., in the radiation algorithms, ice dynamics, ocean coupling) below the threshold, where meaningful sensitivity experiments can be performed.

An alternative approach is to compress the time dimension by taking advantage of the ocean's enormous thermal inertia, thereby transforming a necessity to a virtue. Effectively, one determines the statistical properties of the atmosphere's driving mechanisms of the ocean by an explicit sampling of a detailed atmospheric simulation over a sufficiently long period, such as several years. These statistically determined fluxes of water substance, heat, and momentum are then applied as a boundary condition for driving the ocean over hundreds to a thousand years. This permits a corresponding enhancement of simulated evolution time of the model climatic system by a factor of several hundred. The computational advantage can be gained at little or no sacrifice to the physical complexity needed to describe adequately the model climate system. This approach, in principle, has already been employed in an early attempt to construct a coupled atmosphere–ocean model (Manabe and Bryan, 1969). But much future work is required to assess its shortcomings.

9.7 CONCLUSION

I have tried to expose the physical factors that enter into climatic modeling, the kinds of things that are possible today, their limitations for reliable conclusions at this time, and what the deficiencies in our modeling capacity are.

For the energy question in particular, the most immediately critical outstanding problems are ocean-atmosphere coupling, cryospheric dynamics, cloud feedback, and cloud-aerosol interaction. However, there is still a general requirement for refinement and sophistication of all model elements. Moreover, the transitivity properties of the climatic system must be better understood in order to design meaningful and convincing sensitivity experiments. At the moment, it appears that experiments on the effects of heat sources will be easiest to undertake. CO_2 problems will be more difficult, and those due to particulates the most difficult.

As was pointed out in connection with the earlier discussion of cloud reaction and ocean-continent contrast, the interhemispheric-interseasonal differences are of the magnitude of significant climatic change. A detailed definition and understanding of the contemporary seasonal and interannual-interhemispheric variability is essential.

The details of an orderly program of research to cover all of these needs, as well as those in general for a broad range of climatic sensitivity and stability questions, have been addressed authoritatively by the study conference in Sweden (GPS No. 16, 1975), as was indicated earlier.

The international Joint GARP Organizing Committee meeting in Budapest (JOC X, 1975) initiated a plan to implement accelerated development. It takes the form of a decade of definitive global observation and of modeling research in the 1980's. Between now and then, a great deal of preparatory work will be required.

There seems to be no clear shortcut for a careful and responsible attack on the problem. We should be wary of premature, hasty, and sweeping conclusions at this time.

REFERENCES

Bryan, K., S. Manabe, and R. C. Pacanowski (1975). A global ocean-atmosphere climate model: Part II. The oceanic circulation, *J. Phys. Oceanog. 5,* 30.

GPS No. 16 (1975). *The Physical Basis of Climate and Climate Modeling,* Report of the Study Conference on the Physical Basis of Climate and Climate Modeling, Stockholm, Sweden, 1974, World Meteorological Organization, Geneva, Switzerland.

JOC X (1975). *Report of the Tenth Session of the Joint Organizing Committee for GARP,* Budapest, November 1974, World Meteorological Organization, Geneva, Switzerland.

Mahlman, J. D. (1973). A three-dimensional stratospheric point-source tracer experiment and its implications for dispersion of effluent from a fleet of supersonic aircraft, in *Proceedings of AIAA/AMS International Conference on the Environmental Impact of Aerospace Operations in the High Atmosphere,* Denver, Colo., June 11-13, 1973.

Manabe, S. (1970). Cloudiness and the radiative, convective equilibrium, in *Global Effects of Environmental Pollution,* Proceedings of AAAS Air Pollution Session, Dallas, Tex., Dec. 1968, S. F. Singer, ed., pp. 156-157.

Manabe, S., and K. Bryan (1969). Climate calculations with a combined ocean-atmosphere model, *J. Atmos. Sci. 26,* 786.

Manabe, S., and R. T. Wetherald (1975). The effects of doubling the CO_2 concentration on the climate of a general circulation model, *J. Atmos. Sci. 32,* 3.

Manabe, S., D. G. Hahn, and J. L. Holloway, Jr. (1974). The seasonal variation of the tropical circulation as simulated by a global model of the atmosphere, *J. Atmos. Sci. 31,* 43.

Manabe, S., K. Bryan, and M. J. Spelman (1975). A global ocean-atmosphere climate model: Part I. The atmospheric circulation, *J. Phys. Oceanog. 5,* 3.

Matthews, W. H., W. W. Kellogg, and G. D. Robinson, eds. (1971). *Inadvertent Climate Modification,* Report of the Study of Man's Impact on Climate (SMIC), The MIT Press, Cambridge, Mass.

Namias, J. (1970). Macroscale variations in sea-surface temperatures in the North Pacific, *J. Geophys. Res. 75,* 565.

Newell, R. E. (1970). Stratospheric temperature change from the Mt. Agung volcanic eruption of 1963, *J. Atmos. Sci. 27,* 977.

Smagorinsky, J. (1960). On the dynamical prediction of large-scale condensation by numerical methods, in *Physics of Precipitation,* Monograph No. 5, American Geophysical Union, Washington, D.C., pp. 71-78.

Smagorinsky, J. (1974). Global atmospheric modeling and the numerical simulation of climate, in *Weather and Climate Modification,* W. N. Hess, ed., John Wiley and Sons, Inc., New York, pp. 633-686.

U.S. Domestic Council (1974). *A United States Climate Program,* Environmental Resources Committee, Subcommittee on Climate Change, Dec.

Van Loon, H., J. J. Taljaard, T. Sasamori, J. London, D. V. Hoyt, K. Labitzke, and C. W. Newton (1972). *Meteorology of the Southern Hemisphere,* Meteorological Monographs, Vol. 13, No. 35, Nov.

Vonder Haar, T. H., and A. H. Oort (1973). New estimate of annual poleward energy transport by northern hemisphere oceans, *J. Phys. Oceanog. 3,* 169.

Washington, W. M. (1972). Numerical climatic-change experiments: The effect of man's production of thermal energy, *J. Appl. Meteorol. 11,* 768.

The Carbon Dioxide Cycle and the Biosphere

10

ROGER REVELLE and WALTER MUNK
Scripps Institution of Oceanography, University of California, San Diego

10.1 INTRODUCTION

The industrial revolution and the accompanying rapid rise of human population have resulted in a flux of perhaps more than 200 Gt of carbon dioxide to the atmosphere. Two thirds of this quantity came from fossil-fuel combustion and one third from clearing of forests and other wildlands for agriculture. Only about 40 percent of the carbon dioxide produced in these ways remains in the atmosphere. Two models are developed that account for absorption of the remainder in the biosphere and the oceans. Both models depend on the assumptions that the increase in atmospheric carbon dioxide has caused net primary photosynthetic production to exceed oxidation of organic matter by heterotrophic respiration and fires and that absorption of carbon dioxide by the oceans has been limited up to the present time to about 20 percent of the total carbon dioxide produced because of the buffer mechanism of seawater and the slow exchange of surface and deep ocean waters. In our preferred model, we assume that the biomass that carries out photosynthesis is constant in amount. The equilibrium partition of carbon among the atmosphere, the oceans, and the biosphere is then uniquely determined by the total quantity of carbon in the system. Computations using this model indicate that the atmospheric carbon dioxide content could rise to about 5 times the preindustrial value in the early part of the twenty-second century.

10.2 OBSERVED SECULAR INCREASE IN CARBON DIOXIDE

From 1959 to 1973 inclusive, 51.8 Gt of carbon dioxide in terms of carbon were produced by the worldwide combustion of fossil fuels and cement manufacture (Chapter 4). Accurate measurements of atmospheric carbon dioxide at the South Pole and the Mauna Loa Observatory show that during this 15-year period the atmospheric carbon dioxide content increased by 13.8 parts per million, from 316.2 to 330.0 parts per million, or 4.36 percent, corresponding to an addition of 29.2 Gt of carbon to the atmosphere. The difference of 22.6 Gt between the quantity of carbon released by industrial activity and the quantity remaining in the atmosphere is 43.6 percent of the industrial carbon.

The Carbon Dioxide Cycle and the Biosphere

This difference can be accounted for only by absorption in the oceans and biosphere.

Carbon dioxide produced by industrial activity from 1860 to 1973 inclusive was equivalent to 128 Gt of carbon (Keeling, 1973; Chapter 4). If the proportion between carbon remaining in the air and that produced by fossil-fuel combustion during this 113-year period was the same as during 1959-1973, the carbon dioxide content of the air should have increased by 34 parts per million, from 296 to 330 parts per million, or 11.5 percent of the initial value, corresponding to 72 Gt of carbon. The actual increase may have been significantly greater. Although nineteenth-century data on the atmospheric carbon dioxide content are much less accurate than modern measurements, a best value for the middle of the nineteenth century is 290 parts per million (Chapter 4). The increase in atmospheric carbon dioxide up to 1973 is then 40 parts per million, corresponding to 85 Gt of carbon. This is 66 percent of the carbon added by fossil-fuel combustion and 13.8 percent of the carbon in the atmosphere in 1860. Industrial carbon dioxide production during 1860-1973 corresponded to 60 parts per million, or about 20.7 percent of the nineteenth-century atmospheric carbon dioxide.

As we shall see, carbon dioxide has also been released from part of the biosphere, primarily by clearing of forest lands for agriculture. The total quantity produced between 1860 and 1973 may have been between 70 and 80 Gt of carbon. Thus around 205 Gt of carbon (one third of the original atmospheric content) must be accounted for by partitioning among the atmosphere, the ocean, and the biosphere. In this chapter we present models of the atmosphere-ocean-biosphere system that are consistent with these data and provide a range of projections of future atmospheric carbon dioxide content.

10.3 PROCESSES IN THE TERRESTRIAL BIOSPHERE

The portion of organic matter in the terrestrial biosphere that exchanges carbon with the atmosphere consists of two components: (1) the biomass of living plants and animals, mostly the trunks, branches, roots, and leaves of trees, and (2) litter, detritus, and soil organic matter (humus). Whittaker and Likens (1975) have compiled and evaluated data on the biomass (see Table 10.1). They conclude that 90 percent of the total of around 830 Gt is in the world's forests, which cover nearly 50 million km². Tropical forests, with an area of 24.5 km², contain more than half of the total. Woodland and shrubland, savannas, grasslands, desert and semidesert scrub, swamps and marshes, and cultivated land together contain only 84 Gt, or 10 percent of the total biomass, although they cover nearly 75 million km². Net primary production of organic matter (photosynthesis minus plant respiration) is more evenly divided: forests produce 33 Gt of carbon per year, and all other vegetation produces nearly 20 Gt. These estimates correspond to an average efficiency of photosynthetic conversion of solar

TABLE 10.1 Estimated Standing Crop and Net Photosynthetic Production of the World Biomass[a,b]

Ecosystem Type	Area (10^8 ha)	Mean Net Primary Production (NPP) (tons of C/ha/yr)	Biomass (tons of C/ha)	Total NPP (Gt of C/yr)	Total Biomass (Gt of C)
Tropical rain forest	17.0	9.90	202.5	16.83	344.2
Tropical seasonal forest	7.5	7.20	157.5	5.40	118.1
Temperate evergreen forest	5.0	5.85	157.5	2.92	78.8
Temperate deciduous forest	7.0	5.40	135.0	3.78	94.5
Boreal forest	12.0	3.60	90.0	4.32	108.0
TOTAL FOREST	48.5	6.85	153.3	33.25	743.6
Woodland and shrubland	8.5	3.15	27.0	2.68	23.0
Savanna	15.0	4.05	18.0	6.08	27.0
Temperate grassland	9.0	2.70	7.2	2.43	6.5
Tundra and alpine	8.0	0.63	2.7	0.50	2.2
Desert and semidesert scrub	18.0	0.41	3.1	0.74	5.6
Extreme desert	24.0	0.01	0.1	0.02	0.2
Cultivated land	14.0	2.93	4.5	4.10	6.3
Swamp and marsh	2.0	13.50	67.5	2.70	13.5
Lake and stream	2.0	1.80	0.1	0.36	0.02
TOTAL NONFOREST	100.5			19.61	84.32
TOTAL CONTINENTAL	149	3.55	55.6	52.86	827.9
TOTAL MARINE	361	0.70	0.05	24.75	1.8
TOTAL EARTH	510	1.52	16.3	77.61	829.7

[a] Source: R. H. Whittaker and G. E. Likens (1975).
[b] Carbon estimated at 45 percent of total dry matter.

energy on the earth's land surface of about 0.1 percent. In a steady state, net primary production must be balanced by the oxidative activities of animals and microorganisms (heterotrophic respiration) and fires. Thus about 8 percent of the carbon dioxide content of the atmosphere is turned over each year by terrestrial biological activities, including the oxidation of organic carbon in soils.

As Loomis (1977) has emphasized, both the quantity of organic carbon in the biomass and the rates of turnover are subject to considerable uncertainty. Reiners (1973) has given a range of 37 to 64 Gt yr^{-1} for the rate of carbon turnover between the atmosphere and biosphere.

Bohn (1976) has recently estimated, from the FAO-UNESCO (1971) *Soil Map of the World* and other sources, that the content of organic carbon in the top meter of the world's soils is somewhat less than 3000 Gt, about three times the previously accepted value (Baes et al., 1976). Of this total, about 860 Gt is in peaty materials (dystric and gelic histosols in the FAO nomenclature), covering 4.3 million km^2, which presumably exchange carbon very slowly, if at all, with the atmosphere. The remaining approximately 2000 Gt of carbon in soil organic matter can be assumed to lose carbon to the atmosphere by oxidation at about the same rate as carbon is added by the accumulation of dead plant material. According to Bohn, the annual percentage rates of increase of cultivated area have been estimated by the U.S. Department of Agriculture (1965, 1970; Chugg, 1965; Hertford, 1971; Atkinson, 1969; Hermann, 1972) for various intervals between 1950 and 1970 for a group of Asian and Latin American countries together with two African countries, Sudan and Tanzania. For the Soviet Union, the rate of increase of cultivated area was estimated between 1940 and 1963 inclusive (U.S. Department of Agriculture, 1964). These data are shown in Table 10.2 together with estimates by Hayami and Ruttan (1971) on the increase of cultivated land in Japan and the United States between 1880 and 1940 and 1880 and 1920, respectively. Assuming that the annual rates of increase were constant from 1950 to 1970, we may estimate the total increase in cultivated land over this 20-year period in Asia and Latin America as 84 million hectares and 50 million hectares, respectively (Table 10.3). The data for Africa are obviously insufficient, since the cultivated areas of Sudan and Tanzania are less than 15 percent of the total cultivated area in Africa south of the Sahara in 1970. Nevertheless, assuming that these two countries were typical of the remainder of the sub-Saharan part of the continent, we estimate that the increase in cultivated land over the years from 1950 to 1970 was 74 million hectares.

For the Soviet Union, we estimate that the increase in the 20 years from 1950 to 1970 was 88 million hectares. We have been unable to find data for Australia and New Zealand or China, but we estimate that the 1950–1970 increase in these two regions was small—6 million hectares for Australia and New Zealand and 7 million hectares for China. The total estimated increase over the 20-year period was about 310 million hectares, or 30 percent of the world's cultivated area in 1950.

The world's soils contain on the average 18 kg/m^2 of exchangeable soil organic matter in the top meter, or 180 t/ha (tons per hectare). There is a wide variation in different soil types from 40 to 600 t/ha.

The average residence time of carbon in the terrestrial biosphere is simply the total mass of organic carbon in the part of the biosphere that exchanges with the atmosphere, about 2800 Gt, divided by 53 Gt per year (the rate of heterotrophic respiration plus fires), or 53 years. Variations in global temperature and precipitation should bring about short-term variations in the rates of photosynthesis and oxidation of organic matter in the terrestrial biosphere, but over periods of 10 years or more these changes should be small and tend to balance out. They are likely to be considerably less than the changes brought about by the secular increase in atmospheric carbon dioxide, or by human activities such as clearing of forests, reforestation, and destruction of soil humus.

Several effects of human activities during the past 100 years may have tended to change the size of the terrestrial biosphere and, correspondingly, the content of carbon dioxide in the atmosphere.* Perhaps the most important of these is the clearing of forests and woodlands for agriculture, which is now taking place with the rapid increases of population and food needs in the less-developed countries and occurred in both developed and less developed countries until recent decades. The losses of carbon from agricultural clearing during the past 100 years can be estimated from the probable increase in the area of cultivated land.

Except for the United States and Japan, we have not found any direct information on changes in the world's cultivated area prior to World War II, although careful historical search should yield useful information. However, we may roughly estimate the increases in cultivated land between 1860 and 1950 from the growth of human population in different regions.

During this 90-year period, the earth's total population nearly doubled, growing from somewhat more than 1200 million people to nearly 2400 million people (Table 10.4). The populations of the developed regions increased by 137 percent, and those of the developing regions by 77 percent. After World War II, the rates of increase in the developing regions speeded up and in the developed regions slowed down. Nevertheless, by 1970, the world's population was almost three times what it had been in 1860. Especially in the earlier period, the necessary increase in food production for this expanding population was almost certainly achieved largely by increasing the areas of cultivated land. The demand for food must have risen even more rapidly than population because of the approximate doubling of per capita incomes in the poor countries between 1860 and 1950. In these countries at present, the income elasticity of food demand is about 60 percent, that is, for each dollar increase in per capita income, 60¢ is spent for food. Assuming that

---

*According to our preferred model (see below), the proportions of organic carbon in the biosphere and of carbon dioxide in the atmosphere tend to remain constant if the total amount of carbon in the biosphere plus the atmosphere is unchanged.

The Carbon Dioxide Cycle and the Biosphere

TABLE 10.2 Increase in Cultivated Land in 35 Countries

Region and Country	Rates of Increase in Cultivated Area — % per Year	Rates of Increase in Cultivated Area — Millions of Hectares per Year	Years	Cultivated Land in 1970, Millions of Hectares[a]	Estimated Increase in Cultivated Land 1950-1970, Millions of Hectares
Asia					
Burma	1.9[b]		1954-1967	18.92	5.96
Sri Lanka	1.1[b]		1951-1967	1.98	0.39
Cambodia	2.5[b]		1954-1967	1.84	0.72
India	1.3[b]		1952-1965	165.68	37.93
Indonesia	2.2[b]		1951-1967	18.10	6.44
Iran	2.2[c]		1948-1963	16.15	6.05
Iraq	2.1[b]		1950-1967	10.16	3.48
Malaysia	1.6[b]		1954-1967	3.52	0.96
Pakistan	1.2[b]		1950-1967	19.24	4.10
Philippines	2.7[b]		1950-1967	11.14	4.65
S. Korea	1.5[b]		1950-1967	2.31	0.60
Thailand	2.4[b]		1950-1967	13.94	5.31
Turkey	1.5[b]		1950-1967	27.61	7.16
Laos	2.0[d]		1937-1962	0.95	0.32
Japan		0.2[e]	1880-1940	6.10	—
TOTAL ASIAN COUNTRIES				317.64	84.07
Sudan	2.7[c]		1948-1962	7.10	2.96
Tanzania	3.1[c]		1948-1963	16.24	7.51
Soviet Union		2.96[f]	1940-1963	238	88
United States		2.83[e]	1880-1920	189	—
Central America and Caribbean					
Costa Rica	2.3[b]		1950-1967	0.91	0.34
El Salvador	1.4[b]		1950-1967	0.65	0.16
Guatemala	3.4[b]		1950-1967	1.82	0.90
Jamaica	2.8[b]		1950-1961	0.28	0.12
Mexico	3.2[b]		1950-1965	27.47	12.98
Nicaragua	4.4[b]		1950-1967	1.19	0.70
Panama	3.0[b]		1950-1966	0.54	0.25
South America					
Argentina	1.1[b]		1950-1967	26.03	5.14
Bolivia	2.3[b]		1950-1967	1.51	0.56
Brazil	4.1[b]		1950-1967	34.08	19.07
Chile	0.7[b]		1950-1967	4.80	0.63
Colombia	1.6[b]		1950-1967	5.05	1.38
Ecuador	2.8[b]		1950-1967	3.82	1.64
Peru	2.1[b]		1950-1967	2.87	0.99
Uruguay	−2.4[b]		1950-1967	1.85	−1.14
Venezuela	4.2[b]		1950-1967	7.61	4.32
TOTAL LATIN AMERICA				120.48	48.04

[a] Food and Agricultural Organization (1975).
[b] U.S. Department of Agriculture (1970).
[c] U.S. Department of Agriculture (1965).
[d] Chugg (1965).
[e] Hayami and Ruttan (1971).
[f] U.S. Department of Agriculture (1964).

the ratio between the percentage increase in cultivated area and the percentage increase in population was approximately the same from 1860 to 1950 as it was between 1950 and 1970, we arrive at the estimates for the 1860-1950 increases in cultivated land given in Table 10.3.

During this period, the cultivated land in the presently

TABLE 10.3 Agricultural Land in Different Regions, 1860-1970[a]

Regions and Countries	Agricultural Land, Millions of Hectares			Increase, Millions of Hectares		
	1860	1950	1970	1860-1950	1950-1970	1860-1970
Developed Regions						
Northern America[b]	58	235	235	177	—	177
Europe	100	144	144	44	—	44
Soviet Union	44	150	238	106	88	194
Australia and New Zealand	2	12	18	10	6	16
Japan	5	6	6	1	—	1
TOTALS	209	547	641	338	94	432
Developing Regions						
Africa S. of Sahara[c]	35	91	165	56	74	130
Latin America	18	71	121	53	50	103
China	115	120	127	5	7	12
Other Asia[d]	136	228	312	92	84	76
TOTALS	304	510	725	206	215	421
GLOBAL TOTALS	513	1057	1366	544	309	853

[a] Sources: 1970 areas of agricultural land from FAO (1975), pp. 3-7. 1950 areas extrapolated from last column of Table 10.2. 1860 areas for Africa, Latin America, and other Asia computed from assumed constancy of the ratio between percentage change in land area from 1860 to 1950 and percentage change in population during this period, relative to 1950-1970. For other areas, see text.
[b] United States and Canada.
[c] Includes Sudan.
[d] Omitting 19 countries with less than 3 million inhabitants and also Afghanistan, Hong Kong, North Korea, Taiwan, Nepal, Saudi Arabia, Syria, Yemen, and North Vietnam. These 28 countries had a total population of 105 million people in 1970.

TABLE 10.4 Populations in Different Regions, 1860-1970[a]

Regions and Countries	Population in Millions			Percentage Increase		
	1860	1950	1970	1860-1950	1950-1970	1860-1970
Developed Regions						
Northern America[b]	35	166	226	374	36	546
Europe	209	392	459	88	17	120
Soviet Union	74	180	243	143	35	228
Australia and New Zealand	1	10	15	900	50	1400
Japan	32	84	104	163	57	225
TOTALS	351	832	1047	137	26	198
Developing Regions						
Africa S. of Sahara[c]	76	167	226	120	59	250
Latin America	37	164	283	343	73	665
China	443	558	772	26	39	74
Other Asia[d]	318	661	1050	108	59	230
TOTALS	874	1550	2371	77	53	171
GLOBAL TOTALS	1225	2382	3418	94	43	179

[a] Sources: Populations in 1860 (except for Africa) computed from data given in Zimmerman (1965). Population of Africa south of Sahara in 1860 interpolated from data given by Hauser (1971). Populations in 1950 and 1970 from United Nations (1975).
[b] United States and Canada.
[c] Includes Sudan.
[d] Omitting same 28 countries as in Table 10.3.

TABLE 10.5 Estimated Original Biomass Categories of Cleared Land, 1860–1870[a]

Original Biomass, Millions of Hectares

Region	Tropical Rain Forest	Tropical Seasonal Forest	Temperate Evergreen Forest	Temperate Deciduous Forest	Boreal Forest	Woodland and Scrubland	Savanna	Grassland	Swamp and Marsh	Totals
Northern America	—	—	20	25	10	—	—	122	—	177
Europe	—	—	—	11	—	—	—	11	11	44
Soviet Union	—	—	40	—	20	—	—	134	—	194
Australia and New Zealand	—	—	—	8	—	—	—	8	—	16
Africa S. of Sahara	15	15	—	—	—	50	50	—	—	130
Latin America	15	16	—	—	—	36	36	—	—	103
China	—	—	—	—	—	5	3	2	2	12
Other Asia	15	50	—	—	—	71	40	—	—	176
TOTALS	45	81	60	44	30	173	129	277	13	852
Carbon in biomass, tons/hectare	202.5	157.5	157.5	135	90.0	27.0	18.0	7.2	67.5	
Total biomass carbon, Gt	9.1	12.8	9.4	5.9	2.7	4.7	2.3	2.0	0.9	49.8

[a] It is assumed that the developed regions are all in the temperate zones and that all the developing regions (except China) are in the tropics or subtropics. The cleared areas in forests and nonforest are approximately in proportion to the ratios of these categories indicated in Table 10.1.

developed regions increased by nearly 340 million hectares and in the less-developed regions by over 200 million hectares. Table 10.5 shows the total biomass carbon that could have been transferred to the atmosphere by land clearing. This table has been constructed on the assumption that the areas of forests and other wildlands cleared for agriculture were approximately in proportion to the original ratios of these biomass categories in different regions. It may be seen that 30 percent of the cleared land was originally in forests and 70 percent in woodland and scrubland, savanna, grassland, and swamp and marsh. We note that 50 Gt of carbon were removed by clearing, most of it by the clearing of forests from 1860 to 1870.

There is good reason to believe that the content of soil humus diminishes when forests and other wildlands are cleared for agriculture (Loomis, 1977). Reiners (1973) has given a model in which the total loss of detritus from human activity between 1860 and 1970 was about 200 Gt. On the other hand, Loomis (1977) suggests that in recent decades the content of soil carbon in at least some areas of agricultural land may have increased through the effects of irrigation, legume rotation, use of fertilizers, and soil conservation, all practices that enhance inputs of organic matter to the soil. Bolin (1977) estimates that the total loss of soil carbon to the atmosphere has been between 10 and 40 Gt. We have rather arbitrarily assumed that there was about a 15 percent decrease in the average humus content of recently cleared lands, corresponding to 30 tons per hectare. Combining the figures for the reduction in biomass and the reduction in soil humus, we find that the total loss of carbon from the biosphere to the atmosphere resulting from land clearing was about 75 Gt between 1860 and 1970. If the newly cleared land was immediately planted to agricultural crops, the biomass in the cultivated areas would have been about 4.5 tons per hectare (Table 10.1), and this must be subtracted from the total loss by clearing and destruction of soil humus. We arrive at an estimate of 72 Gt loss from the biosphere to the atmosphere by land clearing over the 110-year period. This is 57 percent of the carbon produced by fossil-fuel combustion. The total addition to the atmosphere from fossil-fuel combustion and land clearing was thus about 200 Gt, roughly a third of the initial atmospheric carbon content.

Populations in the less-developed countries have increased roughly exponentially, and agricultural clearing may well have followed the same course, with an annual growth rate of about 3 percent from 1860 to the present. The total potential arable land on earth is limited, however, to 2600 million hectares (Revelle, 1976a) compared with a presently cultivated area of about 1400 million hectares. Even assuming that all presently uncultivated but potentially arable land is now in forests, the maximum future addition of carbon dioxide from land clearing is only about 240 Gt, less than 5 percent of the total carbon reserve in fossil fuels. Land clearing in the future is likely to follow a logistic curve, similar to that used by Keeling and Bacastow (see Chapter 4) for fossil-fuel carbon dioxide production but with a maximum rate of clearing around A.D. 2010 compared with the maximum carbon dioxide production from fossil fuels near the end of the twenty-first century.

Reforestation, the planting of trees in previously cleared areas, will tend to increase the size of the terrestrial biosphere at the expense of atmospheric carbon dioxide. According to Bolin (1977), reforestation has occurred over 30 million to 60 million hectares in China and about 10 million hectares in other less-developed countries, while in the developed countries it is not certain whether there has been an increase or a decrease in the forested area by as much as 12 million hectares. The increase in net primary production, less oxidation and fires, resulting from reforestation may be as much as 2.5 tons per hectare per year; hence, the annual subtraction of carbon from the atmosphere by reforestation may lie between 70 million and 200 million tons.

Bolin (1977) suggests that wood cut for lumber, paper pulp, and especially firewood may exceed net primary production in existing forests. Wood is the poor man's oil, and in the rural areas of Africa, Latin America, and parts of Asia the average per capita use of firewood (primarily for cooking) is probably much greater than it is today in the developed countries. Table 10.6 gives estimates, based on a variety of sources, of annual fuel wood consumption in the rural areas of 11 countries, with a total rural population of 1400 million people. The average per capita consumption is somewhat less than 700 kg/yr, with a carbon content of 350 kg. Extrapolating to an estimated total rural population of 2500 million people in the developing world gives a total firewood use of 0.875 Gt per year.

This is more than the FAO (1974) estimate of the harvest of roundwood (including fuel wood) from the world's forests. But, even accepting this figure, the total wood harvest is only 5 percent of the estimated net primary production in tropical and subtropical forests (Table 10.7). Thus, it seems unlikely that wood cutting has resulted in a net addition of carbon to the atmosphere.

As Loomis (1977) and G. C. Delwiche (University of California at Davis, personal communication) have pointed out, several other human activities may result in adding or subtracting carbon from the biosphere or the inorganic components of the soil. Cultivation of calcareous soils in arid and semiarid regions, and expansion of areas of irrigated land, may result in an uptake of carbon by the soil as calcium carbonate or organic matter. Drainage of swamps and marshes will liberate organic carbon to the atmosphere, as will some improved forest practices. At present, we are unable to estimate the magnitude of these various effects.

Besides clearing of land for agriculture, the two principal effects of human activities on the biosphere should be an increase in net primary production over oxidation, resulting from the fertilization of the biosphere by carbon dioxide added to the air (Botken et al., 1973; Lemon, 1976; Showcroft et al., 1973), and a similar increase resulting from the excess of nitrogen fixation over denitrification.

According to G. C. Delwiche and G. K. Likens (University of California at Davis, personal communication) the present annual fixation of atmospheric nitrogen from human activities amounts to 68 million tons (40 million tons of nitrogen in fertilizer, 10 million tons in other chemical products, and 18 million tons in combustion of fossil fuels). This is 35 percent of total world nitrogen fixation. The remaining 65 percent (127 million tons) is fixed by terrestrial and marine bacteria—including blue-green algae—and in lightning discharges. Denitrification is estimated at 160 million tons annually. Thus, the excess of nitrogen fixation over denitrification is on the order of 30 million tons per year. If this increment of fixed nitrogen were stored in the terrestrial biomass, principally in the wood of forest trees that have a high carbon-to-nitrogen ratio, the carbon added annually to the biosphere would be 2300 million tons. If it were stored in soil humus, the additional carbon would be only 350 million tons. It seems unlikely that the excess fixed nitrogen has had much effect on the quantity of organic matter in the marine biosphere, which is limited by the flux of phosphorus as well as nitrogen to ocean surface waters (Broecker, 1974).

Any uptake of carbon dioxide by the biosphere requires

TABLE 10.6 Estimated Use of Fuel Wood in Rural Areas of Less-Developed Countries

Country	kg of Fuel Wood per Capita per Year	Rural Population in 1970, Millions	Total Fuel Wood, Metric Tons per Year	Total Carbon, Metric Tons per Year
India[a]	285	440	125	63
Nepal[b]	730	11	8	4
Bangladesh[c]	95	70	7	4
China[d]	830	660	550	275
Thailand[e]	1100	30	33	16
Tanzania[e]	1800	12	22	11
Gambia[e]	1200	1	1	1
Nigeria[d]	1000	50	60	30
Mexico[d]	970	45	44	22
Bolivia[d]	2250	10	22	11
Brazil[f]	1000	70	70	35
TOTALS		1399	942	472

[a] Revelle (1976b).
[b] Revelle and Smith (1977).
[c] Tyers (1977).
[d] Makhijani and Poole (1975).
[e] Openshaw (1974).
[f] Adams et al. (1977).

TABLE 10.7 FAO Estimates of Wood Harvest in Different Regions Compared with Biomass Production in Forests, 1963 and 1974[a]

	Region							
	North and Central America	South America	Europe	Africa	Asia	Oceania	Soviet Union	World Total
Harvest of round wood including fuel wood[a]			(10^8 Tons of Carbon/yr)					
1963	1.23	0.57	0.86	0.71	1.83	0.07	0.78	6.05
1974	1.49	0.67	0.92	0.90	2.06	0.08	1.10	7.22
Forested area			(10^8 Hectares)					
1963	8.26	8.90	1.44	7.10	5.50	0.96	9.10	41.26
Wood harvest per hectare			(Tons of Carbon/ha/yr)					
1963	0.149	0.064	0.597	0.100	0.333	0.073	0.085	0.147
1974	0.180	0.075	0.639	0.127	0.374	0.083	0.121	0.175
Estimated net biomass production in forests			(Tons of Carbon/ha/yr)					
	5.6	7.2	5.6	8.0	7.0	6.5	3.6	6.85
Ratio of wood harvest to net biomass production in 1974			(Percent)					
	3.2	1.0	11.4	1.6	5.3	1.3	2.84	2.6

[a] Source: FAO (1974), and Table 10.1.
[b] Including bark, estimated as 15 percent of wood by weight; original figures are given in cubic meters of wood without bark; assumed density of dry wood = 0.5; assumed carbon content = 50 percent of dry weight of wood.

that the rate of net primary production by photosynthesis be greater than the rate of oxidation by microorganisms, animals, and fires. It has long been known that photosynthetic production increases when crop plants are bathed in a carbon dioxide-rich atmosphere, provided other factors of production are not limiting. In the natural world, however, plant production is limited by the availability of nitrogen, phosphorus, and minor nutrients; the intensity of solar radiation; the adequacy of the supply of water; and environmental stresses, such as low and high temperatures, in addition to the concentration of atmospheric carbon dioxide. Experimental data (Waggoner, 1969; Hardman and Brun, 1971) indicate that in the absence of other limiting factors, net primary production increases with the logarithm of the increase in carbon dioxide. In the heterogeneous conditions of the natural world we may introduce a factor β that takes account of the presence of other limiting factors. In the following section, we present models of the partitioning of carbon dioxide among the ocean, the atmosphere, and the biosphere in which the calculated value of β ranges from 0.05 to 0.3. For these models, we have divided the organic carbon in the biosphere into two components: the portion of living plants that takes part in photosynthesis, designated as y', and the wood of trees, plus organic detritus and soil humus, called y'', which are not directly involved in photosynthesis. We assume that all the biomass material not in forests, amounting to 84 Gt, is required for the net primary production of 19.6 Gt per year by this portion of the biosphere and that a proportionately larger amount of organic material in forests accounts for a net primary production of 33.2 Gt per year. The initial quantity, y_0', is then about 200 Gt (Table 10.1).

10.4 MODELS FOR THE PARTITION OF CARBON DIOXIDE AMONG THE ATMOSPHERE, THE OCEAN, AND THE BIOSPHERE

DEFINITIONS

$x(t)$, Atmospheric CO_2;
$y(t)$, Total organic carbon in biosphere (biomass plus organic detritus in soil), $y = y' + y''$;
$y'(t)$, Biomass that engages in photosynthesis;
$z(t)$, Oceanic carbon, $z = z' + z''$, z' is carbon in mixed layer (assumed 100 m thick);
$w(t)$, CO_2 produced by industrial revolution;
$v(t)$, CO_2 produced by clearing land;
τ, time constant for deep-sea mixing;
τ', time constant for surface layer.

ASSUMED RELATIONSHIPS

$$p = p_0 \left(1 + \beta \log_e \frac{x}{x_0}\right) \frac{y'}{y_0'}$$

Net production of biomass carbon by photosynthesis, where β depends on limitations of photosynthesis by factors other than atmospheric carbon dioxide.

$$q = q' + q'' = a \frac{y'}{y_0'} + b \frac{y''}{y_0''}$$

Consumption of biosphere carbon by respiration and fires; a and b are constants.

MAGNITUDES*

$t = 0$ in 1860.
$x_0 = 617$ Gt.
$y_0 = 2800$ Gt, $y_0' = 200$ Gt, $y_0'' = 2600$ Gt.
$z_0' = 864$ Gt.

$$r = \left(\frac{1}{w}\frac{dw}{dt}\right)_0 = 0.03 \text{ yr}^{-1}, w_0 = 4.5 \text{ Gt}, w_\infty \doteq 5000 \text{ Gt}.$$

$$\hat{r} = \left(\frac{1}{v}\frac{dv}{dt}\right)_0 = 0.03 \text{ yr}^{-1}, v_0 = 3.5 \text{ Gt}, v_\infty \leq 315 \text{ Gt}.$$

$0 \leq \beta \leq 1$. Buffer factor: $R = R_0 + d(x - x_0)/x_0$,
$R_0 = 9, d = 4$.
$P_0 = 53$ Gt yr^{-1}.
Ocean mixing: $\tau = 500$ yr, $\tau' = 12.5$ yr.

*All CO_2 weights in Gt (1 gigaton = 10^9 metric tons) of carbon and in Gt yr^{-1} of carbon.

PRODUCTION OF CARBON DIOXIDE FROM FOSSIL-FUEL COMBUSTION

We model the production of CO_2 during the industrial revolution (see Chapter 4) by

$$\frac{dw}{dt} = rw(1 - w/w_\infty), \quad (10.1)$$

with the solution

$$w = \frac{w_0 w_\infty e^{rt}}{w_\infty - w_0 + w_0 e^{rt}}, \quad (10.2)$$

$$\begin{aligned} &\to 0 && \text{as } t \to -\infty, \\ &= w_0 && t = 0, \\ &\to w_\infty && t \to \infty, \\ &\approx w_0 e^{rt} && w \ll w_\infty. \end{aligned} \quad (10.3)$$

Taking $w_0 = 4.5$ Gt of carbon, $w_\infty = 5000$ Gt, and $r = 0.03$ yr^{-1} gives $\Delta w = w - w_0$ as tabulated below in comparison with observed values.

Years	Δw Computed from Eq. (10.1)	Δw Observed [Keeling (1973) and Chap. 4]
1860–1959	82 Gt of carbon	81 Gt of carbon
1860–1973	126 Gt of carbon	128 Gt of carbon
1959–1973	46 Gt of carbon	52 Gt of carbon

The peak rate of combustion of fossil fuels can be computed by setting the second derivative of Eq. (10.2) equal to zero, from which we find that the peak rate of combustion will occur in A.D. 2094, when $w = w_\infty/2$, or 2500 Gt. The computed annual rate of production of carbon at that time is 39 Gt yr^{-1}, or about 9 times the observed average annual rate between 1969 and 1973.

LAND CLEARING

Let dv/dt be the rate at which CO_2 in the biomass is released to the atmosphere by clearing. We shall use the same formalism as for $w(t)$, so that

$$v = \frac{v_0 v_\infty e^{\hat{r}t}}{v_\infty - v_0 + v_0 e^{\hat{r}t}} \quad (10.4)$$

$$\approx v_0 e^{\hat{r}t} \text{ for } v \ll v_\infty. \quad (10.5)$$

Using $\hat{r} = r = 0.03$ yr^{-1}, $v_0 = 3.5$ Gt, $v_\infty = 315$ Gt, Eq. (10.4) gives 72 Gt for the period 1860–1970, the same as our earlier estimates, based on the world's increase in cultivated area, of 72 Gt. The computed peak rate of release of carbon dioxide to the air from land clearing corresponds to about 2.3 Gt yr^{-1} of carbon. This should occur at or

The Carbon Dioxide Cycle and the Biosphere

near the end of the first decade of the twenty-first century. From Eq. (10.1), the rate of release of fossil-fuel carbon at this time should be approximately 10 Gt yr^{-1}, more than 4 times as great as the biosphere release.

RATES OF CHANGE IN THE BIOSPHERE

We have

$$\frac{dy'}{dt} = p - q' - \phi - k'\frac{dv}{dt}, \quad (10.6)$$

$$\frac{dy''}{dt} = -q'' + \phi - k''\frac{dv}{dt}, \quad (10.7)$$

$$\frac{dy}{dt} = p - q - k\frac{dv}{dt}, \quad y = y' + y'', \quad (10.8)$$

where

$$k' = y'/y, \quad k'' = y''/y, \quad k = 1 \quad (10.9a)$$

for "indiscriminate" clearing, thus removing y', y'' in the ratio of their existing mass. If the deforested area is immediately replanted with a crop having equal photosynthesizing biomass,

$$k' = 0, \quad k'' = y''/y, \quad k = y''/y. \quad (10.9b)$$

Table 10.1 suggests that

$$k' = \tfrac{1}{2} y'/y, \quad k'' = y''/y, \quad k = 1 - \tfrac{1}{2} y'/y. \quad (10.9c)$$

Further

$$w - w_0 = (x - x_0) + (y - y_0) + (z - z_0), \quad (10.10)$$

where z refers to ocean storage and the subscript 0 refers to time 0 (1860). ϕ is the amount of y' going into y''. We assume that

$$\phi = cy'/y_0'. \quad (10.11)$$

Further, we have taken

$$p = p_0 \left(1 + \beta \log \frac{x}{x_0}\right) \frac{y'}{y_0}, \quad (10.12)$$

$$q = q' + q'' = a\frac{y'}{y_0'} + b\frac{y''}{y_0''}. \quad (10.13)$$

INITIAL CONDITIONS IN THE BIOSPHERE

We assume that the conditions are stationary at $t = 0$; strictly this applies to $t = -\infty$, but the error involved is negligible. Equations (10.6) and (10.7) then yield

$$p_0 = q_0' + \phi_0 \text{ or } p_0 = a + c,$$

$$q_0'' = \phi_0 \text{ or } b = c.$$

For further work, let $q' = 0$, or $a = 0$, since p refers to net photosynthetic production of organic matter. Hence,

$$p_0 = b = c,$$

or

$$\phi = p_0(y'/y_0'), \quad q'' = p_0(y''/y_0''). \quad (10.14)$$

We now have

$$\frac{dy'}{dt} = p_0 \beta \frac{y'}{y_0'} \log \frac{x}{x_0} - \frac{y'}{y}\frac{dv}{dt}, \quad (10.15)$$

$$\frac{dy}{dt} = p_0 \left[\left(1 + \beta \log \frac{x}{x_0}\right) \frac{y'}{y_0'} - \frac{y''}{y_0''}\right] - \frac{dv}{dt}. \quad (10.16)$$

THE OCEANS

Let z' refer to the storage in the mixed layer and z'' to the remaining ocean storage. Chemical equilibrium with the mixed layer is assumed to be relatively instantaneous. The buffer mechanism of seawater causes a fractional rise of CO_2 in the mixed layer that is only one ninth of that in the atmosphere (under present conditions). To a good approximation (Bacastow and Keeling, 1973),

$$\frac{(x - x_0)/x_0}{(z' - z_0')/z_0'} = R(x) \approx R_0 + d\frac{x - x_0}{x_0}, \quad (10.17)$$

with

$$R_0 = 9, \quad d = 4.$$

(The buffer factor, as given, is not applicable for $x \ll x_0$.)

Now for the deep ocean, we take a box model:

$$\frac{d(z''/h'')}{dt} = \frac{z'/h' - z''/h''}{\tau} + F, \quad (10.18)$$

where

$$z'/h' \text{ and } z''/h''$$

represent the relative concentrations in the mixed layer (depth h') and remaining ocean, h''. τ is the time constant for diffusing a surface contaminant into the bulk oceans. There have been many ways for estimating τ, and usually these fall between 500 and 1500 years. The simplest conceivable ocean model consisting of a balance between vertical upwelling w (1 cm day^{-1}) and vertical eddy diffusivity κ (1 cm^2 sec^{-1}) leads to exponential distributions with scale height κ/w (~1 km) and scale time κ/w^2 (~300 years) (Munk, 1966).

Here F is associated with various fluxes, such as that resulting from the settling of calcareous skeletons and particles of dead organic matter. Initial balance requires that

$$\frac{z_0'/h' - z_0''/h''}{\tau} + F_0 = 0,$$

so we can write Eq. (10.18) in the form

$$\frac{d(z''/h'')}{dt} = \frac{(z' - z_0')h' - (z'' - z_0'')/h''}{\tau} + F - F_0. \quad (10.19)$$

We assume that F is unaffected by the increase in CO_2,

$$F = F_0. \quad (10.20)$$

Equations (10.17), (10.19), and (10.20) can be combined into the combined ocean effect:

$$\frac{d(z - z_0)}{dt} = \frac{z_0'}{x_0} \frac{d}{dt}\left[\frac{x - x_0}{R(x)}\right] + \frac{z_0'}{x_0} \frac{x - x_0}{\tau' R(x)} - \frac{z - z_0}{\tau}, \quad (10.21)$$

with

$$\tau' = \tau \cdot h'/h, \quad h = h' + h''. \quad (10.22)$$

Observed Tritium Distribution

Suppose an amount z of tritium is initially added to the upper layer (as happened as a result of nuclear weapons tests in the early 1960's) and then diffuses downward, with the total $z = z' + z''$ remaining constant, after allowing for radioactive decay. The solution to Eq. (10.19) for constant z is given by

$$z' = z[1 - (h''/h)(1 - e^{(-t/\tau)})] \approx z e^{-t/\tau'}. \quad (10.23)$$

Thus z' diminishes from z to z/e in a time $\tau' = \tau \cdot h'/h$. If we set $\tau = 500$ years, $h'/h = 1/40$, then $\tau' = 12.5$ years, which is consistent with the observed result (H. G. Östlund, Institute of Marine Science, Miami, Florida, personal communication, 1977) that the tritium has mixed to an average depth of 360 m in 12 years.

Our conclusion is that the box model with $\tau = 500$ years can give a rough idea of the fraction of CO_2 lost to the deep ocean.

We now return to Eqs. (10.19) and (10.20). For an arbitrary $z'(t)$, the solution is

$$z''(t) = z_0'' + \frac{h''}{h'\tau} e^{-t/\tau} \int_0^t (z' - z_0') e^{\tilde{t}/\tau} d\tilde{t}. \quad (10.24)$$

Thus with $x(t)$ given, $z'(t) - z_0'$ is found from Eq. (10.17) and $z'' - z_0''$ from Eq. (10.24).

THE GENERAL PROBLEM

Taking into account fossil fuels and indiscriminate clearing, and the role played by the oceans,

$$(x - x_0) + (y - y_0) + (z' - z_0') + (z'' - z_0'') = w - w_0, \quad (10.25)$$

$$\frac{dy'}{dt} = p - \phi - k'\frac{dv}{dt}, \quad k' = \frac{y'}{y}, \quad (10.26)$$

$$\frac{dy}{dt} = p - q'' - k\frac{dv}{dt}, \quad k = 1, \quad (10.27)$$

$$z' - z_0' = \frac{z_0'(x - x_0)}{R_0 x_0 + d(x - x_0)}, \quad (10.28)$$

$$\frac{d(z'' - z_0'')}{dt} = \frac{z' - z_0'}{h'\tau} - \frac{z'' - z_0''}{h''\tau}, \quad (10.29)$$

with

$$w(t) = \frac{w_0 w_\infty e^{rt}}{w_0(e^{rt} - 1) + w_\infty},$$

$$v(t) = \frac{v_0 v_\infty e^{\hat{r}t}}{v_0(e^{\hat{r}t} - 1) + v_\infty},$$

$$p = p_0[1 + \beta \log(x/x_0)](y'/y_0'), \quad (10.30)$$

$$\phi = p_0(y'/y_0'), \quad (10.31)$$

$$q'' = p_0(y''/y_0''), \quad y'' = y - y'. \quad (10.32)$$

We chose a value for β (=0.05) such that $x - x_0$ has the observed value

$$x - x_0 \approx 85 \text{ Gt from 1860 to 1973}.$$

Numerical solutions are shown graphically in Figures 10.1 and 10.2. In Figure 10.1, the effects of land clearing are omitted [$v(t) = 0$], while they are included in Figure 10.2. The two calculations give almost identical results for the quantities of carbon in the atmosphere and the ocean. The atmospheric carbon content rises to a maximum of about 1400 Gt near the end of the twenty-first century and diminishes to less than the initial value a century later. After the year 2200 the carbon dioxide in the air slowly approaches the initial value in 1860. The content of carbon absorbed by the ocean follows a similar course, but it diminishes more slowly than the carbon in the atmosphere. More than half of the added carbon is absorbed in the biosphere by the end of the twenty-first century, and eventually all of it resides in the biosphere, as we shall now demonstrate.

The Carbon Dioxide Cycle and the Biosphere 151

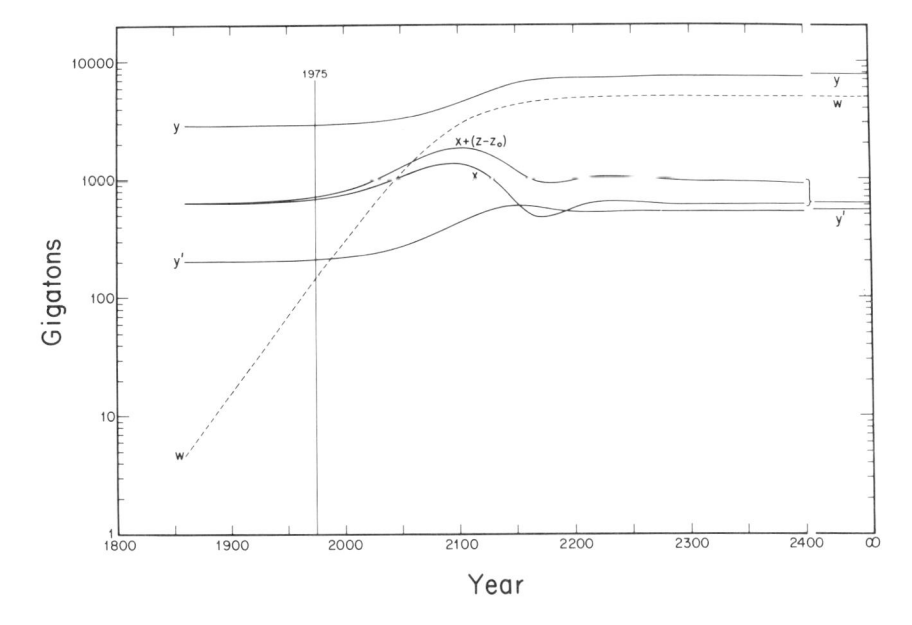

FIGURE 10.1 Partition of carbon dioxide from fossil-fuel combustion (w) among the atmosphere (x), the ocean ($z - z_0$), and the biosphere (y) under the condition that the quantity of carbon in the photosynthesizing portion of the biomass varies with time and that the rates of both net primary photosynthetic production and production of dead organic matter are proportional to the size of the photosynthesizing biomass ($\beta = 0.05$). Note that organic carbon in the atmosphere and the ocean ultimately return to their original value, and all carbon added from fossil-fuel combustion is absorbed in the biosphere.

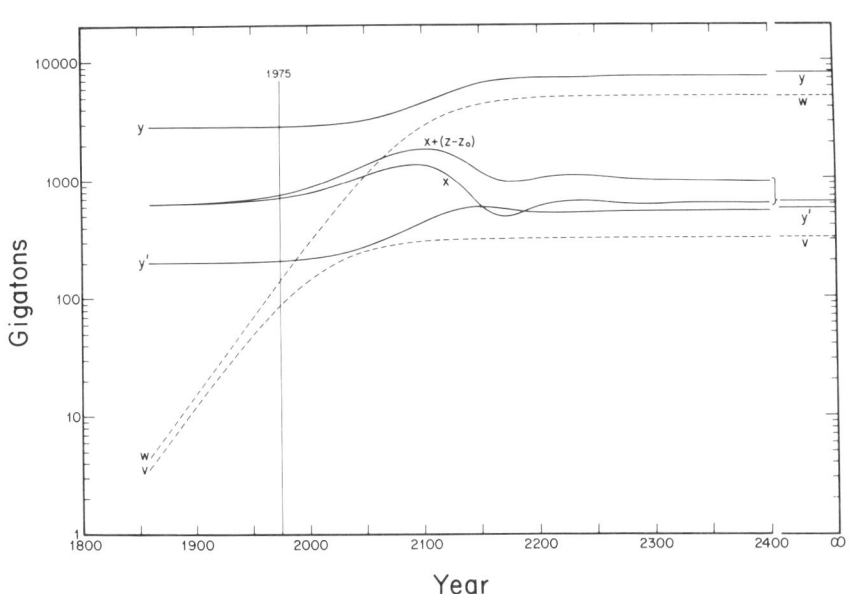

FIGURE 10.2 Partition of carbon dioxide from fossil-fuel combustion (w) and land clearing (v) among the atmosphere (x), the ocean ($z - z_0$), and the biosphere (y) under the condition that the quantity of carbon in the photosynthesizing portion of the biomass varies with time and that the rates of both net primary photosynthetic production and production of dead organic matter are proportional to the size of the photosynthesizing biomass ($\beta = 0.05$). As in the case of Figure 10.1, all carbon added by fossil-fuel combustion and land clearing is eventually absorbed in the biosphere.

THE FINAL BALANCE

A steady balance requires all time derivatives to vanish. From Eqs. (10.26) and (10.27) we then obtain

$$p = \phi = q''. \tag{10.33}$$

Then from Eqs. (10.30), (10.31), and (10.32),

$$1 + \beta \log(x/x_0) = 1 = \frac{y''/y_0''}{y'/y_0'}. \tag{10.34}$$

Equations (10.34) are satisfied for $t = 0$; hence $x = x_0$, $y = y_0'$, $y'' = y_0''$. At the final balance, $x = x_\infty$, and the first Eq. (10.34) implies that

$$x_0 = x_\infty. \tag{10.35}$$

Thus the atmosphere returns to its initial state. This is a consequence of p and ϕ having the same functional dependence on y' (in our case p and ϕ vary linearly with y'). It would not be the case if, for example, $p \sim y'$ and $\phi \sim \sqrt{y'}$ or if $\beta = \beta(y')$.

From Eq. (10.28) we have $z_\infty' = z_0'$, and from Eq. (10.29), $z_\infty'' = z_0''$. Equation (10.25) then yields

$$y_\infty = y_0 + w_\infty - w_0, \tag{10.36}$$

so that the biosphere has absorbed the entire carbon output from the industrial revolution. The effects of clearing are totally gone.

We also have from Eq. (10.34) that $y_\infty''/y_0'' = y_\infty'/y_0'$, which can be combined with Eq. (10.36) to give

$$y_\infty' = y_0' + (w_\infty - w_0)(y_0'/y_0), \qquad (10.37)$$

$$y_\infty'' = y_0'' + (w_\infty - w_0)(y_0''/y_0), \qquad (10.38)$$

so that the two components of the biosphere have gained industrial CO_2 in the proportion of their original fractional mass.

STABILITY OF PREINDUSTRIAL SYSTEM

To illustrate that the atmosphere-biosphere system is stable to small perturbations, ignore the ocean and take the equation

$$\frac{dy}{dt} = p - q'', \qquad (10.39a)$$

$$x + y = x_0 + y_0 = \text{constant}, \qquad (10.39b)$$

$$p = p_0\left(1 + \beta \log \frac{x}{x_0}\right)\frac{y'}{y_0}, \qquad q'' = p_0 \frac{y''}{y_0''}. \qquad (10.40)$$

If we write,

$$\frac{x}{x_0} = 1 + \xi, \quad \frac{y}{y_0} = 1 + \eta, \ldots, \qquad (10.41)$$

then it follows from Eq. (10.39b) that

$$\xi = -(y_0/x_0)\eta. \qquad (10.42)$$

The simplest model is one involving an equal *fractional* increase of y, y' y'', hence $\eta = \eta' = \eta''$. Accordingly, the right-hand side of Eq. (10.39a) is

$$p - q'' = p_0\beta \log(x/x_0)\frac{y'}{y_0'} = -p_0\beta(y_0/x_0)\eta, \qquad (10.43)$$

and so dy/dt is negative and drifts back to y_0. The complete *free* perturbation solution [as distinct from the solution *forced* by $w(t)$ and $v(t)$] is given by

$$\sum_{k=1}^{3} A_k e^{p_k t}, \qquad (10.44)$$

with p_k given by Eq. (10.A.13). The real parts of p_k are all negative; hence the system, for a constant carbon content, is stable to perturbations.

THE SIZE OF THE BIOSPHERE

But this does not come to grips with the question of what determines the magnitude of the biosphere in the first place. Suppose we set

$$\phi = Ay', \ q' = By', \ q'' = Cy'', \ p = Dy' f(x). \qquad (10.45)$$

A, B, C, D are physiological constants that can presumably be determined by laboratory experiments. For example, C is the fractional rate of respiration of dead plant tissue and soil organic matter. The model is quite specific about these constants:

$$A = \frac{P_0}{y_0'} = 0.265 \text{ yr}^{-1}, \ B = 0,$$
$$C = \frac{P_0}{y_0''} = 0.020 \text{ yr}^{-1}, \ D = 0.265 \text{ yr}^{-1}. \qquad (10.46)$$

For $f(x)$ the model involves two empirical constants, β and x_T, where x_T is the threshold CO_2 content of the atmosphere for photosynthesis:

$$f(x) = \beta \log(x/x_T), \qquad (10.47)$$

with

$$\beta = 0.1, \quad x_T = x_0 e^{-1/\beta} = 2.8 \times 10^{-2} \text{ Gt}. \qquad (10.48)$$

Suppose A, B, C, D, β, and x_T were in fact known *a priori*. Then we can compute the equilibrium conditions as follows: from Eq. (10.26), $dy/dt = 0$ implies $p = \phi$, or

$$Dy_0' f(x_0) = Ay_0',$$

hence

$$f(x_0) = A/D. \qquad (10.49)$$

Further, from $dy/dt = 0$, we have $\phi = q''$, hence

$$Ay_0' = Cy_0''. \qquad (10.50)$$

Finally, from the known buffer mechanism, we have a known relation of oceanic to atmospheric CO_2:

$$z_0 = z_0(x_0). \qquad (10.51)$$

Equations (10.48), (10.49), and (10.50), together with

$$x_0 + (y_0' + y_0'') + z_0 = T_0 \qquad (10.52)$$

provide the necessary four relations for computing the equilibrium values x_0, y_0', y_0'', z_0 for a given CO_2 content T_0 of atmosphere, ocean, and biosphere combined.

For illustration, ignore the oceans, and set $T_0 = 3417$ Gt. Then using the specified values of A, D and x_T, β, Eq. (10.49) yields

$$f(x_0) = A/D = 1 = \beta \log(x_0/x_T) \qquad (10.53)$$

or

$$x_0 = x_T e^{1/\beta} = 617 \text{ Gt}. \qquad (10.54)$$

From Eq. (10.52) we compute

$$y_0 = T_0 - x_0 = 2800 \text{ Gt}.$$

Finally from Eq. (10.50),

$$y_0 = y_0' + y_0'' = (1 + C/A)y_0''$$

or

$$y_0'' = \frac{y_0}{1 + C/A} = 2600 \text{ Gt}$$

and

$$y_0' = y_0 - y_0'' = 200 \text{ Gt}.$$

An important result is that the equilibrium content of the atmosphere is not a function of T_0 but depends only on the physiological constants A, D, β, x_T. It follows that if T_0 is increased (as during the industrial revolution), it all ends up in the biosphere. This is a result of having taken p and ϕ proportional to y' and would *not* be the case if, for example, p varied as y' and ϕ as $\sqrt{y'}$ nor if β is a function of y, say. In that case,

$$x_0 = x_T e^{1/\beta(y_0)},$$
$$y_0 = T_0 - x_T e^{1/\beta(y_0)}.$$

A LID ON THE BIOSPHERE

It is difficult to accept a model that suggests that the present biomass is so much smaller than its potential size, given enough CO_2. Other factors, such as the availability of nutrients and water, and ultimately lack of space, must impose an upper limit.

Such considerations led Keeling and Bacastow (see Chapter 4) to clamp an arbitrary lid on the biosphere by setting $\beta = 0$ after A.D. 2025. More generally, we may designate the lid by y_1, lying somewhere between $y_0 = 2800$ Gt and $y_\infty = 7800$ Gt. Let this stage be reached by t_1. Thus for $t < t_1$ we have essentially the previous situation:

$$(x - x_0) + (y - y_0) + (z - z_0) = w - w_0, \quad (10.55)$$

and for $t > t_1$,

$$(x - x_0) + (y_1 - y_0) + (z - z_0) = w - w_0. \quad (10.56)$$

But we can derive the ocean storage from the time history of $x(t)$, and so with $w(t)$ given, $x(t)$ is then determined. As $t \to \infty$, we have from Eq. (10.21)

$$z_\infty - z_0 = \frac{z_0}{x_0 R_\infty}(x_\infty - x_0), \quad z_0 = z_0' h/h', \quad (10.57)$$

$$R_\infty = R_0 + d\xi_\infty, \quad \xi = (x - x_0)/x_0. \quad (10.58)$$

Equations (10.56), (10.57), and (10.58) can be combined into

$$x_0 \xi_\infty + \frac{z_0}{R_0 + d\xi_\infty} \xi_\infty = \Delta = w_\infty - w_0 - (y_1 - y_0),$$

$$\xi_\infty = -\tfrac{1}{2} B + \tfrac{1}{2}\sqrt{B^2 - 4C},$$

where

$$B = \frac{z_0 + R_0 x_0 - d\Delta}{dx_0}, \quad C = -\frac{R_0 \Delta}{dx_0}.$$

Suppose $y_1 = 3500$ Gt; then $\Delta = 4300$, $B = 9.3$, $C = -15.7$, $\xi_\infty = 1.46$, and $z_\infty - z_0 = 3400$ Gt. Most of the carbon dioxide produced by fossil-fuel combustion is finally absorbed in the ocean, and the final amount of carbon dioxide in the atmosphere is 2.5 times the amount in 1860. This does not take into account solution of marine calcareous sediments or weathering processes on land, which would ultimately remove most of the excess carbon dioxide from the atmosphere.

A SECOND MODEL

A more satisfactory model would not only limit the mass of the biosphere to a reasonable maximum value but would also provide a definite relationship within the model itself between the atmospheric carbon dioxide content and the amount of organic carbon in the biosphere. This can be accomplished most simply by assuming that the photosynthesizing biomass y' remains constant. Suppose that prior to t_0 the land *area* was already used by plants as effectively as possible. Thus $y' = y_0'$ and Eq. (10.11) no longer holds. For the final balance,

$$(x_\infty - x_0) + (y_\infty'' - y_0'') + (z_\infty - z_0) = w_\infty - w_0, \quad (10.59)$$

$$p_\infty = p_0[1 + \beta \log(x_\infty/x_0)], \quad (10.60)$$

$$q_\infty'' = p_0 y_\infty''/y_0''.$$

Substituting,

$$z_\infty - z_0 = \frac{z_0}{x_0 R_\infty}(x_\infty - x_0), \quad z_0 = z_0' h/h',$$

gives

$$y_\infty'' - y_0'' = w_\infty - w_0 - (x_\infty - x_0)\left(1 + \frac{z_0}{x_0 R_\infty}\right).$$

For equilibrium $p_\infty = q_\infty''$, and so, with $\xi = (x - x_0)/x_0$,

$$\beta y_0'' \log(1 + \xi_\infty) = w_\infty - w_0(x_0 + z_0/R_\infty)\xi_\infty, \quad (10.61)$$

$$R_\infty = R_0 + d\xi_\infty. \quad (10.62)$$

Numerical solution yields $\xi_\infty = 1.46$, $x_\infty - x_0 = 900$ Gt, $y_\infty - y_0 = 700$ Gt, and $z_\infty - z_0 = 3400$ Gt for $\beta = 0.3$

and storage of carbon in the entire ocean. Equation (10.54) becomes

$$x_0 = x_T \exp\left[\frac{C}{\beta D y_0'}(T_0 - y_0' - x_0)\right].$$

The equilibrium carbon dioxide content of the atmosphere depends on the total amount of carbon in the atmosphere plus that in the biosphere.

If we wish to compute the total time history, then we use the complete set of Eqs. (10.25) to (10.32) (or their perturbation counterpart), where $y' < y_0'$, but write $y' = y_0'$ for subsequent calculations. This takes into account a slight initial dip in y' associated with clearing. A simple case is that associated with $k' = 0$ [Eq. (10.9b)], where there is no initial dip in y_0', so that $y' = y_0'$ holds throughout.

For this case, set

$$\frac{dy'}{dt} = p - \phi - k'\frac{dv}{dt} = 0 \text{ with } k' = 0. \quad (10.63)$$

Then with $y' = y_0' = $ constant,

$$p = p_0(1 + \beta \log x/x_0) = \phi, \quad (10.64)$$

$$q'' = p_0(y''/y_0''), \quad (10.65)$$

and

$$\frac{dy''}{dt} = -q'' + \phi - k''\frac{dv}{dt}, \quad k'' = y''/y$$

$$= -p_0(y''/y_0'') + p_0\left(1 + \beta \log \frac{x}{x_0}\right) - \frac{y''}{y}\frac{dv}{dt}. \quad (10.66)$$

The results are plotted in Figure 10.3 and shown numerically in Table 10.8.

In agreement with Keeling and Bacastow (see Chapter 4), a maximum of atmospheric carbon dioxide is reached in the latter half of the twenty-second century but at a lower level than their estimate, about 4.7 times the original content in 1860. Their computed maximum for the same fuel-consumption pattern is 7 to 8 times the 1860 content, assuming no dissolution of surface ocean carbonate.

The difference is chiefly due to the larger uptake of carbon in the biosphere in our model, over 1000 Gt at the time of maximum atmospheric carbon content. Also, our projection gives a somewhat more rapid uptake of carbon by the deep sea than assumed by Keeling and Bacastow in Chapter 4.

After the middle of the twenty-second century, the combined contents of carbon in the atmosphere and the biosphere slowly diminish, as the ocean absorbs a larger and larger fraction of the total fossil-fuel carbon dioxide. By the beginning of the twenty-fifth century, our projected value for the carbon content of the atmosphere is only 3.3 times the content in 1860.

As would be expected, the biosphere lags somewhat behind the atmosphere, reaching a maximum carbon content 40 percent higher than the assumed 1860 value in the first half of the twenty-third century.

The projected rise in atmospheric carbon during the course of the next 100 years is the most important aspect of this calculation, since beyond that time the uncertainty in the quantity of fossil fuels consumed becomes very large. By the year A.D. 2000, we find that the carbon content of the air should be about 790 Gt, or 370 parts per million of carbon dioxide, nearly 14 percent above the 1970 value and 27.5 percent above that in 1860. The carbon content of the air will be twice the 1970 value by the middle of the twenty-first century, only 75 years from now, and will increase rapidly thereafter to more than 3 times the

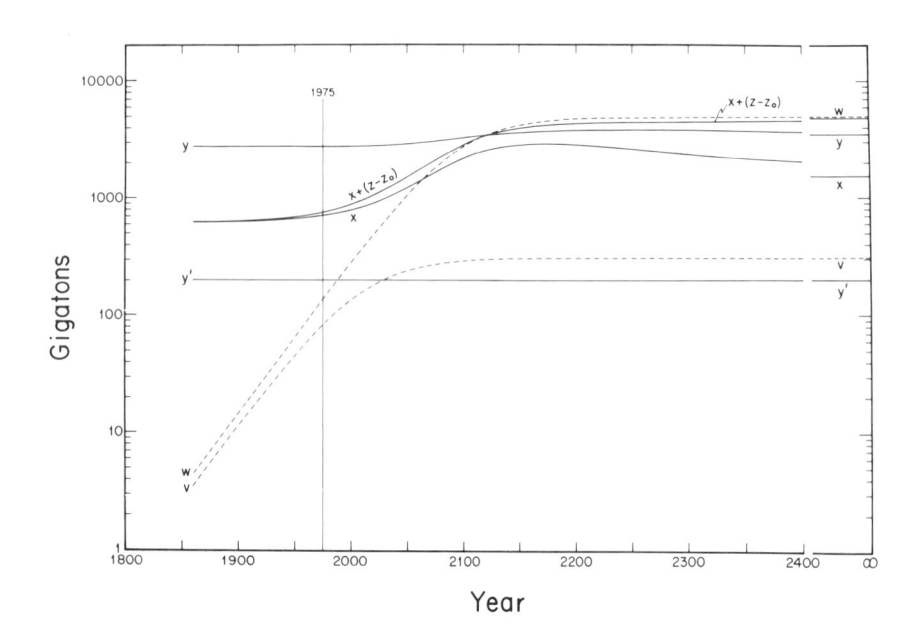

FIGURE 10.3 Partition of carbon dioxide from fossil-fuel combustion (w) and land clearing (v) among the atmosphere (x), the ocean ($z - z_0$), and the biosphere (y) under the condition that the photosynthesizing portion of the biomass (y') remains constant and $\beta = 0.3$. The ultimate quantities of atmospheric carbon, ocean carbon, and biosphere carbon are shown on the right-hand side. Note the absence of oscillations around the initial value of atmospheric carbon dioxide shown in Figures 10.1 and 10.2.

The Carbon Dioxide Cycle and the Biosphere

TABLE 10.8 Partition of Added Carbon Dioxide among the Atmosphere, the Ocean, and Biosphere[a]

Year	Fossil-Fuel Production $(w - w_0)$	Land-Clearing Production $(v - v_0)$	$\begin{pmatrix} v - v_0 \\ + \\ w - w_0 \end{pmatrix}$	Added to Atmosphere $(x - x_0)$	Added to Biosphere $(y - y_0)$	Added to Oceans $(z - z_0)$	$(x - x_0) / \begin{pmatrix} v - v_0 \\ + \\ w - w_0 \end{pmatrix}$	$\begin{pmatrix} v - v_0 \\ + \\ y - y_0 \end{pmatrix} / \begin{pmatrix} v - v_0 \\ + \\ w - w_0 \end{pmatrix}$	$(z - z_0) / \begin{pmatrix} v - v_0 \\ + \\ w - w_0 \end{pmatrix}$	"Airborne Fraction" $\dfrac{(x - x_0)}{(w - w_0)}$
1860	0	0	0	0	0	0	0	0	0	
1950	62	42	103	43	−2.5	21	42.0	37.9	20.2	70.3
1960	84	54	139	58	−1.6	28	41.6	38.0	20.4	68.4
1970	115	70	185	76	+0.4	38	41.3	38.1	20.6	66.5
1980	155	88	243	100	4.2	51	41.3	37.9	20.8	64.7
2000	279	131	410	171	21	87	41.8	37.1	21.1	61.4
2020	489	177	666	286	60	142	43.0	35.6	21.4	58.6
2040	827	220	1047	471	131	229	45.0	33.5	21.5	56.9
2060	1328	253	1581	750	241	338	47.4	31.2	21.4	56.4
2080	1987	271	2262	1123	384	480	49.6	29.1	21.2	56.5
2100	2730	289	3019	1540	546	643	51.0	27.7	21.3	56.4
2120	3432	298	3730	1908	708	817	51.2	27.0	21.9	55.6
2140	3997	302	4299	2154	850	994	50.1	26.8	23.1	53.9
2160	4394	305	4698	2265	963	1165	48.2	27.0	24.8	51.6
2180	4646	307	4952	2272	1043	1330	45.9	27.3	26.9	48.9
2200	4797	308	5105	2216	1095	1486	43.4	27.5	29.1	46.2
2220	4885	308	5193	2130	1122	1633	41.0	27.5	31.4	43.6
2240	4934	308	5242	2032	1132	1770	38.8	27.5	33.8	41.2
2260	4962	308	5270	1936	1128	1898	36.7	27.3	36.0	39.0
2280	4977	308	5285	1844	1116	2016	34.9	27.0	38.2	37.0
2300	4985	308	5294	1760	1099	2176	33.2	26.6	40.2	35.3
2320	4990	309	5298	1683	1079	2228	31.8	26.2	42.0	33.7
2340	4992	309	5301	1613	1057	2323	30.4	25.8	43.8	32.3
2360	4994	309	5302	1549	1034	2410	29.2	25.3	45.4	31.0
2380	4995	309	5303	1492	1012	2491	28.1	24.9	47.0	29.9
2400	4995	309	5304	1439	987	2569	27.1	24.4	48.4	28.8

[a] Assumption: carbon dioxide fertilization factor $\beta = 0.3$; y' = constant = photosynthesizing portion of the biomass.

1970 value by the year 2100. When Keeling and Bacastow (see Chapter 4) assume a faster rate of growth of fossil-fuel consumption (over 4 percent per year until the late 1980's), they find an atmospheric carbon dioxide content of 389 parts per million in A.D. 2000, a doubling of the 1970 content by A.D. 2035, and a fivefold increase by A.D. 2100. Both our values and those of Keeling and Bacastow fall well within the range of possibilities defined by Baes et al. (1977).

Our calculations indicate that up to A.D. 1973 the total quantity of organic material in the biosphere should have remained relatively constant since 1860. That is, the release of carbon to the atmosphere by land clearing has been almost exactly balanced by the increase in the net photosynthetic production and storage of organic material resulting from the higher atmospheric carbon dioxide content. Most of this increased photosynthetic production and storage of organic carbon should have occurred in the forest areas remaining after land clearing. But in 1970, the total increase in production and storage over the previous 110 years amounted to only 4 percent of the original mass of organic carbon in the world's forests, including the humus in forest soils. This is probably not detectable with present methods of estimating forest biomass and soil humus.

The slow rate of uptake of carbon by the oceans is notable. By 1970, only 38 Gt of carbon had been lost to the ocean, or about 21 percent of the total added during the previous 110 years of fossil-fuel combustion and land clearing. According to the model, the fraction of added carbon absorbed in the sea should remain at about 21 percent throughout the twenty-first century.

With the slowing down in the rate of fossil-fuel combustion that we have postulated for the late twenty-second century, the proportion of added carbon dioxide taken up by the ocean rapidly increases, because the deeper waters begin to play a larger and larger role. This effect would be observed at an earlier time and at a much lower concentration of atmospheric carbon dioxide if the rate of fossil-fuel combustion were not allowed to increase significantly after the next few decades. Additional model computations should be made to explore the implications of such a slowing of fossil-fuel use.

ACKNOWLEDGMENT

We are grateful to John Spiessburger for carrying out the numerical calculations shown in the figures and in Table 10.8.

APPENDIX 10A: SMALL TIME PERTURBATIONS

Our calculations of the effects of small time perturbations involve two sets of assumptions:

1. The disturbance in the initial balance of atmosphere, biosphere, and ocean is relatively small:

$$\xi \equiv \frac{x - x_0}{x_0} \ll 1, \quad \eta = \frac{y - y_0}{y_0} \ll 1,$$

$$\zeta = \frac{z - z_0}{z_0} \ll 1, \quad \eta' = \frac{y' - y_0'}{y_0'}, \ldots \quad (10.A.1)$$

2. The release of new CO_2 by fossil fuel and by clearing is still small compared with the total potential release:

$$w \ll w_\infty, \quad v \ll v_\infty. \quad (10.A.2)$$

The first set of assumptions permits the linearization of the model, thus neglecting the squares and products of all perturbation parameters. The second set permits us to use the simple exponential growth rates

$$w = w_0 e^{rt}, \quad v = v_0 e^{\hat{r}t} \quad (10.A.3)$$

in the place of Eqs. (10.2) and (10.4). These limits correspond roughly to the years 2000 and 1975, respectively. The conclusion is that we can justifiably use the perturbation method for the analysis of data up to the present but not for prediction purposes.

We now consider the set of four differential equations, (10.10), (10.15), (10.16), and (10.21) in the independent variables x, y', y, z. The corresponding equations in the perturbation variables are

$$\begin{array}{llll}
x_0\xi & + y_0\eta & + z_0\zeta & = w - w_0 \\
-a\xi & + D\eta' & & = -y_0^{-1}Dv \\
-b\xi & + (D+c)\eta - c\eta' & & = -y_0^{-1}Dv \\
\left(D + \frac{1}{\tau'}\right)\xi & & -R_0\frac{z_0}{z_0'}\left(D + \frac{1}{\tau}\right)\zeta = 0
\end{array}$$

$$(10.A.4)$$

where we have written D for d/dt. Numerical values are

$$a = (p_0/y_0')\beta = 1.33 \times 10^{-2} \text{ yr}^{-1},$$

$$b = (p_0/y_0)\beta = 9.46 \times 10^{-4} \text{ yr}^{-1}, \quad (10.A.5)$$

$$c = p_0/y_0'' = 2.04 \times 10^{-2} \text{ yr}^{-1},$$

based on an estimate $\beta = 0.05$. We now perform the Laplace transform

$$\mathcal{L}(u) \equiv \bar{u} = \int_0^\infty e^{-pt} u(t)\, dt. \quad (10.A.6)$$

In particular, $\mathcal{L}(D\xi) = p\bar{\xi} - \xi_0 = p\bar{\xi}$, since $\xi_0 = 0$, and similarly for η, η', ζ. Further

$$\mathcal{L}(w - w_0) = w_0 \mathcal{L}(e^{rt} - 1)$$

$$= w_0 \left(\frac{1}{p-r} - \frac{1}{p}\right) = \frac{w_0 r}{p(p-r)}, \quad (10.A.7)$$

$$\mathcal{L}(Dv) = \frac{v_0 \hat{r}}{p - \hat{r}}. \quad (10.A.8)$$

The solution can be written $\bar{\xi} = NU/DE$, with

$$DE = \frac{z_0}{z_0'} \begin{vmatrix} x_0 & y_0 & 0 & z_0' \\ -a & 0 & p & 0 \\ -b & p+c & -c & 0 \\ p + 1/\tau' & 0 & 0 & -R_0(p + 1/\tau) \end{vmatrix}; \quad (10.A.9)$$

$$NU = \frac{z_0}{z_0'} \begin{vmatrix} \dfrac{w_0 r}{p(p-r)} & y_0 & 0 & z_0' \\ \dfrac{-v_0\hat{r}}{y_0(p-\hat{r})} & 0 & p & 0 \\ \dfrac{-v_0\hat{r}}{y_0(p-\hat{r})} & p+c & -c & 0 \\ 0 & 0 & 0 & -R_0(p + 1/\tau) \end{vmatrix}.$$

We can write $\qquad(10.A.10)$

$$DE = \frac{z_0}{z_0'}[ep(p + c)(p + f) + g(p + \tau^{-1})(p + h)],$$

$$(10.A.11)$$

where

$$e = z_0' + x_0 R_0 = 6417 \text{ Gt},$$

$$f = \frac{z_0'/\tau' + x_0 R_0/\tau}{e} = 0.0125 \text{ yr}^{-1},$$

$$g = y_0 R_0 b = 23.84 \text{ Gt yr}^{-1},$$

$$h = ac/b = 0.287 \text{ yr}^{-1},$$

The Carbon Dioxide Cycle and the Biosphere

and this can finally be written

$$DE = (z_0/z_0')\, e(p^3 + Ap^2 + Bp + C),$$

where

$A = g\,e^{-1} + c + f = 3.66 \times 10^{-2}\ \text{yr}^{-1},$

$B = g\,e^{-1}(h + \tau^{-1}) + cf = 1.33 \times 10^{-3}\ \text{yr}^{-2},$

$C = gh\,e^{-1}\tau^{-1} = 2.132 \times 10^{-6}\ \text{yr}^{-3},$

$$DE = (z_0/z_0')\, e(p - p_1)(p - p_2)(p - p_3), \quad (10.A.12)$$

where

$p_1 = -0.17 \times 10^{-2}\ \text{yr}^{-1},$

$p_2 = -1.75 \times 10^{-2} + i\,3.11 \times 10^{-2}\ \text{yr}^{-1},$

$p_3 = -1.75 \times 10^{-2} - i\,3.11 \times 10^{-2}\ \text{yr}^{-1}. \quad (10.A.13)$

Similarly

$$NU = \frac{z_0}{z_0'} R_0 (p + \tau^{-1})(p + c)\left(\frac{w_0 r}{p - r} + \frac{v_0 \hat{r}}{p - \hat{r}}\right).$$

$$(10.A.14)$$

From now on we restrict ourselves to the special case $r = \hat{r}$. We then have

$$\bar{\xi} = \frac{R_0 e^{-1}(w_0 + v_0)\,r(p + \tau^{-1})(p + c)}{(p - p_1)(p - p_2)(p - p_3)(p - r)}. \quad (10.A.15)$$

Then by the rules of inverse Laplace transforms,

$$\xi = \frac{NU(p_1)\,e^{p_1 t}}{(p_1 - p_2)(p_1 - p_3)(p_1 - r)}$$

$$+ \frac{NU(p_2)\,e^{p_2 t}}{(p_2 - p_3)(p_2 - r)(p_2 - p_1)} + \ldots, \quad (10.A.16)$$

where the additional terms are obtained from cyclical rotation of subscripts 1, 2, 3, r. This gives

$\xi(t) = [-.05 e^{-0.0017t} + e^{-0.0175t}(-5.27 \cos 0.0311t + 2.74 \sin 0.0311t) + 5.32 e^{0.03t}] \times 10^{-3},$
$\xi(0) = [-0.05 \qquad\qquad -5.27 \qquad\qquad\qquad\qquad\qquad\qquad + 5.32\quad] \times 10^{-3} = 0,$
$\xi(100) = [-0.042 \qquad\qquad +0.92 \qquad\qquad +0.02 \qquad\qquad 106.9\quad] \times 10^{-3} = 0.108,$
$\xi(114) = [-0.041 \qquad\qquad +0.66 \qquad\qquad -0.15 \qquad\qquad 162.6\quad] \times 10^{-3} = 0.163. \quad (10.A.17)$

Thus

1860–1959, $t = 100$, $\zeta = 0.108$, $x - x_0 = 66.5$ Gt (58 Gt observed, Chapter 4)

1860–1973, $t = 114$, $\zeta = 0.163$, $x - x_0 = 100.6$ Gt (83 Gt observed)

1959–1973, $\qquad\qquad\qquad\qquad\qquad\quad x - x_0 = 34.1$ Gt (29 Gt observed). $\qquad\qquad\qquad\qquad (10.A.18)$

The last term predominates after a century. The forecast oscillations have a frequency of $2\pi/0.0311 = 202$ years and decay to e^{-1} in 57 years. The first term decaying in 588 years is presumably associated with ocean coupling.

For the model in which y' is constant, in the place of the four perturbation equations (10.A.4) we now have only three equations:

$$x_0 \xi \qquad\qquad + y_0'' \eta'' \qquad\qquad + z_0 \zeta = w - w_0,$$
$$-(p_0\beta/y_0'')\xi + (D + p_0/y_0'')\eta'' \qquad + 0 = (y_0'')^{-1} Dv,$$
$$(D + 1/\tau')\xi \qquad\qquad + 0 \qquad -R_0 \frac{z_0}{z_0'}(D + \frac{1}{\tau})\zeta = 0.$$

$$(10.A.19)$$

The determinant of the Laplace transform is

$$DE = -(z_0/z_0')\,e(p^2 + Ap + B), \quad (10.A.20)$$

with

$e = z_0' + x_0 R_0 = 6417$ Gt,

$A = \dfrac{P_0}{y_0''} + \dfrac{z_0'/\tau' + x_0 R_0/\tau}{e} + \dfrac{R_0 p_0 \beta}{e} = 5.52 \times 10^{-2}\ \text{yr}^{-1},$

$B = \dfrac{R_0 p_0 \beta}{\tau e} + \dfrac{p_0}{y_0''}\left(\dfrac{z_0'/\tau' + x_0 R_0/\tau}{e}\right) = 2.30 \times 10^{-4}\ \text{yr}^{-1}.$

$$(10.A.21)$$

Note that e and A are the same as in Eq. (10.A.12), except that we are now using $\beta = 0.3$. The roots are

$$p_1 = -4.54 \times 10^{-3}, \quad p_2 = -5.07 \times 10^{-2}. \quad (10.A.22)$$

Similarly

$$NU = -\frac{z_0}{z_0'} R_0 (w_0 + v_0)(p + \tau^{-1}) \frac{r}{p(p - r)}(p + c),$$

$$c = \frac{w_0 p_0}{(w_0 + v_0) y_0''} = 1.15 \times 10^{-2}\ \text{yr}^{-1}. \quad (10.A.23)$$

The solution is of the same form as Eqs. (10.A.15) and (10.A.16), except for the new definition of c (note that w_0 and v_0 now enter not merely as their sum) and the fact that $p_3 = 0$ and that β is now 0.3. The result is

$$\xi(t) = (-0.82 e^{-0.0045t} - 3.40 e^{-0.0507t} - 1.12 + 5.34 e^{0.03t}) \times 10^{-3},$$

$$1860\text{-}1959, \quad t = 100, \quad \zeta = 0.106, \quad x - x_0 = 65.2 \text{ Gt} \quad (58 \text{ Gt observed}),$$
$$1860\text{-}1973, \quad t = 114, \quad \zeta = 0.162, \quad x - x_0 = 99.7 \text{ Gt} \quad (83 \text{ Gt observed}),$$
$$1959\text{-}1973, \qquad\qquad\qquad\qquad\quad x - x_0 = 34.5 \text{ Gt} \quad (29 \text{ Gt observed}). \quad (10.\text{A}.24)$$

REFERENCES

Atkinson, L. F. (1969). *Changes in Agricultural Production and Technology in Colombia*, Foreign Agricultural Economic Rep. No. 52, U.S. Department of Agriculture Economic Research Service, Washington, D.C., p. 55.

Adams, J. A. S., M. S. Matonari, and L. L. Lundell (1977). Wood versus fossil fuel as a source of excess carbon dioxide in the atmosphere, *Science 196*, 54.

Bacastow, R. B., and C. D. Keeling (1973). Atmospheric carbon dioxide and radio-carbon in the natural carbon cycle: Changes from A.D. 1700 to 2070 as deduced from a geochemical model, in *Carbon and the Biosphere*, G. M. Woodwell and E. V. Pecan, eds., U.S. Atomic Energy Commission, Washington, D.C., pp. 86-135.

Baes, C. F., Jr., H. E. Goeller, J. S. Olson, and R. M. Rotty (1976). *The Global Carbon Dioxide Problem*, Oak Ridge National Laboratory Rep. ORNL-5194.

Baes, C. F., Jr., H. E. Goeller, J. S. Olson, and R. M. Rotty (1977). Carbon dioxide and climate: the uncontrolled experiment, *Am. Scientist 65*, 310.

Bohn, H. L. (1976). Estimate of organic carbon in world soils, *Soil Sci. Soc. Am. J. 40*, 468.

Bolin, B. (1977). Changes of land biota and their importance for the carbon cycle, *Science 196*, 613.

Botkin, D. B., J. F. Janak, and J. R. Wallis (1973). Estimating the effects of carbon fertilization on forest composition by ecosystem simulation, in *Carbon and the Biosphere*, Proceedings of the 24th Brookhaven Symposium on Biology, G. M. Woodwell and E. V. Pecan, eds., Technical Information Center, Office of Information Services, U.S. Atomic Energy Commission, Washington, D.C.

Broecker, W. S. (1974). *Chemical Oceanography*, Harcourt Brace Jovanovich, Inc., New York.

Chugg, B. A. (1965). *Agriculture in the Southeast Asia Rice Bowl and Its Relation to U.S. Farm Exports*, Foreign Agricultural Economic Rep. No. 26, U.S. Department of Agricultural Economic Research Service, Washington, D.C., pp. 15, 17, 65.

FAO-UNESCO (1971). Food and Agriculture Organization of the United Nations *Soil Map of the World, 1:5,000,000*, UNESCO, Paris.

FAO (1974). *FAO Yearbook of Forest Products*, Food and Agriculture Organization of the United Nations, Paris, pp. 3, 4, 17, and 35.

FAO (1975). *Production Yearbook 1974*, Food and Agricultural Organization, United Nations, Rome, pp. 3-7.

Hardman, L. L., and W. A. Brun (1971). Effects of atmosphere enriched with carbon dioxide on different developmental stages from growth and yield components of soybeans, *Crop Sci. 11*, 886.

Hauser, P. H. (1971). in *Rapid Population Growth*, National Academy of Sciences, Johns Hopkins Press, Baltimore, Md., p. 106.

Hayami, N., and V. W. Ruttan (1971). *Agricultural Development, An International Perspective*. Johns Hopkins Press, Baltimore, Md., p. 113.

Hermann, L. F. (1972). *Changes in Agricultural Production in Brazil, 1947-65*, U.S. Department of Agriculture, Economic Research Service, Washington, D.C., pp. 23-26.

Hertford, R. (1971). *Sources of Change in Mexican Agricultural Production*, Foreign Agricultural Economic Rep. No. 73, U.S. Department of Agricultural Economic Research Service, Washington, D.C., pp. 25-27.

Keeling, C. D. (1973). Industrial production of carbon dioxide from fossil fuels and limestone, *Tellus 25*, 174.

Lemon, E. R. (1976). The land's response to more carbon dioxide, in *The Fate of Fossil-Fuel CO₂*, U.S. Office of Naval Research Conference, Honolulu, Hawaii, in press.

Loomis, R. S. (1977). CO₂ and the biosphere, manuscript prepared for ERDA Miami workshop on Environmental Effects of Carbon Dioxide from Fossil Fuel Combustion, March 7-11, 1977, Miami Beach, Florida, 25 pp. (in press).

Makhijani, A., and A. Poole (1975). *Energy and Agriculture in the Third World*, Ballinger Publishing Co., Cambridge, Mass. pp. 46, 52, 54.

Munk, W. H. (1966). Abyssal recipes, *Deep-Sea Res. 13*, 707.

Openshaw, K. (1974). Wood fuels the developing world, *New Scientist 61*, 271.

Reiners, W. A. (1973). Terrestrial detritus and the carbon cycle, in *Carbon and the Biosphere*, Proceedings of the 24th Brookhaven Symposium on Biology, G. M. Woodwell and E. V. Pecan, eds., Technical Information Center, Office of Information Services U.S. Atomic Energy Commission, Washington, D.C.

Revelle, R. (1976a). The resources available for agriculture, *Sci. Am. 235*, 165.

Revelle, R. (1976b). Energy use in rural India, *Science 192*, 969.

Revelle, R., and D. Smith (1977). Personal observations, March.

Showcroft, R. W., E. R. Lemon, and B. W. Stewart (1973). Estimation of internal crop water states from meteorological and climatic parameters, in *Plant Response to Climatic Factors*, R. O. Slater, ed., UNESCO, Paris, pp. 449-459.

Tyers, R. (1977). Harvard Center for Population Studies, personal communication.

United Nations (1975). *Demographic Yearbook, 1974*, United Nations, New York.

U.S. Department of Agriculture (1964). *Soviet Agriculture Today*, Foreign Agricultural Economic Rep. No. 13, Washington, D.C., pp. 4-6.

U.S. Department of Agriculture (1965). *Changes in Agriculture in 26 Developing Nations*, Foreign Agricultural Economic Research Service Rep. No. 27, Washington, D.C., pp. 15-16.

U.S. Department of Agriculture (1970). *Economic Progress of Agriculture in Developing Nations, 1950-1968*. Economic Research Service, Foreign Agricultural Economics Rep. No. 59, Washington, D.C., pp. 9, 16.

Waggoner, P. E. (1969). Environmental manipulation for higher yields, in *Physiological Aspects of Crop Yield*, Am. Soc. Agron. and CSSA, Madison, Wisconsin.

Whittaker, R. H., and G. E. Likens (1975). The biosphere and man, in *Primary Productivity of the Biosphere*, H. Lieth and R. H. Whittaker, eds., Springer Verlag, New York.

Zimmerman, L. J. (1965). *Rich Lands, Poor Lands, the Widening Gap*, Random House, New York, pp. 34-37.